Corporate Social Responsibility in the Construction Industry

The construction process, right through from planning and design to use and demolition, has a major impact on society. Initially, concern was focused on its environmental impact and the quest for sustainability, but this has now extended into the wider remit of corporate social responsibility (CSR). This involves businesses acknowledging the expectations of society and acting on behalf of all stakeholders. It essentially means that businesses must act (voluntarily) in a socially ethical manner by developing policy that encompasses the core principles enshrined by CSR. Unlike several other EU countries (notably France, Denmark and the Netherlands), the UK has yet to pass laws requiring firms to report on social and environmental issues. Yet some firms in the UK have grasped the challenge.

This textbook examines the impact of the construction industry on society and incorporates expert contributions on subjects including government intervention, human and employee rights, community involvement, corruption within the procurement process, and environmental damage. It is argumentative in nature and exposes competing views that seek to justify support for or ignorance of CSR principles.

This book is designed to be essential reading for all built environment undergraduate and postgraduate courses. CEOs and senior managers within construction businesses who are likely to embark on developing a CSR strategy should also read it.

Mike Murray is Course Director of the Construction Management MSc at the University of Strathclyde, UK.

Andrew Dainty is Professor of Construction Sociology at Loughborough University, UK.

Also available from Taylor & Francis

Ethics for the Built Environment
P. Fewings

Hb: ISBN 978–0–415–42982–5
Pb: ISBN 978–0–415–42983–2

Construction Cost Management
K. Potts

Hb: ISBN 978–0–415–44286–2
Pb: ISBN 978–0–415–44287–9

Project Management Demystified, 3rd edition
G. Reiss

Pb: ISBN 978–0–415–42163–8

Re-thinking IT in Construction and Engineering: Organisational Readiness
M. Alshawi

Hb: ISBN 978–0–415–43053–1

Procurement Systems
D. Walker and S. Rowlinson

Hb: ISBN 978–0–415–41605–4
Pb: ISBN 978–0–415–41606–1

Innovation in Small Construction Firms
P. Barrett *et al.*

Hb: ISBN 978–0–415–39390–4

Economics for the Modern Built Environment
L. Ruddock (ed.)

Hb: ISBN 978–0–415–45424–7
Pb: ISBN 978–0–415–45425–4

Information and ordering details

For price availability and ordering visit our website
www.tandfbuiltenvironment.com
Alternatively our books are available from all good bookshops.

Corporate Social Responsibility in the Construction Industry

Edited by

Mike Murray and Andrew Dainty

Taylor & Francis
Taylor & Francis Group

LONDON AND NEW YORK

First published 2009
by Taylor & Francis
2 Park Square, Milton Park, Abingdon, Oxon OX14 4RN

Simultaneously published in the USA and Canada
by Taylor & Francis
270 Madison Ave, New York, NY 10016

*Taylor & Francis is an imprint of the Taylor & Francis Group,
an informa business*

Typeset in Sabon by
Integra Software Services Pvt. Ltd, Pondicherry, India
Printed and bound in Great Britain by
TJ International Ltd, Padstow, Cornwall

The publisher makes no representation, express or implied, with regard
to the accuracy of the information contained in this book and cannot
accept any legal responsibility or liability for any efforts or
omissions that may be made.

British Library Cataloguing in Publication Data
A catalogue record for this book is available from the British Library

Library of Congress Cataloging in Publication Data
Murray, Mike, 1964-
Corporate social responsibility in the construction industry / Michael Murray
and Andrew Dainty.
p. cm.
Includes bibliographical references and index.
1. Construction industry--Management. 2. Construction industry--Moral and
ethical aspects. 3. Corporate social responsibility--Cross-cultural studies.
4. Business ethics--Cross-cultural studies. 5. Construction industry--
Cross-cultural studies. I. Dainty, Andrew. II. Title.
HD9715.A2M87 2008
624.068′4--dc22
2008003867

ISBN10: 0–415–36207–5 (hbk)
ISBN10: 0–415–36208–3 (pbk)
ISBN10: 0–203–01233–X (ebk)

ISBN13: 978–0–415–36207–8 (hbk)
ISBN13: 978–0–415–36208–5 (pbk)
ISBN13: 978–0–203–01233–8 (ebk)

Contents

List of figures   viii
List of tables   ix
List of contributors   x

PART I
**Evolution of corporate social responsibility
in the construction industry**   1

1   Introduction: Corporate social responsibility: challenging
    the construction industry   3
    MIKE MURRAY AND ANDREW DAINTY

2   The evolution of corporate social responsibility in
    construction: defining the parameters   24
    STUART GREEN

3   A business case for developing a corporate social
    responsibility policy   54
    LOUISE RANDLES AND ANDREW PRICE

PART II
Impact of construction on communities   71

4   The role of construction and infrastructure
    development in the mitigation of poverty and disaster
    vulnerability in developing countries   73
    GIRMA ZAWDIE AND MIKE MURRAY

5   Community interaction in the construction industry   98
    KRISEN MOODLEY AND CHRIS PREECE

PART III
Prevalence and nature of corrupt practices 119

6 Corruption in the UK construction industry: current
  and future effects 121
  JOHN TOOKEY AND DALE CHALMERS

7 Corruption within international engineering-construction
  projects 141
  MIKE MURRAY AND MOHAMED RAFIK MEGHJI

8 Cartels in the construction supply chain 165
  STEVE MALE

PART IV
Sustainable development 189

9 The evolution of sustainable development 191
  MARTIN SEXTON, PETER BARRETT AND SHU-LING LU

10 The alternative eco-building movement and its impact
   on mainstream construction 214
   TOM WOOLLEY

11 Corporate social responsibility and the UK
   housebuilding industry 235
   DAVID ADAMS, SARAH PAYNE AND CRAIG WATKINS

PART V
International perspectives on corporate social
responsibility in construction 259

12 Occupational health and safety (OH&S) and
   corporate social responsibility 261
   JOHN SMALLWOOD AND HELEN LINGARD

13 Corporate social responsibility in the American continental
   construction industry 287
   AGUINALDO DOS SANTOS AND FAUSTO AMADIGI

14 Corporate social responsibility and public sector procurement
   in the South African construction industry 304
   PAUL BOWEN, PETER EDWARDS AND DAVID ROOT

15 Corporate social responsibility in the Hong Kong and Asia
Pacific construction industry   327
STEVE ROWLINSON

16 Corporate social responsibility in the Australian construction
industry   351
HELEN LINGARD, NICK BLISMAS AND PETER STEWART

Index   381

Figures

2.1   Social responsibility categories   35
5.1   Developing a community policy   103
7.1   Flow of funds from international companies to LHWA
      chief executive   156
8.1   Construction demand and supply chain system   173
9.1   Model of interaction between ecological and social systems   192
9.2   Worldview framework   203
9.3   Systemic nesting of scales   207
9.4   Basic rearrangement of PSR framework   207
9.5   Framework for change   209
11.1 A typical CSR management system in housebuilding   246
12.1 The holistic role of CSR in OH&S and the role of OH&S
      in overall performance   280
13.1 A ready-to-assemble product for roof structures of
      low-income houses   299
16.1 Australia's population trends projected to 2051   355
16.2 Hours worked per week by managers and professionals
      in the Australian construction industry   373

Tables

4.1	Millennium development goals and role of construction	77
6.1	Fraud and corruption within construction projects	126
9.1	Gaps in knowledge and understanding and their implications	209
11.1	Key delivery objectives for CSR in the UK housebuilding industry	244
11.2	Top ten housebuilders' SAP ratings and brownfield completions in 2004	248
15.1	GDP growth rate in Singapore	329
15.2	GDP in Hong Kong	331
15.3	GDP in Macau	333
15.4	Macau construction expenditure	334
15.5	GDP and construction statistics in China	335
16.1	Proportion of construction workers with benefits	372
16.2	Cross-occupational comparison of mean burnout scores	374

Contributors

David Adams holds the Ian Mactaggart Chair of Property and Urban Studies at the University of Glasgow. His main researchinterests are in urban policy and regeneration, land and property development, and state–market relations in policy planning and implementation. He has researched and published widely in these fields, most notably as author of *Urban Planning and the Development Process* (1994), coauthor with Craig Watkins of *Greenfields, Brownfields and Housing Development* (2002) and as coeditor with Craig Watkins and Michael White of *Planning, Public Policy and Property Markets* (2005). His research has been funded by the Economic and Social Research Council, the Office of the Deputy Prime Minister, UK, and other public and private sector organisations.

Fausto Amadigi graduated in History from the Universidade Federal do Paraná (1999) and in Law from Pontifícia Universidade Católica do Paraná (2002). He is a specialist in Environmental Law at Centro Universitário Positivo – Unicenp (2006). Since 2003 he has worked within the area of corporate social responsibility, and between 2005 and 2007 he had also served as the Municipal Environment Advice President of Rio Negrinho/SC.

Peter Barrett is Professor and Pro-Vice-Chancellor for Research and Graduate Studies at the University of Salford. He is the Chairman of the Salford Centre for Research and Innovation in the Built Environment, Salford's EPSRC-funded Innovative Manufacturing Research Centre, and is the President of the International Council of Research and Innovation in Building and Construction (CIB) where he is championing a worldwide 'Revaluing Construction' initiative. CIB is the premier international research network linking 6000 construction researchers in over 50 countries.

Nick Blismas has ten years' experience as a researcher in construction management, both in Australia and in the United Kingdom. His strong industry-based research and publications have spanned such areas as

OH&S, multi-project management and off-site manufacture. Nick is currently Senior Lecturer in the School of Property, Construction & Project Management at RMIT University in Melbourne, where he also coordinates all research and postgraduate affairs. Prior to his academic career, Nick spent three years in industry as a project manager. Nick holds a PhD from Loughborough University, UK, and two Bachelor's degrees from the University of the Witwatersrand South Africa, in Building and Natural Sciences.

Paul Bowen is Professor and former Head of the Department of Construction Economics and Management at the University of Cape Town. He is a nationally rated researcher with the National Research Foundation and a member of the Academy of Science of South Africa. His professional and academic experience relating to the construction and property industries has resulted in more than 130 refereed publications. His research interests lie in the fields of building procurement, client briefing, process management and value management in the attainment of best value.

Andrew Dainty is Professor of Construction Sociology in the Department of Civil and Building Engineering at Loughborough University, UK. His research focuses on human social action within construction and other project-based sectors and, in particular, the social rules and processes that affect people working within the industry. He has published widely in both academic and industry journals. He holds a number of research grants from the EPSRC, ESRC and various UK government and European agencies, as well as serving in the role of advisor on human and organisational issues for a wide range of contracting and consultancy firms. He is coauthor of *HRM in Construction Projects* (2003) and *Communication in Construction: Theory and Practice* (2006) and is coeditor of *People and Culture in Construction: A Reader*, all of which were published by Taylor & Francis. He is also coeditor of the journal *Construction Management and Economics*.

Peter Edwards has lived and worked in England, Scotland, South Africa, Australia and the United States. Trained as a quantity surveyor, he holds an MSc from the University of Natal, a PhD from the University of Cape Town, and has jointly authored over 100 academic papers and a book, all of which focus on topics related to construction and project management. Recently retired from a full-time post at RMIT University, he lives in Melbourne and continues to be active in teaching, research, writing, peer review and consultancy.

Stuart Green is Professor of Construction Management in the School of Construction Management and Engineering at the University of Reading. He is Director of the EPSRC-funded Innovative Construction Research Centre (ICRC) where he is responsible for a multi-million-pound research programme in collaboration with industry. Previously he worked in contracting for several years before gaining design experience with an engineering consultancy. He is a chartered civil engineer and a chartered builder. Stuart has published widely in a range of international journals and has extensive experience of working with industry to improve practice. He is well known as a long-standing critic of 'lean thinking' and other approaches that fail to take account of social, economic and environment sustainability.

Helen Lingard, after completing a PhD in the field of occupational health and safety in the construction industry, worked as Area Safety Advisor for Costain Building and Civil Engineering (Hong Kong). Helen has lectured in occupational health and safety and human resource management at RMIT and Melbourne universities and has provided consultancy services to corporate clients in the mining, construction and telecommunications sectors. She has researched and published extensively in the areas of occupational health and safety, work–life balance and human resource management. She has coauthored two books, *Human Resource Management in Construction Projects* and *Occupational Health and Safety in Construction Project Management*. She is writing a third book on the subject of *Managing Work–Life Balance in the Construction Industry*, published in 2008. Helen is Associate Professor (Construction Management) in the School of Property, Construction and Project Management, RMIT University.

Shu-Ling Lu (PhD) is Research Fellow within the School of the Built Environment at the University of Salford. Her research interests are construction innovation and the organisation and management of small construction firms and professional service firms. She is Joint Coordinator of the CIB Task Group 65 in the Management of Small Construction Firms. Dr Lu has published one book and two book-chapters, and over 40 conference and journal papers.

Steve Male (PhD) has undertaken extensive industrial research and consultancy. He was awarded his doctorate in 1984, became a full-time lecturer the same year, was made Senior Lecturer in 1990, a Professor in 1993 at the age of 41, and Head of School of Civil Engineering at the University of Leeds in 1995. Major themes of his work in industrial contracts have involved knowledge and technology transfer, and include: value-for-money studies with a range of blue-chip and Government clients on major projects or with significant refurbishment programmes; studies for clients or consortia wishing to develop long-term partnering and supply chain arrangements; organisational change and restructuring; implementing information technology systems or business process reengineering of projects and teams. More recently he has led teams of academics and industrialists to develop a national asset management framework for managing the £23bn replacement value asset base for the Environment Agency, and for the United Kingdom's Office of Government Commerce research study into improving property asset management in the central civil government estate, worth approximately £220bn.

Mohamed Rafik Meghji is Consulting Engineer/Managing Director of M-Konsult Ltd, a Tanzanian Consulting Engineering company operating regionally in Africa. He was the President of The Institution of Engineers Tanzania (IET) (2001–2003), Executive Committee member of FIDIC (International Federation of Consulting Engineers) (1999–2003), Vice President (Africa) of Comtech committee of the World Federation of Engineering Organizations (WFEO) (2002–2006), Board Member (representative for Africa) of the World Council of Civil Engineers (WCCE) (2006–2010), as well as Board Member of International Federation of Housing and Planning (IFHP) (2004 onwards). He had chaired FIDIC's Capacity-Building Committee, which updated the World Bank 'Training Manual' *Guide to Practice: The Business of a Professional Services Firm.*

Krisen Moodley is Lecturer in Construction Management in the School of Civil Engineering, University of Leeds, and Director of Postgraduate Programmes at Leeds. His first academic appointment was at Heriot-Watt University where he spent four years before joining Leeds in 1994. His research interests include business management, and business and professional ethics. His recent books include *Engineering, Business and Professional Ethic* (2007); *Corporate Communications in Construction* (2006) and *Construction Business Development and Meeting New Challenges, Seeking Opportunity* (2003).

Mike Murray is Lecturer in Construction Management in the Department of Civil Engineering at the University of Strathclyde. He completed his PhD research in June 2003 and also holds a 1st Class Honours degree and MSc in Construction Management. He has lectured at three Scottish universities (The Robert Gordon University, Heriot-Watt and currently Strathclyde) and has developed a pragmatic approach to both research and lecturing. Mike has delivered research papers to academics and practitioners at UK and in overseas symposiums and workshops. He began his career in the construction industry with an apprenticeship in the building services sector and was later to lecture in this topic at several Further Education colleges. Mike is a member of the Chartered Institute of Building (MCIOB) and is coeditor of several textbooks: *Construction Industry Reports 1944–1998* (2003); *The RIBA Handbook of Construction Project Management* (2004); and coauthor of *Communications in Construction: Theory and Practice* (2006).

Sarah Payne is currently a doctoral student in the Department of Urban Studies at the University of Glasgow. Her PhD research focuses on the UK speculative housebuilding industry and the challenge of brownfield development, which is funded by the ESRC and Jenkins & Marr as a CASE studentship. Her other research interests include institutional analysis in British property research, urban planning and governance, the residential development process and the sustainable urban form.

Chris Preece is Fellow of the Chartered Institute of Building, a Member of the Chartered Institute of Marketing and Fellow of the Higher Education Academy. Since 1994 he has been Programme Leader of the MEng Civil Engineering with Construction Management (International) course in the School of Civil Engineering, University of Leeds. Chris was external assessor for the construction management degree programme at Glasgow Caledonian University and recently has been visiting professor at the International Islamic University, Kuala Lumpur, Malaysia. Chris's research interests include marketing, business development, and procurement and project management in the construction sector. He has published more than 40 books and journal papers nationally and internationally.

Andrew Price, Professor of Project Management, Director of Postgraduate Studies and Academic Director of the Health and Care Infrastructure Research and Innovation Centre, has over 25 years' design, construction and research experience. His current research focuses on improving construction performance, innovative design and construction solutions for health and care infrastructures, whole life value, continuous improvement, and sustainable urban environments.

Louise Randles began her career as a building surveyor before becoming research engineer with Loughborough University, where earlier research in social risks within urban regeneration developed into a passion for CSR. Currently Louise is working as a Project Manager helping to deliver sustainable regeneration in Liverpool.

David Root is Senior Lecturer in the Department of Construction Economics and Management at the University of CapeTown. A chartered surveyor and chartered builder by training, David worked in industry for a number of years for a variety of public and private sector organisations before working as a researcher at the Universities of Bath and Loughborough. He is currently active in research in the areas of procurement and project management with a specific interest in the response of the construction sector to the implementation of preferential procurement and Black economic empowerment in South Africa.

Steve Rowlinson is Professor in the Department of Real Estate and Construction at Hong Kong University, and he is involved actively in research and doctoral supervision in the areas of procurement systems, construction management, occupational health and safety and ICT. He has been coordinator of the CIBW 092 working commission on Procurement Systems for over ten years now and has coorganised numerous conferences and symposia in this capacity. Steve has authored and coauthored more than ten books and over 100 peer-reviewed papers. He is Adjunct Professor at Queensland University of Technology where he has a particular interest in sustainability, international project management and construction innovation. Steve has acted as a consultant to, *inter alia*, Hong Kong Works Bureau, Hong Kong Housing Authority and Queensland Department of Main Roads, as well as numerous consultancy reports. He has produced over 100 expert reports in relation to construction site accidents and construction disputes over the past 20 years in Hong Kong. He is a member of the Institution of Engineers (Hong Kong), the Institution of Civil Engineers (United Kingdom) and Fellow of the Royal Institution of Chartered Surveyors.

Aguinaldo dos Santos is a civil engineer who graduated in 1992 from the Federal University of Paraná (UFPR), and obtained an MSc in Civil Engineering from the Federal University of Rio Grande do Sul (1993–1995) and a PhD in Operations Management from the University of Salford (1996–1999). He is currently Professor at the Design Department at UFPR, supervising research projects within UFPR's Design Postgraduate Program and Civil Construction Postgraduate Program. He coordinates the Design & Sustainability Research Center (www.design.ufpr.br/nucleo) where his main research interests are lean production and sustainable design, with a focus on low-income families. He has been a visiting scholar at the Glasgow Caledonian University, Scotland, and Escuela Militar del Ecuador, and has worked within a research fellowship at the University of Salford. In 2006, he was the Chair of the 7th Brazilian Congress on Design Research & Development (www.design.ufpr.br/ped2006) and in 2007 he was the Chair of the International Symposium on Sustainable Design (www.design.ufpr.br/issd). He also coordinates the cooperation agreements between UFPR and Köln International School of Design, Germany, and Politecnico di Milano, Italy.

Martin Sexton is Professor and Associate Head of the School of the Built Environment at the University of Salford. His research interests focus on innovation management in project-based organisations, change management and knowledge management. He is Joint Coordinator of the CIB Working Commission in the Organisation and Management of Construction, with membership from over 45 countries. Martin is currently engaged in international and national funded projects to the value of over £6m, and has published over 150 journal and conference papers.

John Smallwood is Professor and Head of the Department of Construction Management, Programme Director of the MSc (Built Environment) Programme, and a director of two research, technology and innovation entities at the Nelson Mandela Metropolitan University. Both his MSc and PhD (Construction Management) addressed Health and Safety. He is also a National Research Foundation-rated researcher specialising in construction-related issues such as H&S. Twenty years of H&S experience includes contracts management, research, publishing, course/programme/seminar/workshop development, lecturing, conference organising, judging, and chairing of regional and national forums and committees. He is Fellow of the Chartered Institute of Building (CIOB), and a member of the Association of Construction Project Managers (ACPM), Ergonomics Society of South Africa (ESSA), and Institute of Safety Management (IoSM), and a member and councillor of the South African Council for the Project and Construction Management Professions (SACPCMP).

Peter Stewart has more than 30 years' experience in the construction and education sectors, and is currently the General Manager of Infrastructure Division of the Department of Education in Victoria. In this role he has responsibility for the capital works programs which include new schools, modernisations and regeneration projects, emergency services, provision planning and property management. Prior to joining the department, Peter was Associate Professor and Head of the School of Property, Construction and Project Management at RMIT University. Earlier in his career he worked for Fluor Daniel, and with the Victorian Public Works Department and Ministry of Education. Peter holds a doctorate from RMIT University, a Masters from the University of Newcastle, and an Honours degree from the University of Melbourne.

John Tookey took up a senior lectureship in Construction Management in the Department of Civil and Environmental Engineering at the University of Auckland, New Zealand, in 2006, having previously worked at Glasgow Caledonian University from 1998 to 2006. He graduated in 1993 with a BSc in Industrial Technology and subsequently obtained his PhD in Industrial Engineering from the University of Bradford in 1998 before moving into built environment research. He has published widely in the last decade within his main research interest areas, which include supply chain management (SCM), logistics and procurement in a construction context.

Craig Watkins is Reader in the Department of Town and Regional Planning at the University of Sheffield. His research interests include housing economics and policy, and urban regeneration and commercial property market analysis. He is coauthor with David Adams of *Greenfields, Brownfields and Housing Development* (2002) and coeditor with David Adams and Michael White of *Planning, Public Policy and Property Markets* (2005). His recent research has been funded by ESRC, RICS, Investment Property Forum and several local and central government departments.

He is currently researching 'Outcome Indicators for Spatial Planning in England' for CLG and the RTPI.

Tom Woolley (BArch, PhD) is an architect and environmental researcher living in County Down, Northern Ireland. He has been Professor of Architecture at Queen's University Belfast since 1991 but now works for Rachel Bevan Architects where he concentrates on sustainable design and consultancy. He has worked at the Architectural Association, the University of Strathclyde, and Hull School of Architecture. He is a visiting professor at UITM, Malaysia, and at the Centre for Alternative Technology. He was editor of the *Green Building Handbook* and author of *Natural Building*. He has contributed to many other books and international conferences and is Chairman of the UK Hemp Lime Construction Products Association.

Girma Zawdie (PhD) is a lecturer in the David Livingstone Centre for Sustainability at the University of Strathclyde, Glasgow. He is an economist with over 30 years of teaching and research experience gained from work in the United Kingdom and overseas. Areas of his research interest cover technology and innovation management, policy issues relating to environmental management, world poverty, and sustainable development. He has published papers widely in peer-reviewed journals, conference proceedings, books and monographs. He is cofounder and coeditor of the *International Journal of Technology Management and Sustainable Development*, which is published in collaboration with the Association of Commonwealth Universities.

He has lived and worked in developing countries as an academic, a consultant and freelance journalist. He has at various times in the past undertaken research and consultancy work for UN agencies, ILO, DFID, the British Council and different NGOs, covering a wide range of development issues. He has also at various times played key roles in the organisation and management of international conferences.

Part I

Evolution of corporate social responsibility in the construction industry

Introduction

Corporate social responsibility: challenging the construction industry

Mike Murray and Andrew Dainty

Introduction

Corporate social responsibility (CSR) represents one of the most widely debated contemporary business issues. In broad terms, CSR is the responsibility an organisation takes for the impact of its corporate activities on the various stakeholders with whom it interfaces and whom it affects (e.g. employees, customers, and communities) and on the environment. Thus, the concept holds that corporate emphases extend well beyond legal and statutory obligation to encompass a real and sustained responsibility to those whom the business affects. The emergence and growth of the concept reflects a deepening societal interest in the ethical behaviour of businesses. Stakeholders are increasingly prone to criticise unethical practices that damage the environment, exploit the workers or damage communities. Accordingly, in recent years, the nature and meaning of the concept has evolved from one of corporate philanthropy to a key business driver that influences organisations' strategic priorities. This has seen the formal recognition of CSR as a key driver and metric for business performance. Thus, although assertions of a positive relationship between social and financial performance are by no means uncontested, the need for effective approaches to CSR is beginning to have purchase within many industries and sectors.

Given the impact of construction activity on society, the economy and the environment, and its significance as an employer and provider of work, the construction industry has more reasons to focus on its CSR than most others. Indeed, in recent years construction has found itself under more scrutiny than ever before, particularly given the media's focus on transgressions in social responsibility (Herridge 2003). If CSR is viewed as a journey (see Business in the Community 2003), then the industry's evolution in CSR most probably began many decades ago with largely unrecognised philanthropic business activities that emphasised a concern for people and local environments. However, the UK construction sector as a whole is yet to embrace fully the CSR agenda (DTI 2006), and this suggests that it has made slow progress in comparison to other industries and sectors. Thus, many commentators have begun to call for CSR to begin to shape the way the industry conducts its business.

The aim of this book is to provide a broad introduction to CSR and its relevance and application to the construction industry. It shares common ground with Crowther's and Rayman-Bacchus's (2004) book – *Perspectives on Corporate Social Responsibility* – in that the chapters present a plurality of definitions and perspectives on the concept. No attempt has been made to knit them into a single overarching theme or to anchor the perspectives of the individual contributions to a common, all-encompassing definition. Rather, it should be recognised that the contributions originate from a variety of scholars from different countries, cultures and contexts, each of whom views the concept of CSR through a different theoretical lens. This is appropriate given that this textbook is the first to consider CSR in depth within the specific context of the construction industry. This is not to suggest that it is not a well-researched topic within the construction research literature. Other texts and papers abound that have examined topics such as ethical behaviour within the built environment (e.g. Bowen *et al.* 2007; Fox 2000; Liu *et al.* 2004; Poon 2004; Sassi 2004; Suen *et al.* 2007) and engineering and environmental ethics (e.g. Wilcox and Theodore 1998). Construction management researchers have also given CSR considerable attention in recent years (see Barry 2003; Barthorpe *et al.* 2004; Herridge 2003; Jones *et al.* 2006, Leão-Aguiar *et al.* 2006; Rameezdeen 2007; Wilkinson *et al.* 2004). However, this work has tended to focus on distinct elements of the concept. Although a few of the authors have attempted to provide a more holistic overview of the CSR concept in relation to the industry, a multifaceted consideration of the application of CSR and what it means to be socially responsible in construction is arguably long overdue.

This opening chapter has a dual purpose in that it attempts to problematise the application of CSR to the contemporary construction sector and to place the individual contributions contained in this book within this broader context. It provides a broad overview of the CSR concept and its application to construction, discussing its relevance and importance to the industry's organisations and stakeholders. In doing so, it sets out some of the contested and controversial issues that pervade the CSR literature. This provides a backcloth to the more in-depth treatment that specific CSR topics receive in the ensuing chapters and also a point of departure for those seeking to make connection between the perspectives presented within this text. It is, however, important to recognise that this book is devoid of normative or instrumental advice on 'how to' take account of CSR in corporate governance. It is incumbent on the readers to assimilate these different worldviews and perspectives and to relate them to their personal contexts when seeking to address the considerable challenges that CSR presents to construction organisations.

Towards a definition of CSR: a case for localised conceptions?

Before examining how the UK construction industry has responded to the CSR challenge, it is important first to construct a cogent definition of the concept to act as a point of reference for the remainder of the chapter. The growing prominence of CSR is considered to be an important 'trend' that requires fresh ways of thinking about the issues, both internal and external, which confront organisations (Jonker *et al.* 2004). However, despite the considerable development in both research and practice of CSR in recent years, there exists no single unifying definition of the concept (Ward and Smith 2006). Indeed, as Herridge (2003) notes, there is no definitive answer to the question of what CSR actually is. The term has become problematic as it has been appropriated and applied in different ways by organisations seeking to support their own particular agendas and perspectives. Thus, it is necessary to explore the central principles of the concept, as revealed in the various definitions used to describe it, in order to construct a definition that accords with the sectoral context within which it is to be applied in this book.

As with many other business concepts, CSR has evolved over several decades. The historical roots and the development of CSR are more fully described by Green in Chapter 2 in relation to the wider socio-political context, but as a precursor to reviewing the industry's approach towards CSR in recent years, it is worth reflecting briefly on how the contemporary meaning of the concept has emerged. According to *Corporate Watch* (2007), the phrase CSR was coined in 1953 to coincide with the publication of a book authored by H.R. Bowen, *Social Responsibilities of the Businessman* (Bowen 1953).

Bowen's definition acknowledged concerns about the pervasive powers of corporations and how they affected those within society. The 'obligations' of those with responsibilities for running businesses have since become better understood and codified, and there now exist 'standards' against which organisations can measure their CSR performance. Thus, as Tepper Marlin and Tepper Marlin (2003) suggest, the evolution of the concept can be conceptualised in terms of the ways in which it has been *reported*. They suggest that although the term CSR was in common use in the early 1970s, the first phase of reporting really only comprised advertisements and annual-report sections and had little relationship with corporate performance. A second phase began in the late 1980s with the emergence of *Stakeholders Reports* to describe social responsibility performance. A third phase of reporting introduced certification by third party bodies that are accredited to standards of social or environmental performance. The onset of auditable standards has necessitated more precise definitions in order that CSR metrics can be better understood. A range of definitions and terms has therefore emerged to embody aspects of the CSR concept including sustainable development/business practices, business ethics, corporate responsibility/citizenship and the ubiquitous 'triple-bottom-line'.

In seeking greater definitional clarity on the term, it is useful to compare some popular definitions rooted within a particular socio-economic context. Four definitions of CSR, popular in the UK, are provided below. The first of these is attributed to the current UK Prime Minister Gordon Brown and the second, to a UK Government CSR website. The other two definitions are taken from leading UK construction-industry-related sources.

> Today, corporate social responsibility goes far beyond the old philanthropy of the past – donating money to good causes at the end of the financial year – and is instead an all year round responsibility that companies accept for the environment around them, for the best working practices, for their engagement in their local communities and for their recognition that brand names depend not only on quality, price and uniqueness but on how, cumulatively, they interact with companies' workforce, community and environment. Now we need to move towards a challenging measure of corporate responsibility, where we judge results not just by the input but by its outcomes: the difference we make to the world in which we live, and the contributions we make to poverty reduction.
>
> (Brown 2004: 2)

> The Government sees CSR as the business contribution to our sustainable development goals. Essentially it is about how business takes account of its economic, social and environmental impacts in the way it operates – maximising the benefits and minimising the downsides. Specifically, we see CSR as the voluntary actions that business can take, over and above compliance with minimum legal requirements, to address both its own competitive interests and the interests of wider society.
>
> (csr.gov.uk 2007)

> The commitment to integrate socially responsible values and concerns of stakeholders into their operations in a manner that fulfils and exceeds current legal and commercial expectations.
>
> (Constructing Excellence 2004)

> Approaching your business aims responsibly, with an awareness of your surrounding social and environmental needs, in order to fulfil stakeholder demands, achieve a profit and remain competitive. There are several areas that construction companies should be addressing in order to become more socially responsible. Many of the issues that are being grouped under the heading of CSR are already part of the daily work of construction professionals, but are not being recognised specifically

as social obligations. These include Ethics, Human Rights, Community, Environment and Employee Relations.

(CIOB 2007)

Whilst there are some notable differences in emphasis, common themes undergird all of these definitions. For example, they all allude to the need for respect for communities and/or stakeholders, the need for awareness of social, economic and environmental impacts and they all emphasise the alignment of business objectives with societal goals in a way that goes above and beyond minimum legislative requirements. These components will be returned to later in this chapter and throughout this book.

It is important to note that the definitions listed above do not attempt to impose any framework on the way in which the concept should be interpreted and enacted within a particular context. As Ward and Smith (2006) suggest, CSR needs to be localised at both conceptual and operational levels so that it becomes more manageable and embedded within an organisation. This will inevitably mean that it is still better if CSR is shaped by the interests of stakeholders closest to a company's operations on the ground. As Porter and Kramer (2003) note, adopting a context-focused perspective makes a company's philanthropic practices far more effective, even though it goes against the grain of mainstream practice. In accepting this perspective, the individual authors have been encouraged to establish and operationalise their own definitions which reflect the issues on and the contexts within which their work is grounded. Such an approach resonates with other CSR textbooks that present a plurality of CSR definitions appropriate to the organisations to which they relate (e.g. Crowther and Rayman-Bacchus 2004).

Problematising the CSR concept: corporate philanthropy or revenue opportunity?

It is important to recognise that even if the motives for CSR are rooted in social conscience, its enactment by businesses may represent a different appropriation of this intention, and so CSR is not without its detractors. This is due in part to the language associated with CSR lending itself to rhetorical statements and grandiose claims which seem disconnected from the realities of corporate life. Corporate Watch (2006) argues that CSR is, in fact, a contradiction in terms given that companies are legally bound to maximise profits for their shareholders. As such, wider social good can ever be only incidental to the interest in making profit. This perspective asserts that social justice is a business opportunity fundamentally flawed. This sceptical view is exemplified by the mocking attempts to amend the CSR acronym (e.g. Complete Sidelining of Reality, Companies Spouting Rubbish and Corporate Slippery Rhetoric). Thus, although CSR has transformed from voluntary philanthropic actions to a key strategic activity for leading organisations (see Carroll 1999), the extent to which a firm can attend to both people and profits in the course of its business has been widely disputed (see Margolis and Walsh 2001).

Taking this argument a stage further, there is a clear distinction that can be made with regard to the drivers for embracing the CSR agenda: a philanthropic desire to contribute towards societal well-being, and a business desire to embrace social responsibility for reasons of financial performance. These contrasting perspectives were recently debated by leading writers in a collection of essays published by the *Harvard Business Review* (HBR 2003). Several contributors argued that corporate financial performance and CSR are not mutually exclusive, but are, in fact, inexorably linked. In other words, as Smith (2003) argues, philanthropic and business units have joined forces to generate powerful competitive positions through philanthropic actions, particularly in international markets. However, although these 'business case' arguments for CSR are attractive, a distinction needs to be drawn between what has been defined as 'cause-related philanthropy' and true 'strategic philanthropy'. Porter and Kramer (2003) argue that 'strategic philanthropy' has been widely misused to refer to acts that are aimed at generating good publicity and marketing rather than genuine social impact. As Handy (2003) states, '[m]ore corporate democracy and better corporate behaviour will go a long way to improve the current business culture in the eyes of the public, but unless these changes are accompanied by a new vision of the purpose of business, they will be seen as mere palliatives'. Thus, to take hold, the language and measures of business must be reversed; businesses should be seen as wealth-creating communities, not as pieces of property.

Margolis and Walsh (2001) reviewed the numerous studies linking social and financial performance. They dichotomise these studies into those which take a narrow economic perspective on the firm (where socially responsible practices squander a firm's resources) and those which take a broader perspective which sees positive returns from social performance as evidence that such policies do nothing to jeopardise the financial performance of the firm. Overall, their review presents a positive perspective on the bottom-line benefits of socially responsible business. However, they also warn that such an assertion must be treated with caution given the weaknesses in the measures used for social performance and the diversity of financial performance metrics; the complexity of such concepts do not lend themselves to simple correlations. The jury is still out, therefore, on whether the responsible CSR really is a catalyst for improved financial performance.

The growing prominence of CSR in the construction industry

As the succeeding chapters of this book will reveal, CSR is growing in prominence as a core issue confronting the construction industry and its organisations. This is related in part to its size and significance as an employer. The global construction industry is said to account for around ten per cent of the world's gross domestic product (GDP) and is the largest industrial employer, accounting for seven per cent of worldwide employment (International Federation of Building and Wood Workers 2004). However, it is the impact that the industry's products have on societal well-being that provides the most compelling reasons for raising the profile of CSR. The built environment is instrumental in influencing human health and social behaviour as well as cultural identity and civic pride (Pearce 2003a). Moreover, prominent architects and engineers can leave a socio-economic legacy that can extend for centuries. For example, Ford (2007) highlights Thomas Telford's road-building programme in Scotland in 1803–1821. His projects provided a transport link between the fishing ports on the west coast and the Caledonian Canal and promoted commerce, assisted in reducing emigration and, in doing so, supported many of the aspirations of sustainable development long before it became fashionable.

Of course, not all construction projects are embraced by those whom they impact upon. For example, many stakeholders are deeply concerned about the impact that building and infrastructural development has on their environment and well-being. Determining which projects are truly beneficial to society and eliminating those flavoured by 'optimism bias' (Mott MacDonald 2002) or 'engulfing illusions' (Miller and Lessard 2000: 45) are necessary to prevent the misuse of the taxpayer's money on public projects that have no clear benefit for society. This issue has been explored by CEE bankwatch network and Friends of the Earth Europe (2006). They argue that plans for 22 new construction projects, ranging from motorways, dams and canals to waste incinerators, are 'environmentally damaging, economically unjustified and socially controversial'. Miller and Lessard (2000) note that 'effective' projects must combine economic performance, technical functionality, social as well as environmental acceptability, political legitimacy and economic development. Thus, whilst project sponsors and the construction industry supply-side may actively support a project, local activists and pressure groups may see little long-term benefit in it. Indeed, large engineering projects present complex dilemmas that will most certainly involve trade-offs. The new BAA T5 project at Heathrow is one such example. Whilst the project won environmental awards during construction, one could question whether society gains by sanctioning the ultimate increase in air travel brought about through its construction. The economic counter-argument – that trade would go elsewhere in Europe and that the United Kingdom would lose the associated benefits derived from increased air travel through Heathrow – ignores the legitimate CSR challenge to the project's overall benefit to society.

The economic benefit versus environmental impact debate is just one of the hundreds of CSR dilemmas that confront the industry on a daily basis. For example, according to an estimate by Hampton (2003), the global construction industry could require between two and eight planets' worth of resources (e.g. steel, concrete, energy and aluminium) in just 50 years or so. Thus, the rate at which the industry is consuming resources is not sustainable. Moreover, given that built assets act against nature by rendering an area of soil incapable of producing those natural resources, they can be considered parasitic (Curwell and Cooper 1998, cited in Guy and Farmer 2000). The challenge is, of course, not restricted to environmental issues. The current structure of the industry with its reliance on very small firms may reward entrepreneurship, but is not necessarily conducive to embedding principles of CSR throughout the supply chain. Pearce (2003b) suggested that as the 'triple-bottom-line' of profitability and environmental and social responsibility comes at such a cost to small firms they are unlikely to be affordable. The fragmented nature of the industry also leads to informality and the possibility of corrupt business practices. For example, the Office of Fair Trading (OFT) has recently completed a two-year investigation into the English construction industry, which has revealed some £3 billion worth of tenders that had been unfairly bid for (Building 2007). The specific practices used by contractors involved 'cover pricing', where a firm arranges for a rival to enter a bid that is too high to win but low enough to give the appearance of competition. Compensation payments (or 'bungs') are then passed between competitors in exchange for the cover price. In a similar vein, in November 2002 the European Commission imposed fines totalling: £250 million on companies who had been operating a cartel in the plasterboard sector (Building 2007: 33). This was based on a secret information-exchange system designed to avoid over-aggressive competition and a price war in the sector. The examples examined in Part III of this book present much more evidence of the industry's susceptibility to corruption.

The evolution of the CSR concept over the past 30–40 years reflects a shift in the ways in which not only the corporations communicate with their stakeholders, but also how society has begun to question organisations as to their social responsibility. The Business in the Community framework (2006: 7) summarises society's expectation of business in relation to environmental improvements over the last four decades. It summarises these phases as: 'trust you' (1970s), 'tell me' (1980s), 'show me' (1990s) and 'prove to me' (2000s). The report outlines how, in spite of the establishment of the environmental agenda in the 1970s, trust in the government and big business fell in the 1980s following some major accidents such as Bhopal and Chernobyl. This led to a series of regulatory measures being initiated in the 1990s and eventually to environmental audits, the results of which tend to be disclosed to the investor community. As such, the societal expectations of business have been influenced by the socio-political context over this period.

Viewing the industry through this lens, a similar developmental trajectory can be seen to have been played out within the construction industry over this period in respect of its wider CSR performance. In the 1970s, the landmark

Health and Safety at Work Act (1974) revolutionised the policy and proced- ures adopted in all businesses within the United Kingdom. The construction industry had a particular need to improve its health and safety (H&S) per- formance due to rising numbers of workplace accidents and fatalities. It could be argued that the implementation of the legislation has promoted an increased level of 'trust' between employee and employer. Additional legisla- tion, such as The Reporting of Injuries, Diseases and Dangerous Occurrences Regulations 1995 (RIDDOR), has reinforced this by placing a legal duty on employers to report accidents and ill health in the workplace. Such report- ing is very much in line with the 'prove to me' ethos of the 1990s. In the 1980s, a 'tell me' analogy seemed relevant to the way in which the term 'sustainable development' was conceived in *The Brundtland Report* (1987) as 'development that meets the needs of the present without compromising the ability of future generations to meet their own needs'. However, despite Brundtland raising awareness of sustainability in the latter part of this dec- ade, the industry has largely operated without any widespread evidence of CSR being embraced in a way that reflects the growing concerns of society to 'tell me'.

The boom-to-bust economy and financial collapse of some construction conglomerates (see Hillebrandt *et al.* 1995), as well as many insolvencies within the SME community, reflect an industry motivated by turnover, profit and returns to shareholders. However, the brutal trading environment exper- ienced by suppliers during the 1980s, particularly with respect to cash-flow problems resulting from late payment and non-payment, was to provide Michael Latham (1993) with the ammunition for his interim report on the UK construction industry – 'Trust and Money'. Constructor's annual reports from this decade show little sign of the CSR references that are now common- place within stock market listed companies' reports. However, even though society was expected to 'trust' both the demand and supply sides of the industry to undertake development in a sustainable manner, the concept of sustainable construction was born in the 1980s.

In the 1990s the impact of one particular UK project, the M3 motorway through Twyford Down, was arguably a contributory catalyst for construc- tion's journey into CSR proper. The campaign to stop this project spread from a local campaign to receive national and eventually international cov- erage, and culminated in environmental protestors disrupting the major contractors annual meeting in 1994. This prompted a recognition of the importance of the environment as a business issue and led to the formation of an independent advisory panel, culminating in the publication of an envir- onmental report (*First Report: Tarmac in the Environment*) in 1995. The emphasis here appears to have been societal pressure on construction busi- nesses to 'show me' that they care and the industry dialect (i.e. sustainable construction) would perhaps suggest a new CSR paradigm. Further evidence of the need to provide 'proof' can be found in the ethical procurement of building materials and the requirement to provide documentary assurance that timber for use on projects has been procured through an accredited and sustainable source such as the Forest Stewardship Council (FSC 2007). Furthermore, new legislation introduced in England and Wales in April 2008 (*Site Waste Management Plans – SWMP*) requires constructors' to identify the amount and type of waste that will be produced on a new construction project and demonstrate how it will be reused, recycled or disposed of (Defra 2007). Such national initiatives are taking place under a wider international concern for CSR whereby the International Standards Organization intend to publish Guidance on Social Responsibility (ISO 2600) in 2009 (ISO 2007). The construction industry has assisted in developing the content of this document through representation from Swedish contractor Skanska (2005: 43).

The 'business case' for CSR in construction

Given the inexorable relationship between economic growth and construction development, there are now new pressures from consumers and users of the built environment for greater social responsibility from the sector. Construction clients and investors, supported by a media hunger for corporate transgressions, increasingly demand that companies comply with standards linked to social and environmental performance (Herridge 2003). Companies that develop and initiate such policies are seen as providing themselves with market advantage over their competitors (CIRIA 2003). Thus, it is arguably the 'business case' arguments which have emerged as the primary driver for socially responsible business practices within many construction companies.

A first set of business case drivers is rooted in the desire of the construction firm to project a positive and responsible image to its customers. According to Constructing Excellence (2004), effective CSR practices can secure strategic advantage (e.g. by securing goodwill of the local community and facilitating planning permission); improve reputation (e.g. by developing a brand for doing good business); reduce costs (e.g. by enhancing productivity by avoiding conflicts with community groups) and minimise risks (e.g. relating to corporate image or project programme). To this end, the government encourages socially responsible practices in relation to both the treatment of people and sustainable construction. However, as Pearce (2003b) argues, this 'soft regulation' encourages corporations to self-regulate rather than having legislation imposed on them. The dilemma for many organisations, as Barry (2003) notes, is 'what do companies do when the ethical option is not the most profitable one?' It is all too easy for organisations to de-prioritise CSR practices in favour of commercial imperatives, or for CSR to become little more than rhetorical exhortations rather than meaningful actions. Some employees expose companies that transgress CSR and legislative requirements. This 'whistle-blowing' is protected through The Public

Interest Disclosure Act 1998 (HMSO 1998). However, although many construction companies devote web space to this issue, Vinten's (2004) study revealed that the British character has been ambivalent to such practice.

Given the inherent weaknesses of voluntary approaches to CSR, construction firms are increasingly being compelled to demonstrate socially responsible practices in order to win work. According to Margolis and Walsh (2001) social trends effectively enforce a reconsideration of the role and responsibilities of the firm and what is expected of them. Indeed, the call for constructors to be 'considerate' whilst building has been reviewed by Glass and Simmonds (2007). In this respect, companies are increasingly called upon to resolve problems to which they have contributed or from which they stand to benefit (e.g. the negative impact on the local environment resulting from development). A second driver is where firms find themselves in specific situations that force them to respond in a socially responsible manner. For example, a contractor may operate in a country where Western welfare institutions may not exist. A condition of operating in such an environment might be that they contribute to the creation of such welfare provision. Thus, this perspective sees CSR as a necessary and desirable component of securing market position.

In many countries, public sector procurement is used by governments as a tool with which to embed socially responsible and sustainable principles within the construction activity. In the United Kingdom, for example, sustainable development outcomes are often enshrined within public-sector contracts. However, the extent to which contemporary procurement arrangements support such agreements remains somewhat questionable. For example, whilst the Private Finance Initiative (PFI) acts as an economic multiplier and provider of new facilities, critics now point to 'saddling of debt' to future generations. Kelly and Hunter (2005) examined the legacy of PFI in Scotland and concluded that local authorities and health boards have largely ignored their CSR responsibilities in favour of satisfying the HM Treasury's six primary drivers for value for money. The National Audit Office has also reported on mixed fortunes of PFI projects, and reports from end users often comment on performance failures. This has led to the public sector union Unison (2002) to suggest that CSR and PFI are incompatible. Thus, the ability of such measures to provide community benefits remains questionable.

Fresh perspectives on CSR in construction

In recent years, there has been an increasing interest in CSR from within the built environment research community. One of the first academic papers on CSR was Hamil's (1993) book chapter 'Building on ethics: Ove Arup and corporate social responsibility'. This provided an overview of the document that came to be known as 'The Key Speech' delivered by Sir Ove Arup in

1970. Following this, early perspectives on CSR tended to focus on environmental issues, before broadening out to encompass more general concerns for businesses' stakeholders (see Barthorpe *et al.* 2004). Most recently there has been a group of papers that have examined ethical behaviour within the built environment (Bowen *et al.* 2007; Fox 2000; Liu *et al.* 2004; Poon 2004; Sassi 2004; Suen *et al.* 2007) and engineering and environmental ethics (Wilcox and Theodore 1998). Other construction researchers have also given CSR attention as part of other studies (e.g. Barry 2003; Barthorpe *et al.* 2004; Jones *et al.* 2006; Leão-Aguiar *et al.* 2006; Rameezdeen 2007; Wilkinson *et al.* 2004). The growing prominence of CSR as a research topic in construction is also reflected in the institutional interest in the topic. Key institutions amongst these within the United Kingdom have been the industry improvement organisation Constructing Excellence (2004) and professional bodies such as The Chartered Institute of Building (CIOB 2007), both of which have published fact sheets on CSR in the construction industry.

Despite the growing research and industry interest in CSR, however, this text is the first that examines, within a single volume, the various facets of the concept and their significance for the construction sector. The ensuing chapters present an eclectic range of perspectives that traverse the entire concept, but share the common objective of examining their relevance and application to the construction sector. Each offers a unique perspective on how various aspects of CSR are being addressed by construction organisations. The contributions have been grouped into five broad sections, which examine (I) the CSR concept and its evolution; (II) the impact of construction on society; (III) corruption and illegal practices; (IV) sustainability; and (V) international perspectives on the CSR debate. The contributions have been clustered in this way to help the reader to find synergies and connections between the perspectives provided, although each of the chapters has resonances and interconnections with the others presented in this book.

Part I of the book contains three chapters charting the evolution of CSR in the construction industry. Following this chapter's introduction of the CSR concept, in Chapter 2 (The evolution of CSR in construction), Green places current debates about CSR in the broader context of social and political changes. This provides a more detailed analysis of the evolution of CSR as a foundation for the chapters that follow. In charting the evolution of the concept, Chapter 2 traces its origins back to the 1930s and describes the emergence of social responsibility in business, from systems perspectives in the 1960s through stakeholder theory in the 1970s to the enterprise culture and globalisation of today. In discussing the way that CSR has been influenced by the prevailing economic–political doctrine, Green's analysis reveals the tension between business and society, whilst acknowledging that both the construction worker and the construction workplace have been subject to politically motivated interference. He uses the change movement within the UK construction industry (see also Adamson and Pollington 2006) to reveal the relationships between CSR and a diverse range of construction industry issues, several of which (e.g. Health & Safety) are discussed in later chapters. Green argues that enlightened self-interest has become the dominant justification for CSR, as even if CSR cannot be proven to contribute to short-term profit, it must still seemingly be justified in terms of the long-term economic interests of the firm. This baton is picked up by Randles and Price in Chapter 3 (A business case for developing a corporate social responsibility policy). The tensions between business and society noted in Green's chapter are further elaborated here in relation to the perspectives of shareholders in relation to that of other stakeholders. The authors discuss the need for businesses to move from purely profit-centred strategies to those where ethical considerations are given equal prominence in decision-making. They suggest that the development of a 'social conscience' is not readily associated with construction companies, but that corporate governance is now being used as a yardstick for company performance. With this realisation has come an acknowledgement that is positive for staff recruitment and retention, the loyalty of stakeholders and shareholders, and the minimisation of risks from negative working practices, as well as for ensuring that the business stays ahead of future legislation and competition. Thus, it would seem that CSR now offers a route to sustained competitive advantage for many construction firms.

In Part II, issues concerning the impact of construction on communities are explored. Two chapters provide accounts of both the positive and the negative impact of construction activities in a diverse range of communities. In Chapter 4 (The role of construction and infrastructure development in the mitigation of poverty and disaster vulnerability in developing countries), Zawdie and Murray note the call for construction firms to have operational objectives compliant with the UN Millennium Development Goals (MDGs) and underpinned by poverty reduction objectives. They present several case studies to show that post-disaster recovery and poverty reduction are more successful when construction consults those stakeholders who have been afflicted. This emphasises the importance of delivering sustainable infrastructure in sympathy with the respective social, cultural, economic, political and environmental context within which such development takes place. In Chapter 5 (Community interaction in the construction industry), Moodley and Preece examine the strategies that are often adopted by the construction industry to mitigate the impact of its operations during construction activity. A central plank of their argument is for construction to be more 'considerate' to neighbours through avoiding/reducing potential activities that would cause environmental pollution to local neighbourhoods. From a more strategic perspective, they also discuss the manner in which construction companies can demonstrate their philanthropic credentials through active participation within communities and financial donations to charitable foundations. Their chapter demonstrates the benefits of corporate involvement in the community both for enhancing the reputation of those responsible for construction development and for ensuring political and social support for development through the goodwill of communities. This has clear resonances with the issues discussed in Chapter 3 mentioned above.

The darker side of construction activities is examined in Part III in which the prevalence and nature of corrupt practices are revealed. In Chapter 6 (Corruption in the UK construction industry), Tookey and Chalmers uncover the unsavoury behaviour of bribery and corruption within the UK construction sector, reporting that the UK Office of Fair Trading (OFT) has a special interest in the industry and is intent on exposing unfair business practices such as price-fixing. However, despite legislation outlawing corrupt practices, construction industry professionals often consider some unethical practices to be an acceptable business custom inbuilt in the procurement process. Indeed, in spite of the obvious possibilities for anti-competitive and unethical behaviour, there appears to be no coherent and unified approach adopted to tackle the problem. In Chapter 7 (Corruption within international engineering-construction projects), Murray and Meghji move the debate overseas by drawing upon research evidence from Transparency International to highlight the economic and social consequences that arise from corrupt practices in large engineering-construction projects. Murray and Meghji find that constructors emanating from developed countries are particularly prone to offering bribes to public officials whilst undertaking work in developing countries. The chapter includes a case study of a large construction engineering project exposed to corruption and explores the involvement of funding institutions and export credit agencies. The authors conclude their chapter by reminding readers that anti-corruption initiatives are foremost a matter of personal morality and ethical behaviour. This section of the book closes with a chapter by Male (Cartels in the construction supply chain) who takes an in-depth look at the manner in which cartels operate in construction, which act against the normal operations of competitive behaviour. He views this problem through a theoretical and investigative commentator approach, raising and debating issues surrounding the presence of such cartels and the resultant effect on competition within the industry. He argues that the characteristics of construction projects both nationally and internationally provide opportunities for corrupt practices to occur, often involving bribery and/or fraud, to the extent that they are regarded by some as normal business in the industry. The collusive practices uncovered in this chapter will be of particular concern to those with a vested stake in improving the popular image of the industry.

The three chapters in Part IV (Sustainable development) are opened by Sexton, Barrett and Lu (The evolution of sustainable development) who provide a theoretical exploration of the concept of sustainable development in its broadest sense. They encourage researchers and practitioners to locate and carry out their CSR work within a robust 'sustainable-development' framework and reinforce the message that all construction activities, whether

CSR compliant or not, fall within a wider scope of sustainable development. They argue that the CSR agenda is located predominantly within a neo-classical context which emphasises short-term profit generation and hedonistic client satisfaction, a position which will inevitably constrain the motivation and capability of construction stakeholders to bring about sustained, meaningful CSR. They present a dynamic Pressure/State/Response (PSR) model as a potentially fruitful framework to develop appropriate CSR strategies. In Chapter 10 (The alternative eco-building movement and its impact on mainstream construction), Woolley investigates 'green-building' and warns that terms such as 'sustainable' and 'green' are increasingly being applied to almost every branch of conventional construction practice. He argues that while corporate statements from large companies make claims of ethical behaviour, the more radical alternative movement provides a deeper green model that challenges the *status quo*. He therefore prefers the use of the term 'ecological' and uses this to distinguish genuinely environmentally responsible approaches to construction. The chapter concludes by challenging construction to fully embrace the implications of climate change and environmental issues without indulging in 'greenwash'. This radical position demands a 100 per cent acceptance of social and environmental responsibility grounded in ethical principles, thereby rendering many design and building solutions currently in use as not ethically acceptable. This section of the book closes with a chapter by Adams, Payne and Watkins (Corporate social responsibility and the UK housebuilding industry) in which they argue that the rapid dissemination of CSR as an embodiment of a new business ethic presents the speculative housebuilding industry with particular challenges. In their review of the top UK housebuilders, they find that at a national level, at least, most of the major developers are beginning to take CSR seriously, with many having adopted policies or strategies that indicate a desire to take a much broader view of their business activities. However, they argue that the real commitment of the major housebuilders to CSR will be tested at a local level, which would need to be supported by a significant policy shift. They call for new research to examine the local implementation of CSR policies by major housebuilders.

The final section of the book explores international perspectives on CSR in construction. Part V opens with a chapter by Smallwood and Lingard (Occupational health and safety (OH&S) and corporate social responsibility) who argue that construction organisations need to develop organisational cultures that are ethical and supportive of OH&S, a requirement that demands much more than the establishment of formal OH&S management systems. In common with other chapters in this book, they contend that stakeholder theory offers a significant opportunity for improving the construction industry's OH&S performance and introducing greater equity into the allocation of risk within construction firms. Smallwood and Lingard call for construction firms involved in international contracting to adopt better

practices with regard to OH&S irrespective of local standards and practices of their host country. They also acknowledge, however, that this will
demand improved collaboration between industry and project stakeholder
groups, an acknowledgement and acceptance of OH&S responsibility by
all parties and creative approaches to better embedding OH&S through the
construction supply chain. Chapter 13 (Corporate social responsibility in
the American continental construction industry) moves across the continents to South America, where Santos and Amadigi centre their discussion on
the Brazilian construction industry. Here they reveal the prevalence of child
labour among the poorly educated, who find it hard to break out of the
sector. A cause of this situation is the shortage of quality housing for low-
income families throughout the American continent. Santos and Amadigi
note that various initiatives are taking place to alleviate this burden on the
continent's poorest citizens and that this provides ample opportunity for
the construction organisations to align their business objectives with CSR
through more socially responsible practices.

In Chapter 14 (Corporate social responsibility and public sector procurement in the South African construction industry), Bowen *et al.* explore the
challenges of adapting to the new socio-economic and political realities of
post-apartheid South Africa. They note the importance of the South African
government's first attempt to improve the quality of life of the previously
disadvantaged segments of South African society which was through the
Reconstruction and Development Programme and that this has been followed by further improvement initiatives. Bowen *et al.* argue that the concept
of *ubuntu* provides a way forward in respect of CSR as it epitomises the
African philosophy of respect and human dignity. Fortunately, government
legislation, public sector procurement, social and economic development
and the consensus forged over ten years of democracy are combining to
support the adoption of *ubuntu* as an African conceptualisation of CSR.
In Chapter 15 (Corporate social responsibility in the Hong Kong and Asia
Pacific construction industry), Rowlinson argues that it is very important
to include the cultural dimension when examining CSR, in terms of both
the cultural values and beliefs and the stage and pace of economic development, before making judgements as to its applicability. He examines the
evolution of CSR in the Asia Pacific region by using examples taken from
China, Hong Kong, Singapore, Macao and Malaysia. These countries and
city-states have very different cultures with all but one placed in the global
top 20 in terms of GDP. Rowlinson's analysis reveals that, whilst CSR has
taken hold in the region, the reasons for this are very different from those
in the West and relate more to religious ideals and cultural values. Thus,
the systems and approaches that are adopted in the West to drive CSR are
inappropriate in the Asian context. Considerable challenges remain, therefore, in implementing CSR in a culturally sensitive manner. The final chapter
in this book (Corporate social responsibility in the Australian construction
industry) turns to Australia where Lingard *et al.* test the 'ethical climate'

of the construction industry. They describe the CSR profile of Australian business in general, and the Australian construction industry in particular, summarising recent reports that highlight the detrimental social impact of misconduct within the industry. They argue that addressing the social irresponsibility of the industry will demand cultural, societal, moral and ethical change, if it is to be successful. A range of CSR indices are presented as tools for benchmarking the industry's performance, which could provide a starting point for making improvements in the future. However, the scale of the challenge confronting the Australian sector resonates with that of the other countries discussed within this book. It would seem that the size of the task of embedding CSR across the global construction sector is considerable.

Conclusion: CSR – challenging the industry

As will become apparent as the reader moves through the chapters summarised above, it is not the intention of the editors or the individual chapter authors to present a 'how to' book on CSR for the construction industry. Generic business guides are available elsewhere, and are usually rooted in the assumption that CSR is a positive contributor to the achievement of business objectives whilst satisfying societal needs. In this book, the application of the CSR concept is juxtaposed against the complex and problematic realities that the construction sector presents. Each author has drawn from his or her own experience and research insights that have been gained in a number of different socio-political contexts. They view the CSR concept through a range of different theoretical lenses to provide a rich understanding of the scale of the challenge confronting the industry and some possible solutions to address them. The challenge for the readers is to assimilate these different worldviews and to relate them to their personal context. We do not lay claim to the cumulative insights provided within this book as having found any kind of panacea for overcoming the challenges that achieving socially responsible construction presents, but the chapters do point to the dilemmas faced and the directions that researchers and practitioners must take if they are to address these issues in the future. Its aims will be validated if students, researchers and practitioners use the ideas within this book to make their own contribution to the CSR discourse.

References

Adamson, D.M. and Pollington, T. (2006) *Change in the UK Construction Industry: An Account of the UK Construction Industry Reform Movement 1993–2003*, Oxon: Routledge.

Arup, O. (1970) *The Key Speech*. Online. Available at: www.arup.com/_assets/_down load/download5.pdf (accessed 10 June 2007).

Barry, M. (2003) Corporate social responsibility – unworkable paradox or sustainable paradigm? *Proceedings of the ICE, Engineering Sustainability*, Issue ES3, September, 129–130.

Barthorpe, S., James, R. and Taylor, S. (2004) Corporate social responsibility: An imperative or imposition upon the UK construction industry? Toronto: CIB World Congress.

Bowen, H.R. (1953) *Social Responsibilities of the Businessman*, New York: Harper & Row.

Bowen , P., Akintola, A., Robert, P. and Edwards, P. (2007) Ethical behaviour in the South African construction industry, *Construction Management and Economics*, 25: 631–664.

Brundtland, H. (1987) *Our Common Future: A Report of the World Commission on Environment and Development*, World Commission on Environment and Development, 1987. Published as Annex to General Assembly document A/42/427, http://www.un-documents.net/wced-ocf.htm (accessed 08/06/08).

Brown, G. (2004) Cited in, Department of Trade and Industry (2004) *Corporate Social Responsibility: A Government Update*. Online. Available at: www.csr.gov.uk/pdf/dti_csr_final.pdf (accessed 24 July 2007), p. 2.

Building (2007) OFT may use criminal powers in cartels investigation, issue 48, http://www.building.co.uk/story.asp?sectioncode=29&storycode=31012549 (accessed 8 June 2008).

Business in the Community (2003) *The Business Case for Corporate Responsibility*. Online. Available at: www.bitc.org.uk/resources/publications/cr_business_case. html (accessed 11 November 2006).

Business in the Community (2006) *Looking Back Moving Forward: Building the Business Case for Environmental Improvement*. Online. Available at: www.bitc. org.uk/resources/publications/looking_back_moving.html (accessed 05 June 2007).

Carroll, A.B. (1999) Corporate social responsibility: The evolution of a definitional construct, *Business and Society*, 38: 268–295.

CEE Bankwatch and FOEE (2006) *EU Funds in Central and Eastern Europe: 6 Billion Euros for Damaging Projects*. Online. Available at: www.bankwatch.org/ newsroom/release.shtml?x=1601568 (accessed 12 March 2006).

Chartered Institute of Building (2007) *Corporate Social Responsibility and Construction*, CIOB Information and Guidance Series. Online. Available at: http://support.freecpd.net/support_data/1/Reports/csr.pdf (accessed 13 June 2007).

Construction Industry Research and Information Association (2003) *Social Responsibility and the Business of Building*, London: CIRIA.

Constructing Excellence (2004) *Corporate Social Responsibility Fact Sheet*. Online. Available at: www.constructingexcellence.org.uk/resourcecentre/publications/docu ment.jsp?documentID = 116258 (accessed 10 October 2004).

Corporate Watch (2006) *What's Wrong with Corporate Social Responsibility*, Oxford: Corporate Watch.

Crowther, D. and Rayman-Bacchus, L. (eds) (2004) *Perspectives on Corporate Social Responsibility*, England: Ashgate Publishing Limited.

Crowther, D. (2004) Corporate social reporting: Genuine action or window dressing, in Crowther, D. and Rayman-Bacchus, L. (eds), *Perspectives on Corporate Social Responsibility*, England: Ashgate Publishing Limited, 140–160.

Csr.gov.UK (2007) *The UK Government Gateway to Corporate Social Responsibility*. Online. Available at: www.csr.gov.uk (accessed 12 August 2007).

Curwell, S. and Cooper, I. (1998) The implications of urban sustainability, *Building Research and Information*, 26 (1): 17–28.

Defra (2007) *Consultation on Site Waste Management Plans for the Construction Industry*, Department for Environment Food and Rural Affairs. Online. Available at: www.defra.gov.uk/corporate/consult/construction-sitewaste/consultation.pdf (accessed 10 October 2007).

Department of Trade and Industry (2006) *Review of Sustainable Construction*.

Ford, C.R. (2007) Telford's highland roads – a new way of life for Scotland, *Proceedings of ICE, Civil Engineering*, 160 May: 36–42.

Forest Stewardship Council (2007) *Because Forests Matter*. Online. Available at: www.fsc.org/en/ (accessed 10 February 2007).

Fox, E. (2000) *Ethics and the Built Environment*, Routledge, Oxon.

Glass, J. and Simmonds, M. (2007) Considerate construction: case studies of current practice, *Engineering Construction and Architectural Management*, 14 (2): 131–149.

Guy, S. and Farmer, G. (2000) Contested constructions: The competing logics of green buildings, in Fox, W. (ed.), *The Ethics of the Built Environment*, Oxon: Routledge, 73–87.

Hamil, S. (1993) Building on ethics: Ove Arup and corporate social responsibility, in *Good Business? Case Studies in Corporate Social Responsibility*, Bristol: SAUS Publications,

Hampton, D. (2003) *Common Sense Reality and the Built Environment*, Authentic Business Articles. Online. Available at: http://www.authenticbusiness.co.uk/archive/buildenviron/ (accessed 17 June 2007).

Handy, C. (2003) What's a Business For? In Harvard Business Review (HBR) (2003) *Harvard Business Review on Corporate Responsibility*. Boston MA: Harvard Business School Press, 65–82.

Harvard Business Review (HBR) (2003) *Harvard Business Review on Corporate Responsibility*. Boston, MA: Harvard Business School Press.

Herridge, J. (2003) The potential benefits of corporate social responsibility in the construction industry, *CIOB Construction Information Quarterly*, 5 (3): 12–16.

Hillebrandt, P.M., Cannon, J. and Lansley, P. (1995) *The Construction Company In and Out of Recession*, UK: Palgrave Macmillan.

HMSO (1998) *Public Interest Disclosure Act 1998*, Office of Public Sector Information. Online. Available at: http://www.opsi.gov.uk/acts/acts1998/19980023.htm (accessed 10 October 2007).

International Federation of Building and Wood Workers (2004) *Improving Working and Living Conditions in Construction: Addressing Needs through International Labour Standards in World Bank Procurement*, http://siteresources.worldbank.org/INTABCDESLO2007/Resources/Paper2.pdf (accessed 9 July 2008).

International Standards Organisation (2007) *ISO 2600 Social Responsibility Information*. Online. Available at: http://isotc.iso.org/livelink/livelink/fetch (accessed 30 October 2007).

Jones, P., Comfort, D. and Hillier, D. (2006) Corporate social responsibility and the UK construction industry, *Journal of Corporate Real Estate*, 8 (3): 134–150.

Jonker, J., Cramer, J. and Van Der Heijden, A. (2004) *Developing Meaning in Action: (Re)constructing the Process of Embedding Corporate Social Responsibility (CSR) in Companies*, ICCSR Research Paper Series, no. 16. Online. Available at: http://www.nottingham.ac.uk/business/iccsr/pdf/researchpdfs/16-2004.pdf (accessed 9 January 2007).

Kelly, J. and Hunter, K. (2005) Never in the field of construction procurement has so much been won by so few – an analysis of PFI in Scotland, *Proceedings of the Queensland University of Technology Research Week International Conference*, 4–8 July 2005, Brisbane, Australia. Online. Available at: www.rics.org/NR/rdonlyres/DA5D8B2A-2B43-4936-B86F-03ED98D22F78/0/ Never_in_the_field20051124.pdf (accessed 12 December 2006).

Latham, M. (1993) Trust and money, *Interim Report of the Joint Review of Procurement and Contractual Arrangements in the UK Construction Industry*, London: HMSO.

Leão-Aguiar, L., Ferreira, E.A.M. and Marinho, M.M.O. (2006) Integrating CSR, ethics and sustainable development principles into the construction industry, *CIB W107 Construction in Developing Countries International Symposium 'Construction in Developing Economies: New Issues and Challenges'*, 18–20 January 2006, Santiago, Chile. Online. Available at: www.irbdirekt.de/daten/iconda/CIB1927.pdf (accessed 15 June 2007).

Liu, A., Fellows, R. and Ng, J. (2004) Surveyors perspectives on ethics in organisational culture, engineering, *Construction and Architectural Management*, 11, 6: 438–444.

Margolis, D.M. and Walsh, J.P. (2001) *People and Profits? The Search for a Link Between a Company's Social and Financial Performance*, New Jersey: Lawrence Erlbaum.

Miller, R. and Lessard, D.R. (2000) *The Strategic Management of Large Engineering Projects: Shaping Institutions, Risks, and Governance*, Massachusetts Institute of Technology, USA.

Mott Macdonald (2002) *Review of Large Public Procurement in the UK*, A Report Commissioned by HM Treasury, London, UK.

Pearce, D. (2003a) The social and economic value of construction: the construction industry's contribution to sustainable development, *New Construction Research and Innovation Strategy Panel (nCrisp) Report*, London: nCrisp, http//www.cidb.co.zo/documents/knowledgecentre/sev-construction.pdf (accessed 8 June 2008).

Pearce, D. (2003b) Environment and business: Socially responsible but privately profitable, in Hirst, J., (ed.) *The Challenge of Change: Fifty Years of Business Economics*, London: Profile Books, 54–65.

Poon, J. (2004) An investigation of the differences in ethical perceptions amongst construction managers and their peers, *Proceedings of 12th ARCOM Conference*, 1–3 September 2004, Edinburgh: Heriot Watt University, 2: 985–993.

Porter, M.E. and Kramer, M.R. (2003) The competitive advantage of corporate philanthropy, in *Harvard Business Review* (HBR) *Harvard Business Review on Corporate Responsibility*, Boston, MA: Harvard Business School Press, 27–64.

Rameezdeen, R. (2007) Image of the construction industry, in M. Sexton, Kahkonen, K. and Lu, S.-L. (eds), *CIB Priority Theme – Revaluing Construction: A W065 Organisation and Management of Construction Perspective*, CIB Publication 313, May 2007: 76–87.

Sassi, P. (2004) Professional responsibility: sustainability, ethics and architectural practice, *Proceedings of Corporate Social Responsibility & Environmental Conference*, University of Nottingham, 28–29 June 2004; 264–274.

Skanska (2005) *Annual Report 2005*. Online. Available at: www.skanska.com/files/documents/investor_relations/2005/Skanska_Annual_Report_2005.pdf (accessed 30 October 2007).

Smith, C. (2003) The new corporate philanthropy, in *Harvard Business Review* (*HBR*) (2003) *Harvard Business Review on Corporate Responsibility*, Boston, MA: Harvard Business School Press, 157–188.

Suen, H., Cheung, S.O. and Mondejar, R. (2007) Managing ethical behaviour in construction organizations in Asia: how do the teachings of Confucianism, Taoism and Buddhism and globalization influence ethics management? *International Journal of Project Management*, 25: 257–265.

Tarmac (1995) *Tarmac in the Environment: First Report of the Independent Advisory Panel*, Wolverhampton, UK.

Tepper Marlin, A. and Tepper Marlin, J. (2003) A brief history of social reporting, *Business Respect*, 51, 9 March 2003. Online. Available at http://www.mallenbaker.net/csr/CSRfiles/page.php?Story_ID=857 (accessed in September 2007).

Unison (2002) *Employment and Corporate Social Responsibility: UNISON Scotland's Response to the Scottish Parliament's European Committee's Inquiry on Employment and Corporate Social Responsibility*. Online. Available at: http://www.unison-scotland.org.uk/response/empcsr.html (accessed 10 June 2007).

Vinten, G. (2004) Whistleblowing: The UK experience, Part 2, *Management Decision*, 42 (1): 139–151.

Ward, H. and Smith, C. (2006) *Corporate Social Responsibility at a Crossroads: Futures for CSR in the UK to 2015*, London: International Institute for Environment & Development.

Wilcox, J.R. and Theodore, L. (1998) *Engineering and Environmental Ethics: A Case Study Approach*, New York: John Wiley & Sons.

Wilkinson, S.J., Pinder, J. and Franks, A. (2004) Conceptual understanding of corporate social responsibility in the UK construction and property sectors, *16th CIB World Congress, Toronto, Building for the Future*. Online. Available at: www.irbdirekt.de/daten/iconda/CIB1764.pdf (accessed 20 June 2007).

Chapter 2

The evolution of corporate social responsibility in construction

Defining the parameters

Stuart Green

Introduction

The purpose of this chapter is to place current debates about corporate social responsibility (CSR) in a broader context. The adopted perspective is underpinned by a belief that concepts such as CSR are grounded in broader processes of social and political change. The chapter primarily focuses on the evolution of CSR within the specific context of the UK construction sector. Nevertheless, many of the trends described have global resonance; readers from other countries will therefore be able to make their own comparisons.

The concept of CSR was initially advocated in the 1950s in the aftermath of World War II. Memories of the Great Depression still loomed large in the public consciousness, and the emphasis lay on 'taming the excesses of business'. CSR then matured in accordance with the social idealism of the 1960s, deriving its academic legitimacy from systems thinking and stakeholder theory. The parameters of the debate were significantly reshaped subsequently as a result of the shift to the political right during the 1980s. The social-democratic consensus of the post-war Beveridge settlement was abandoned in favour of the enterprise culture. This shift in the prevailing political climate questioned the very existence of 'society' and emphasised the responsibility of firms to make a profit. The discourse of the enterprise culture had a significant and lasting effect on the construction sector. Trade union activists were blacklisted and government acted to incentivise self-employment. Even within the sphere of health and safety, there was an increasing reluctance to impose regulation on business lest it should impede entrepreneurial spirit. The enterprise culture further served to undermine long-established notions of professionalism in the construction sector in favour of managerialism and reliance upon instrumental tool kits. The change in the prevailing political climate initiated a wave of studies that sought to establish a link between CSR and profitability – most of which were inconclusive. It is against this background that the prevailing approach to CSR is reviewed, with particular emphasis on the guidance documentation published by the Department of Trade and Industry (DTI). Consideration is also given to recent government-sponsored reports specific to the construction sector and to the promotion of CSR through procurement.

The evolution of corporate social responsibility

Origins of the concept

In common with many other business-related concepts, there is no universally accepted definition of CSR. Interpretations differ across time and space. If current debates about CSR are to be understood, it is necessary to paint out the broad historical context. Early sources such as Clark (1939) and Kreps (1940) tended to refer to social responsibility (SR) rather than CSR, thereby reflecting an era that pre-dated the iconic status of the corporation. They further tended to emphasise the need for the 'social control of business'. Such ideas were undoubtedly a hangover from the Great Depression of the 1930s. Memories of widespread unemployment and the collapse of the stock market had undermined the doctrine of Adam Smith (1937). It was therefore no longer taken for granted that profit-maximising business units would inevitably benefit society at large.

The decline of *laissez-faire* in no small way rested on the experience of massive government intervention during World War II, which was instrumental in defeating fascism and overcoming the malaise of the Great Depression. The organisation of resources necessary for a successful war effort was clearly beyond the means of the private sector. The interventionist philosophy prevailed beyond the war by means of the Beveridge settlement in the United Kingdom and Roosevelt New Deal in the United States of America. The post-war social consensus on both sides of the Atlantic derived its theoretical support from Keynesian economics, which questioned the validity of the classical economic doctrine of *laissez-faire* and emphasised the need for government intervention.

The above provides the necessary background to understand the historical starting point of CSR in the 1950s. The modern concept of CSR is frequently credited to Bowen's (1953) seminal *Social Responsibilities of the Businessman*. Leaving aside the omission of businesswomen (who seemingly did not exist in 1953), Bowen's contribution was reflective of continuing social concern about the power of corporations and their intrusive effect upon the lives of individuals. This was a theme that continued throughout the 1960s and the 1970s as the literature on SR expanded dramatically. Davis (1960: 70–76) was influential in proposing his so-called 'Iron Law of Responsibility', which contended that the 'social responsibilities of businessmen [*sic*] should be commensurate with their social power'. McGuire (1963) further suggested that 'the corporation has not only economic and legal obligations but also certain responsibilities to society that extend beyond these obligations'. Such sources set the agenda for much of the subsequent debate about CSR, which centred on whether businesses should be encouraged to take their social responsibilities seriously, or whether they should be regulated.

The idealistic 1960s and the emergence of systems thinking

During the late 1960s 'systems thinking' came into vogue across a wide range of academic disciplines, and CSR was no exception. Numerous sources started to conceptualise businesses as open systems that engage in a dynamic interaction with the broader environment. According to Davis and Blomstrom (1966), the notion of social responsibility 'refers to a person's obligations to consider the effects of his [sic] decisions and actions on the whole social system'. Also emergent during this period was the notion of a 'stakeholder approach', whereby managers are seen to be responsible for balancing the needs of different interest groups. Johnson (1971), amongst others, contended that business takes place within a 'socio-cultural system', and suggested that:

> [a] socially responsible firm is one whose managerial staff balances a multiplicity of interests. Instead of striving only for larger profits for its stockholders, a responsible enterprise also takes into account employees, suppliers, dealers, local communities, and the nation.
>
> (Johnson 1971)

Systems thinking, and the need to adapt to the prevailing social environment, underpinned much of the prevailing management literature during the late 1960s and the early 1970s (cf. Berrien 1976; Burns and Stalker 1961; Rice 1963). Lawrence and Lorsch (1967) were especially influential in arguing that organisations engage in a dynamic interaction with their broader environment. Kast and Rosenzweig (1970) pulled much of this thinking together to focus on the relationship between businesses and the societal environment within which they operate:

> [o]rganizations are subsystems of a broader suprasystem – the environment. They have identifiable but permeable boundaries that separate them from their environment. They receive inputs across these boundaries, transform them, and return outputs. As society becomes more and more complex and dynamic, organisations need to devote increasing attention to environmental forces.
>
> (Kast and Rosenzweig 1970)

At the time, the above message resonated with the prevailing social idealism. Managers were not expected to confine their attention to narrowly construed notions of profit maximisation. The previously dominant premise of 'rational economic man' was seen to be based on 'unrealistic, closed-systems' assumptions. In academic circles, to label someone as a 'closed-systems thinker' became a rhetorical put-down. Closed-systems thinking was considered unsustainable in the long term, and was commonly associated with the optimisation algorithms of Taylorism. Both were seen to share the same myopic perspective.

Stakeholder theory

The 1970s saw the development of stakeholder theory. During the pre-war highpoint of economic liberalism, corporations were responsible solely to their shareholders. However, the emergence of CSR was underpinned by the development of stakeholder theory (Freeman 1984; Freeman and Reed 1983: 88–106). Stakeholders were seen to be groups or individuals who have a stake in the activities of the corporation, i.e. they provide support, and in some way, the corporation is seen to be responsible for their welfare. Freeman and Reed (1983: 88–106) defined stakeholders in a broad and a narrow sense. In the wide sense, stakeholders were seen as 'any identifiable group or individual who can affect the achievement of an organisation's objectives or who is affected by the achievement of an organisation's objectives'. In the narrow sense, they saw stakeholders as those groups or individuals that the organisation depends upon for its continued survival. This distinction is key to understanding the ongoing debate about CSR. The narrow definition of stakeholders is directly compatible with an organisation's economic self-interest, although it does serve to remind managers that any short-term exploitation of stakeholders may not be in the organisation's long-term interests. In this respect, the issue is one of enlightened self-interest. The wider definition of stakeholders resonates more with the idealism of systems theory. But such ideas have always been difficult to sell to practising managers who, even in the 1970s, found their lives increasingly dominated by short-term performance measures. What is clear is that any debate about CSR must take into account the dynamics of the broader socio-political environment.

A shift to the right

As the 1970s drew to a close, the wider conceptualisation of stakeholders continued to lose ground in the adopted discourse of practising managers. This gradual shift in emphasis was in no small way a reflection of the changing political climate in the Anglo-American world. The Keynesian economic doctrine that had prevailed after World War II progressively gave way to a revised form of neo-liberalism proposed by the followers of the American economist Milton Friedman. In the United States of America, Ronald Reagan advocated the benefits of 'trickle-down' economics, whilst Margaret Thatcher in the United Kingdom infamously proposed that there 'is no such thing as society'. And if there is no such thing as society, CSR will inevitably struggle to find a place within the business lexicon.

The enterprise culture

The changing political landscape presented a direct challenge to the social-democratic consensus of the post-war era. The declared task of the UK government throughout the 1980s was to reenergise Britain by encouraging the development of an 'enterprise culture' (Legge 1995). The Conservative Government elected in 1979 sought to abandon the corporatism of the Beveridge settlement in favour of the neo-liberal doctrine of trickle-down economics. Whilst the rhetoric was consistently harsher than applied policy, the government sought to dismantle the social structures that underpinned the welfare state. Specific targets included high levels of taxation, militant trade unionism, high levels of welfare benefit and the regulation of business practices by the state. The advocated policies sought to extend the domain of the free market throughout the economy in the cause of competition. Competitiveness in the global economy became the new mantra. Key policy dimensions included privatisation, deregulation, reduction of trade union power, removal of council houses from public ownership and the lowering of direct taxes. The prevailing political climate emphasised the 'survival of the fittest'. In consequence, the government was much less willing to rescue 'lame ducks' than had been the case in previous decades. State intervention was seen to impede economic efficiency and undermine the enterprise culture by promoting welfare dependency. The decimation of the traditional manufacturing base and the associated rise in unemployment were seen to be a price worth paying in the cause of competitiveness. The needs of big business were considered paramount in the belief that the benefits would 'trickle down' to other levels of society. There was also a marked growth in part-time and temporary employment with a widespread reduction of employment protection. The accepted euphemism for such trends was the 'flexible economy'. The combination of policy, legislation and rhetoric has characterised the 'enterprise discourse' (du Gay and Salaman 1992: 615–633; Keat and Abercombie 1991). In this sense, the notion of discourse includes a complex web of ideas, linguistic expressions, policies, social institutions and material practices. And the discourse of the enterprise culture had a significant and lasting impact on the UK economy and society at large. It shaped business education and practice to such an extent that it became accepted 'common sense'.

Globalisation

It is, of course, easy to demonise Margaret Thatcher, but it must be remembered that she was supported by a sustained electoral coalition that enabled her to achieve three successive electoral victories. Many continue to argue that she imposed a much-needed modernisation on a structurally misaligned economy. It must further be recalled that Keynesianism had already been abandoned by the preceding Labour government of James Callaghan at the bidding of the International Monetary Fund (Gray 1998). The winds of change were already blowing, and Thatcher's policies were in no small way driven by external forces. The shift to the political right was in part precipitated by the onset of globalisation, driven by time–space compression owing to technological change (cf. Harvey 1989: 188–209). The managed economy of the post-war period would almost certainly have expired irrespective of the political opportunism of Margaret Thatcher. The Thatcher years were also instrumental in the transformation of the British Labour Party, which progressively abandoned its socialist heritage in favour of 'New Labour' managerialism. The epochal event was the 1984 Miners' Strike and the defeat of the National Union of Mineworkers (NUM); the face of British industrial relations was changed forever. It is especially notable that the New Labour government elected in 1997 failed to repeal any of the industrial relations legislation their predecessors had so roundly condemned. Indeed, the New Labour government embraced the discourse of the enterprise culture and perhaps even extended it to new heights.

Social responsibility condemned

The shift in the prevailing political climate caused many to question the legitimacy of CSR. During an era when economic externalities were of minor concern, it was inevitable that notions of corporate responsibility became marginalised. The dominant discourse focused on the need to reduce regulation with the aim of encouraging entrepreneurial activity. Indeed, Milton Friedman, the guru of the espoused economic theory of the Thatcher government, famously condemned the 'doctrine of social responsibility':

> few trends could so thoroughly undermine the very foundations of our free society as the acceptance by corporate officials of a social responsibility other than to make as much money for their stockholders as possible.
>
> (Friedman 1962)

Friedman further argued that if managers used corporate resources for any cause other than profit maximisation, it would constitute a form of theft (Snell 1999: 507–526). Whilst such views may appear extreme, they must be understood within the context of the cold war and the associated excesses of McCarthyism. For Friedman, CSR seemingly constituted a concession in the direction of communism. At the time it was by no means unusual for American commentators to conflate economic policy with notions of individual freedom. The implication was that American executives not only had a contractual obligation to serve the interests of the firm's owners, but also a democratic duty to maximise profit in the face of the 'red peril' of communism.

The enterprise culture in construction

The discourse of the enterprise culture had a fundamental impact on the UK construction industry. From the mid-1970s to the current era the sector has experienced significant structural change. The discourse of the enterprise culture has further acted to transform the self-identities of the workforce. The most obvious manifestation of the enterprise culture was the active encouragement of self-employment.

Trouble and strife

The commencement of the enterprise culture in the construction sector can be traced back to the national building strike of 1972, which was successful in securing higher wages for many building operatives. However, the strike left a lasting resentment and many employers felt bitter at having been 'held to ransom'. This legacy of bitterness led to widespread blacklisting of 'militant' trade unionists through a private enterprise blacklisting organisation known as the 'Economic League' (Hollingsworth and Tremayne 1989). The league was originally established in 1919 to fight 'bolshevism' and remained shrouded in secrecy until subjected to scrutiny by a House of Commons Select Committee in 1989. It frequently intervened in British industrial relations throughout the 1970s and the 1980s, before being eventually disbanded in 1994. The League attracted numerous complaints that it illegally held information on individuals who had no means of redressal (Hencke 2000). It is of significance that the largest single group of workers on the Economic League's register comprised those in the construction industry. The majority of the major contracting firms subscribed to the Economic League, and used its services for screening potential employees. Individuals found themselves blacklisted even for entirely legal involvement in trade union affairs. Concerns about civil liberties did not prevent employers from colluding in the blacklisting of active trade unionists.

Economic efficiency

The end of the 1970s also saw a major political campaign to open up the local authority Direct Labour Organisations (DLOs) to private sector competition. DLOs were originally created at the end of the nineteenth century to offset private sector problems of profiteering through collusion, price rings, labour casualisation and inadequate training provision (Langford 1982). In 1980, the newly elected Conservative government introduced legislation that made significant changes in the way DLOs were organised and accounted for their activities (Kirkham and Loft 2000). The legislation required significant changes to accounting practices with a view to emphasise economic efficiency over social responsibilities. The social responsibilities of DLOs were no longer to stand in the way of the more narrowly defined economic efficiency. This legislation was the forerunner of much subsequent legislation that introduced 'market testing' into the delivery of public services. At their peak, DLOs employed in excess of 200,000 building workers. However, by 1995, their role, at least in new construction, was virtually over (Harvey 2003: 188–209). The demise of the DLOs demonstrated that the social costs of labour casualisation and inadequate training had not been allowed to stand in the way of economic efficiency.

Incentivisation of self-employment

Notwithstanding the activities of the Economic League and the erosion of the DLOs, the most important aspect of the manifestation of the enterprise culture on the construction sector lay with the growth of self-employment. The period 1980–1995 witnessed the institutionalised incentivisation of self-employment through the tax and insurance system. In consequence, the percentage of self-employed operatives grew from under 30 per cent in 1980 to over 60 per cent in 1995 (Harvey 2001; ILO 2001). The partial reimposition of barriers to self-employment since 1997 has seen some success in shifting 185,000–210,000 workers back into direct employment (Harvey 2003: 188–209). The institutionalised incentivisation of self-employment is, of course, only part of the story. The discourse of the enterprise culture embraced 'Essex Man'[1], who willingly traded his trade union membership card for a mobile phone or a 714 certificate. The growth of self-employment also undermined the demand for the blacklisting services of the Economic League, which ultimately became a victim of its own success. Self-employed operatives were rarely trade union members and were not subject to the same degree of employment protection as directly employed workers. Hence, concerns about employing 'trouble makers' were progressively alleviated by changing patterns of employment. Winch (1998: 531–42) also links the growth of self-employment to the dominant strategy amongst construction firms of adopting flexibility as the key means of achieving competitive advantage. The need to adapt in the face of fluctuating demand cycles remains central to the industry recipe of the construction sector, with an associated reliance on outsourcing and labour-only subcontracting. Such trends saw the retreat of most major contractors from direct involvement in training, the implications of which were exacerbated by the decline of the DLOs. An inevitable consequence was the decline of the apprentice system and an increasing reliance on migrant workers supplied through agencies. Collectively, such trends add up to a freeloading mentality whereby UK construction firms largely rely on others to invest in the skills of the industry's workforce.

Health and safety

A further relevant theme concerns the interplay between the discourse of the enterprise culture and the construction sector's health and safety record. During the years of the Social Contract (1974–1979), Government and organised labour combined to place health and safety at the centre of the political agenda, although the Confederation of British Industry (CBI) continued to lobby against 'restrictive' legislation (Beck and Woolfson 2000: 35–49). The discourse of the enterprise culture railed against government regulation of business practices, preferring to emphasise the importance of self-regulation. This applied to health and safety issues alongside all other business practices. Indeed, immediately upon assuming power in 1979, the Thatcher government instructed the Health and Safety Commission (HSC) to consider the 'economic implications' of any proposed new regulations (Dawson *et al.* 1988: 35–49). Two subsequent white papers, 'Lifting the Burden' (Department of the Environment 1985) and 'Building Businesses... Not Barriers' (Department of Employment 1986), further emphasised the need for a reduction in the regulatory burden on business (Beck and Woolfson 2000: 35–49). Countless other policy initiatives sought to ensure that health and safety inspectors were sympathetic to the needs of business. Morale within the Health and Safety Executive (HSE) was further undermined by 15 million pounds' worth of budget cuts instigated between 1994 and 1996. The desire to ensure that business is not burdened by over-regulation continued to prevail after the election of the Labour Government in 1997. As was the case with industrial relations legislation, there was little desire to impose measures that could be construed as 'anti-business'. The parameters of the debate on CSR had seemingly changed forever. Regulation to control the 'excesses of business' became unthinkable; the ideological climate had shifted in such a way that governmental 'encouragement' of CSR became the only acceptable policy direction.

Notwithstanding the above, it is difficult to prove any causal connection between the growth of self-employment and the industry's safety record. Statistics regarding health and safety in the construction industry are notoriously unreliable as a result of systemic under-reporting (Gyi *et al.* 1999: 197–204). The reporting rate for reportable (non-fatal) injuries has been estimated to be as low as 40 per cent (HSC 2003). Reporting rates are especially low amongst the self-employed who have little incentive to comply with statutory procedures. Furthermore, there is a general consensus that prevalent subcontracting arrangements have negative implications for health and safety as a result of blurred responsibility demarcations (Gyi *et al.* 1999: 197–204; Haslam *et al.* 2005: 401–415). Even more starkly, Clarke (2003: 40–57) argues that the possibility of a positive safety culture is seriously undermined by workplaces that comprise a reduced number of permanent employees supplemented by contract and contingent workers. The high percentage of non-standard forms of employment in the construction sector therefore stands as a direct barrier to an improved safety culture. But

Government, of course, would not want to implement restrictive legislation that would impede the enterprise culture. It is further notable that the offence of 'corporate manslaughter', first mooted by the Law Commission in 1996, is yet to reach the statute book despite repeated promises.

The decline of professionalism

The final issue that deserves attention is the association between CSR and the concept of professionalism. The accepted orthodoxy throughout the 1960s and the 1970s implied that there is a close relationship between ethical behaviour, social responsiveness and the notion of professionalism (cf. Kast and Rosenzweig 1985). However, professionals also suffered at the hands of the advocates of the enterprise culture. Professional groups were repeatedly cast in the role of self-serving monopolies with detrimental outcomes for the economy and society (Flynn 2002: 18–36). Paradoxically, professionals also came under attack during the 1970s from those on the political left, who contended that they disempowered citizens whilst facilitating social control on behalf of elite groups. Notwithstanding harsh criticisms from both extremes of the political spectrum, it is difficult to establish any universal definition of a 'professional'. It is similarly difficult to distinguish professional associations from other occupational groups.

Professionals are perhaps distinguished by the expectation that they should have an underpinning body of knowledge that is accumulated through a prolonged period of training. Entry qualifications are frequently governed by professional associations, which also establish unique social configurations specific to individual professions (cf. Vollmer and Mills 1966). This specialised knowledge accords professionals with some degree of authority in the eyes of their clients; however, the exercise of their authority remains subject to broad social sanction. Professionals are also frequently governed by codes of ethics, thereby making an explicit connection between CSR and professional firms.

Professionals – and professional firms – have always been strong in the construction sector, especially within the domains of architecture and engineering. The self-identities of practising architects and engineers have long since been shaped by notions of serving society at large. Professional associations have also had an important role in maintaining the distinction between professionals who take account of broader societal concerns and those who engage in commercial trade for the purposes of profit alone. Whilst the distinction was initially forged within the parameters of the British class system (cf. Bowley 1966), the battle lines have been more recently drawn between the conflicting doctrines of professionalism and managerialism (Exworthy and Halford 2002; Fournier 2000). Of particular concern is the extent to which creeping managerialism involves the rationalisation and codification of professional knowledge. The proliferation of 'key performance indicators' arguably acts to remove professional discretion from decision-making.

Furthermore, the systemic instrumental focus on prescribed targets tends to eradicate actions that lack any specific short-term economic justification. The decline of professionalism should therefore be of direct concern to those who advocate CSR.

Corporate social responsibility reconstituted

Shifting terrain

The shift to the political right in the United States and the United Kingdom during the 1980s initiated a gradual reconstitution of CSR. The evolving political climate obliged authors such as Carroll (1979: 497–505) to stress that the primary responsibility of businesses is to produce goods and services that society wants and to sell them at a profit. Profits were seen to be the natural reward for organisational efficiency and effectiveness. Management gurus such as Druker (1984: 53–63) also entered the arena to reemphasise the importance of business making a profit. According to Druker, the primary purpose of social responsibility is

> to tame the dragon, that is to turn a social problem into economic opportunity and economic benefit, into productive capacity, into human competence, into well-paid jobs, and into wealth.
>
> (Druker 1984: 53–63)

Society is seemingly something that has 'problems', whereas business is something that has 'economic opportunities'. Such a construction reflects directly the spirit of the enterprise culture.

An important contribution to the definitional debate was provided by Carroll:

> the social responsibility of business encompasses the economic, legal, ethical and discretionary expectations that society has of organizations at a given point of time.
>
> (Carroll 1979: 497–505)

The above definition attributes four parts to social responsibility, as illustrated in Figure 2.1. The argument is that businesses have a responsibility to make a profit, to obey the law and then to 'go beyond' these activities. The ethical responsibilities relate to what society expects over and above the requirements of the law. The final category of responsibility in Carroll's model relates to the discretionary responsibilities. In this case, the expectations of society are often unclear. Much is left to the judgment of individual managers and company directors and their desire to engage in social issues beyond the expectations of ethical business behaviour. Examples would include sponsorship of local sporting clubs and community programmes,

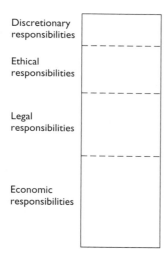

Figure 2.1   Social responsibility categories.
Source: Carroll, 1979.

although it should be noted that activities motivated by a 'business case' in terms of marketing benefits would not qualify.

Aupperle *et al.* (1985: 446–463) subsequently revisited Carroll's (1979: 497–505) four-part taxonomy: relabelled the 'economic' as 'concern for economic performance' and grouped together the 'legal, ethical and discretionary' as 'concern for society'. But the prevailing tendency was to argue that a company's first and foremost 'social responsibility' was to ensure wealth creation through improved economic performance. Few with memories of the stagflation of the 1970s would argue with the importance of wealth creation; but if carried through to their extreme, such arguments accord too easily with the 'greed is good' philosophy immortalised by Gordon Gekko in the 1987 film *Wall Street*.

Linking CSR to profitability

The evolving political environment provided the context within which several studies sought to establish an empirical link between social responsibility and profitability (Aupperle *et al.* 1985: 446–463; Cochran and Wood 1984: 42–56). The quest to prove the business case for CSR was a theme that continued throughout the 1990s with further studies by Balabanis *et al.* (1998: 25–44), Griffin and Mahon (1997: 5–31), McWilliams and Siegal (2001: 117–127) and Ruf *et al.* (1998: 119–133). Despite such efforts, the evidence in favour of a causal link between CSR and financial performance remained doggedly inconclusive. Though the claim that CSR is 'good for business' has obvious rhetorical appeal, it is difficult to find convincing supporting evidence in the research literature.

Despite this lack of supporting evidence, Reeves (2003: 33) argues that the 'need for a business case' in support of CSR has become part of the new orthodoxy. He contends that issues such as 'diversity' and 'respect for people' can no longer be advocated on the basis that they are worthwhile in their own right; they must be supported by a business case. Certainly the expression 'business case' has crystallised into a new dogma. Thirty years ago, supporters of the free market were often amused by the requirement for Soviet planners to state how their policies were compatible with Marxist/Leninist doctrine. But it now seems that thinking in the West is similarly constrained to conform with the omnipotent 'business case'. It must further be recognised that exhortations in support of the 'business case' are invariably based on the organisation's current model of operation. Rarely is the case made on the basis of how the business might operate in ten years. The orthodoxy of insisting on a business case is therefore deeply conservative and inevitably discourages innovation. Beyond the strictures of the need for a business case lies justification founded on enlightened self-interest.

Enlightened self-interest

The notion of enlightened self-interest reflects the idea that the socially responsive actions may not be justified in respect of short-term profit, but are nevertheless held to be in the long-term interests of the firm (Kast and Rosenzweig 1985). Such arguments take CSR beyond the realm of operational managers into the domain of corporate strategy. One commonly cited strategic benefit to adopting voluntary socially responsible policies is that it offsets the likelihood of imposed regulation. From this perspective adherence to social standards such as Social Accountability 8000 and European Corporate Sustainability Framework can be regarded as a strategy for mitigating risk (cf. Rosthnor 2000: 9–19; van Marrewijk and Hardjono 2003: 121–132).

Perhaps the most obvious justification for the development of a CSR policy lies within the domain of marketing. Customers and employees increasingly expect organisations to possess CSR policies, and such policies are often issued as addendums to formal company reports. It is notable that the majority of UK contractors possess formal CSR statements. Good corporate citizenship can impart a significant benefit to an organisation's reputation, thereby generating increased turnover. Other benefits may accrue through better recruitment and improved access to capital (Carroll 1999). However, in common with any other espoused strategy, an instrumental link between CSR policies and operational practice cannot be taken for granted. But nor should it be assumed that such practices comprise cynical marketing practices and nothing more. It cannot be argued that the discourse of the enterprise culture has had a significant impact on business practice, without conceding the same possibility to the discourse of CSR. In both cases, generalisations regarding predictable and uniform responses are likely to be confounded by localised constraints and complexities. Further, it is important to recognise that such discourses are likely to be continuously contested within the context of micro power struggles. However, the fragmented structure of the construction sector coupled with the continued tendency towards multi-tiered subcontracting can only serve to dissipate espoused CSR policies.

Voluntary standards and soft regulation

The main thrust of the current DTI policy is to encourage business to be socially responsible, primarily on the basis of enlightened self-interest. The approach is primarily characterised by a regime of 'soft regulation', whereby firms are encouraged to comply with voluntary social audits and standards.

Business and society

Notions of regulation clearly remain anathema within the corridors of the DTI. Their promotional materials resound with slogans about how 'in a world of change and competition the DTI is working to create the conditions for business success and help the UK respond to the challenge of globalisation'. There is certainly little hint of the need for the 'social control of business' in the way the DTI promotes CSR. The argument is that businesses already take CSR seriously as a major component of their 'intangible assets' (DTI 2001). The United Kingdom is held to be a global leader in the measurement and reporting of CSR activities. The DTI's emphasis lies in the promotion of the business case in support of CSR, despite the inconclusiveness of the supporting research. The argument in support of the business case seemingly rests on three foundations: (i) reputation and standing, (ii) competitiveness and (iii) risk management. In essence, it is argued that a failure to adhere to 'corporate citizenship' will damage a company's reputation amongst customers and potential employees. CSR is also seen as a risk-mitigation strategy to offset the chances of consumers boycotting a company's products.

Included amongst several exemplary case studies is Laing Partnership Housing, which has been commended for setting up a not-for-profit company to provide training for local people from disadvantaged backgrounds. If widely replicated, such initiatives may begin to redress years of neglect vis-à-vis training for the construction industry. The quoted 'business benefits' include additional publicity and extra business from clients who wish to partner companies who can also deliver effective social and economic programmes. Perhaps most importantly, the initiative promotes a discourse that companies thrive through close community relations. Certainly, within social housing, such local engagement would seem to be an essential condition for operating within the sector. Indeed, without such engagement, it could hardly be described as 'social' housing. Whilst not wishing to denigrate such initiatives, which hold much promise within local contexts, it is suggested that they must be set against the broader and changing context of housing provision within the United Kingdom, namely, the withdrawal of the state from its role as a provider of mass social housing. Government encouragement of such local initiatives must therefore be tempered by the bigger picture. The case remains that social housing is much less 'social' than it was 25 years ago. The big picture is that the state has retreated from what was once seen to be one of its core responsibilities, and training issues that were once central components of public procurement policy are now left to the private sector (provided the 'business case' can be demonstrated). Similar arguments can be applied to the demise of the DLOs.

Winning with integrity

Many of the DTI's (2001) arguments are drawn from 'Winning with Integrity', the report of the Business Interest Task Force established by the Prince of Wales (Business in the Community 2000). 'Winning with Integrity' argues that social responsibility can help business succeed in three areas: building sales, building the workforce and building trust in the company as a whole. The report further exhorts,

> [i]t's vital to appreciate that this is all about delivering benefits to both business and society. It's a win-win, creating value for everybody. This will always involve setting priorities among competing demands. There will be trade-offs and dilemmas. But the key point is that behaving in a socially responsible way is not only the right thing to do but makes good business sense, for both large and small businesses.
>
> (Business in the Community 2000)

Clearly, to argue that CSR is the 'right thing to do' is not in itself a sufficiently compelling argument. The endorsement of the report by the Prince of Wales, notorious for his woolly-headedness, arguably does little to encourage business leaders to take the proffered advice seriously. Kim Howells, UK Government Minister for Consumer and Corporate Affairs, was at least able to derive some legitimacy from the fact that he was democratically elected – which is more than what can be said for the Prince of Wales. Milton Friedman would no doubt be relieved that the government has retreated from any orientation towards regulation, but he would also undoubtedly be amused to see government departments offering advice to business. Business leaders are unlikely to credit the DTI with any more expertise in ensuring 'bottom-line' profitability than they do with Prince Charles. It has also become increasingly fashionable to question the continued need for the DTI. In 2004 the

Conservative Party unveiled plans 'to cut the DTI down to size' as part of its campaign to slash back Whitehall's bloated bureaucracy. They estimated their proposals would save taxpayers £750 million a year. Such a course of action would presumably result in the promotion of CSR being entirely outsourced to industry task forces under the umbrella of quangos such as 'Business in the Community'. And, of course, business leaders are especially pleased to serve if they get to meet such luminaries as the Prince of Wales. Though they may not value his advice on business, they do undoubtedly enjoy garden parties at the Palace and the resultant opportunities to fraternise with the higher echelons of British society.

Matters of principle

The message of 'soft regulation' is further reinforced by the DTI in their *Corporate Social Responsibility Report* of 2002 (DTI 2002). The foreword by Douglas Alexander, then Minister of State for E-commerce and Competitiveness, opined that 'many businesses accept the need for responsible behaviour as a matter of principle; but they also report that CSR can help build brand value, foster customer loyalty, motivate their staff, and contribute to a good reputation among a wide range of stakeholders'. The policy of soft regulation was subsequently made especially clear by Stephen Timms, who succeeded Alexander as the minister responsible for CSR in 2002:

> I am well aware of the many and increasing calls for more regulation of company behaviour. And I agree that Government has a responsibility to ensure minimum legal standards. I remain convinced that the main focus of CSR should be a voluntary one.
>
> (DTI 2004)

The argument promoted by the DTI is therefore at least superficially consistent; CSR is the 'right thing to do', and it brings business advantages. But none of the benefits listed would seem to extend beyond Carroll's (1979: 268–295) category of 'economic responsibilities', i.e. they can all be justified on the basis of enlightened self-interest. Alexander's foreword to the *Corporate Social Responsibility Report* of 2002 emphasises the need to be realistic:

> [T]o be sustainable in a competitive world, CSR needs to have a genuine economic foundation. It must assist, not compromise, performance; and it needs to be guided by experience, not dogma.
>
> (DTI 2002)

And, of course, if 'doing the right thing' clashes with business interests (as it often does), it is especially important to be pragmatic rather than dogmatic.

It is also necessary to be reminded that research seeking to demonstrate a link between CSR and economic performance remains stubbornly inconclusive. Precisely what constitutes 'dogma' also deserves some degree of discussion. In the absence of any evidence to support the link between performance and CSR, it is surely dogmatic to continue to claim that it does. Any Keynesian economist from the 1960s would be immediately struck by the narrow dogma of the 'business case'. It is further necessary to question whether businesses can accept the need for CSR as a 'matter of principle'. The claim seems to conflate two units of analysis. Individuals are certainly capable of holding points of principle, and many managers and CEOs in the construction industry undoubtedly accord to strong personal codes of moral and ethical behaviour. But to contend that businesses are capable of acting in accordance with matters of principle, or will 'do the right thing', is much more contentious. And though individuals may strive to act in accordance with deeply held principles, their actions are inevitably mediated by the context within which they operate. The difference between individuals and the businesses for which they work is well illustrated by the following discussion, taken from John Steinbeck's novel *The Grapes of Wrath*, between representatives of a land-owning bank and the tenants they were trying to evict:

'We're sorry. It's not us. It's the monster. The bank isn't like a man.'

'Yes, but the bank is only made of men.'

'No you're wrong there – quite wrong there. The bank is something else than men. It happens that every man in the bank hates what the bank does, and yet the bank does it. The bank is something more than men. I tell you. It's the monster. Men made it, but they can't control it.'
                    (Steinbeck 1939, quoted in Maclagan 1995: 159–177)

Herein lies the dilemma for those who seek to apply CSR as a point of principle. The point is that whilst 'men' [*sic*] may act on principle, this does not necessarily extend to businesses *per se*.

Echoes in a canyon

In addition to the juxtaposition of 'matters of principle' and the 'business case', the *Corporate Social Responsibility Report* (DTI 2002) echoes many other themes covered in the academic literature. The report follows Carroll (1979: 268–295) in suggesting businesses must first of all adhere to legal requirements relating to social obligations, such as equal opportunities and Health and Safety, as well as environmental requirements. But DTI (2002) also emphasises that CSR goes beyond legal compliance. A responsible organisation is claimed to do three things:

1. It recognises that its activities have a wider impact on society.
2. It takes account of the economic, social, environment and human rights impact of its activities across the world.
3. It seeks to achieve benefits by working in partnership with other groups and organisations.

But the overriding tendency is to downgrade the previous focus on 'matters of principle' in favour of 'real business benefits'. The ethical and discretionary responsibilities included in Carroll's (1979: 268–295) model are noticeably absent.

The *Corporate Social Responsibility Report* (DTI 2002) is also notable for its international outlook and the plea for international codes that support responsible behaviour. Support is pledged for the International Labour Organisation's (ILO) 1998 Declaration on Fundamental Principles and Rights at Work. Support is also promised for the Guidelines for Multinational Enterprises produced by the Organisation for Economic Cooperation and Development (OECD). The emphasis is very much on encouraging businesses to comply with voluntary standards of behaviour. Soft regulation remains the dominant philosophy. Certainly there is little appetite for 'taming the excesses of international capitalism' amongst Western governments. Such views would now be considered extreme, and outdated. Only anti-capitalist demonstrators are expected to voice such opinions, and these are seemingly accorded the same social status (and policing techniques) as football hooligans.

The UK Government is, of course, wise enough to avoid the moral high ground when preaching CSR in the international arena. Upon election in 1997, the current New Labour administration pledged its commitment to an 'ethical foreign policy'. This commitment was soon jettisoned in the face of international *realpolitik* and the importance of the global arms trade. The Blair government is also likely to be judged harshly for its 'ethical' commitment to Iraq and its associated human rights record.

However, DTI policy on CSR comprises (slightly) more than encouraging words. The government apparently recognises that fiscal incentives may be necessary to encourage social engagement, but they are careful to warn of the dangers of creating 'artificial dependencies that ultimately do more harm than good'. Certainly, there is little danger of supporting economic lame ducks in the cause of CSR. It is further interesting that the DTI (2002) cites payroll giving as the most obvious example of fiscal backing for responsible behaviour. The emphasis of the scheme seems to lie on enabling employees to make pre-tax donations to charity. As such, it seems to be encouraging *employee* social responsibility rather than stimulating *corporate* social responsibility. In either case, the sums are tiny when compared with the sums generated by the National Lottery, which is increasingly funding core government services through encouraging low-income groups to participate in gambling. Once again, the ethical high ground is a difficult place to occupy. Nevertheless, such comments should not deter corporates from promising (albeit embryonic) initiatives such as the Community Investment Tax Credit and Community Development Venture Fund.

Rethinking construction and beyond

Up until this point the discussion has focused on defining the parameters within which the CSR debate is conducted. Emphasis has been given to the evolution of the concept over time, and the way that it has been influenced by the prevailing economic-political doctrine. The discussion has also included the enactment of the enterprise culture in the construction sector and the associated structural adjustments over the last 30 years. But the picture would be incomplete without considering the impact of the pivotal *Egan Report, Rethinking Construction* (DETR 1998), together with its follow-up *Accelerating Change* (Strategic Forum 2002). Also of relevance are the associated *Respect for People* initiative (Rethinking Construction 2002) and the most recent report to be chaired by Sir John Egan, *Skills for Sustainable Communities* (ODPM 2004). Consideration will further extend to the increasing promotion of CSR through procurement policy.

The Eganites

Since its publication in 1998, the *Egan Report* has dominated the debate on improvement for the construction sector. The primary focus concerned the application of 'lean thinking', as derived from Toyota Manufacturing System. The *Egan Report* comprised the recommendations of the Construction Task Force (CTF) appointed by the Deputy Prime Minister, John Prescott, upon the election of the New Labour Government in 1997. The CTF was tasked with improving quality and efficiency in the construction sector and was notable for its emphasis on annual improvement targets. The report lacks any explicit references to CSR, which had clearly not impinged itself upon the thinking of the authors. There was no inclusion of any environmental targets, and the only target with a conceivable social dimension was one relating to a 20 per cent reduction in the number of reportable accidents, annually. But even here the inference was that a reduction in accidents would be accompanied by 'consequent reductions in project costs'. There was certainly no appeal to the moral imperative to reduce accidents, or to the need to speed up the moribund legislation on corporate manslaughter. It seems even the target of reducing accidents must be accompanied by the omnipotent business case.

With its ruthless focus on performance measurement, *Rethinking Construction* represents the epitome of managerialism. The authors display a touching faith in instrumental improvement recipes such as lean thinking, partnering, value management and benchmarking. It is notable that all of these are seen in terms of their contribution to efficiency; there is little that strays beyond the short-term imperatives of the business case. The *Egan Report* is further notable for promoting the cause of standardisation and pre-assembly. Once again, possible 'solutions' are evaluated in terms of their contribution to the narrowly defined efficiency. There is no recognition of the extent to which pre-assembly deskills local communities. *Rethinking Construction* is notable especially for its emphasis on meeting client needs, thereby reflecting the dominant discourse of customer responsiveness. Such an emphasis is directly reflective of the enterprise culture and acts further to undermine long-established notions of professionalism.

Had the *Egan Report* sunk quickly into obscurity, along the lines of many other government-sponsored reports, its neglect of CSR would have been of only passing concern. However, the *Egan Report* had a significant impact on the discourse of industry improvement. Many clients continue to evaluate contractors on the extent to which they are 'Egan compliant'. The *Rethinking Construction* report also spawned the Movement for Innovation (M4i) which sought to raise awareness of the need for change in the construction industry.

In summary, the *Egan Report* did much to promote a managerialist discourse of improvement within the construction industry. Not only was CSR so notably ignored, but it was also crowded out from the lexicon of industry improvement. To argue in favour of CSR would not have been 'Egan compliant'. The impact of *Rethinking Construction* is summed up by Murray (2003: 178–195) as the 'Eganisation of construction'. Murray further describes how the disciples of *Rethinking Construction* became know as 'Eganites', displaying many attributes of an evangelical cult. The *Egan Report* popularised the concept of Key Performance Indicators (KPIs), thereby reinforcing the industry's pre-dilection for instrumental management techniques.

Accelerating Change

Accelerating Change (Strategic Forum 2002) sought to build on and reaffirm the principles of *Rethinking Construction*. The Strategic Forum report was once again published under the chairmanship of Sir John Egan and followed similar lines of argument. The main thrust of the report was the extent to which the targets for improvement outlined in *Rethinking Construction* were being achieved in practice. *Accelerating Change* therefore reconfirmed the culture of performance measurement, justified on the basis of, 'if you do not measure how can you demonstrate improvement?'. The seven original targets were expanded to twelve with the addition of client satisfaction measures, thereby exacerbating the primacy of the client amongst stakeholders. There are brief references to the M4i's *Respect for People* initiative, as well as to the proposed Environmental Performance Indicators and Design Quality Indicators. There is also a fleeting reference to a 'triple bottom line' case study thereby hinting briefly that firms have social and environmental responsibilities, as well as economic ones. However, the report also emphasises the importance of 'proving and selling the business case for change' – this argument is applied to clients, integrated supply chains and 'respect for people' issues. There is seemingly only one basis upon which to justify action: the omnipotent business case.

Accelerating culture change in people issues

Accelerating Change (Strategic Forum 2002) devotes an entire chapter to 'people issues'. The authors express a determination to reverse the long-term decline in the industry's ability to attract and retain a quality workforce. The discussion once again lacks any explicit reference to CSR. The dominant instrumentalism is indicated by the way the workforce is described as the industry's most valuable 'asset': as if it were something to be owned and exploited. However, in stark contrast to other parts of the report, the argument noticeably strays beyond the boundaries of the immediate business case. For example, when addressing health and safety, practitioners are exhorted to be more aware of their responsibilities under the CDM regulations. Support is also pledged for the *Construction Skills Certification Scheme* (CSCS 2007), without feeling the need to emphasise the benefits in terms of short-term performance. Employers are further exhorted to address non-uniform pay and conditions in the sector. They are even asked to ensure that all operatives are embraced by the industry's new stakeholder pension scheme. However, it is noticeable that no KPIs are promoted to measure the extent to which firms are improving their performance against these criteria. Pleas are also made in support of *Investors in People* (IIP 2004) and greater workforce diversity. The underlying philosophy would seem to accord with that of enlightened self-interest. Such practices may not be justified in respect of short-term profit, but are nevertheless held to be in the long-term interests of the individual firm. Supportive comments are made even in respect of the role of professional institutions and the Construction Industry Training Board (CITB). The chapter on people issues goes as far as to criticise the informal construction economy and the associated high numbers of falsely self-employed operatives. However, the inference is that small and medium enterprises (SMEs) are to blame; strangely, there is no mention of government complicity in the systemic encouragement of self-employment. The report is on more familiar ground when suggesting that SMEs are not active in looking at 'people culture' issues because they do not see a good supporting business case. Given the broader socio-historical context, it would seem somewhat disingenuous to lay the blame at the feet of SMEs.

Respect for People

Accelerating Change was soon followed by the publication of a dedicated report by the *Respect for People Working Group* (Rethinking Construction

2002). The report claims to have established a clear business case for *Respect for People* in terms of improved performance for companies and projects. It also emphasises that *Respect for People* is not altruism.

The report draws its legitimacy from the trial of a tool kit in support of the espoused principles. The trial resulted in five consolidated themes: (i) equality and diversity, (ii) working environment and conditions, (iii) health, (iv) safety and (v) career development and life-long learning. Success amongst the trialists was attributed to four common elements, the first of which concerns the need to promote the business case. The argument here centres on the supposed relationship between better pay and conditions for employees and more satisfied clients. No recognition is made of the institutional barriers to the implementation of the required policies, or of the difficulty of establishing causality between implemented approaches and desired outcomes. Paradoxically, having previously claimed to have established a business case for *Respect for People*, the report subsequently backtracks to suggest that the business case to encourage the industry to improve performance in the 'people' area is at present inadequate. The reader is then directed towards unspecified studies in other sectors to develop the business case for construction.

The second element of the change process relates to the need to start measuring performance. Apparently there was a slow start in this respect on the trial projects, leading to the recommendation that organisations should consider the appointment of a 'champion'. The third element of the advocated change model relates to the need for integrated reporting systems, and the fourth relates to the need to support management teams involved in implementing the advocated change towards *Respect for People*. The report displays a level of naivety common to many prescriptive organisational change programmes, coupled with a failure to consider the institutional context within which managers operate. In summary, the *Respect for People* report contributes little to CSR. Whereas the chapter on people in *Accelerating Change* inches towards a position of enlightened self-interest, the subsequent full report retreats back to advocating instrumental tool kits. The response from industry was at best lukewarm, and tended in some quarters towards the hostile. Once again, the question is one of perceived legitimacy. Had the report focused more on the ethical arguments in support of progressive approaches to people management rather than agonise about the business case, it might have been better received.

Skills for Sustainable Communities

It has already been suggested that the Eganisation of construction has been detrimental to the cause of CSR. However, Egan's most recent report *Skills for Sustainable Communities* (ODPM 2004) breaks decisively with the ethos of *Rethinking Construction*.

The *Skills* report addresses the skills that are necessary to support the government's vision of *Sustainable Communities* (ODPM 2003), which receives a hearty endorsement by Sir John Egan – despite his previous obsession with the narrowly-framed efficiency improvement. Milton Friedman would undoubtedly be suspicious of Egan's unexpected conversion on the road to Damascus (and would perhaps have denounced him as a communist). All of a sudden, the discourse of globalisation is displaced by an emphasis on localisation. Sustainable communities are defined as those which

> meet the diverse needs of existing and future residents, their children and other users, contribute to a high quality of life and provide opportunity and choice. They achieve this in ways that make effective use of natural resources, enhance the environment, promote social cohesion and inclusion and strengthen economic prosperity.
>
> ODPM (2004)

The report goes on to express admiration for the visionary leadership of local authority chief executives. Developers are seen as key partners in the strategy to deliver sustainable communities, but they are exhorted to adopt a new approach:

> [r]ather than using the financial bottom line as an excuse to deliver mediocre design and mediocre building quality – numerous examples of isolated 'placeless' estates across the country are testament to how to do it wrong – developers must buy in to the common goal, commit to delivering high quality attractive places for people to live....
>
> (ODPM 2004)

This, indeed, is a significant deviation from the accepted script. It seems that the 'business case' is now just an excuse for inaction, rather than an essential pre-requisite. The inference here is that the financial bottom line should not stand in the way of 'doing the right thing'. Lean thinking, it seems, is not quite the panacea we were led to believe. Although the *Skills* report is not about CSR as such, it still provides comfort for those who wish to find signs of optimism for the future. But there is little evidence that the Eganites will fall into line beyond *Skills for Sustainable Communities* as they did with *Rethinking Construction*. Nevertheless, if the Deputy Prime Minister and Sir John Egan have recognised the limitations of narrowly defined efficiency, there is hope that construction firms may also reconnect with the communities within which they operate. But, in truth, the vast majority of local firms have always operated in close connectivity with their respective host communities, irrespective of the vicissitude of industry visionaries and the government.

Promoting CSR through procurement

Notwithstanding the DTI's continued commitment to soft regulation, there are several government departments that are actively promoting CSR through their construction procurement policies. Recent years have seen a plethora of reports urging public sector clients 'to drive sustainability through construction procurement' (e.g. DEFRA 2006; GCCP 2000; NAO 2005). It is further interesting to note an increasing tendency to conflate CSR with notions of sustainable development:

> Corporate Social Responsibility is about the behaviour of private sector organisations and their contribution to sustainable development goals. But the approach and values of good citizenship are also important to other organisations including Government.
>
> (DTI 2004)

So, the circle is finally squared. Far from seeking to regulate the 'excesses of business', government departments are now exhorted to emulate the enlightened practices of private sector firms. Certainly, the credibility of government in the eyes of the private sector can only be enhanced if it is seen to practice what it preaches. Most government departments have taken up the clarion call of CSR to some extent – at least in words, if not always in actions. But it is the Welsh Assembly Government and the Scottish Executive which have been at the forefront of promoting CSR through procurement. Both have generally adopted a more interventionist policy than the one that continues to emanate from Westminster. Local and regional government agencies are also acting to promote CSR, often with support from the regional development agencies. Public sector procurement policy remains fragmented in the absence of consistency across government departments. Nevertheless, the sum total of these initiatives undoubtedly represents a significant shift from the mechanistic focus on efficiency exemplified by *Rethinking Construction*. Of course, it is not difficult to find cynics. Many private sector firms forced to pre-qualify against CSR criteria see the process as a needless bureaucratic exercise. Ironically, this would be especially true for those firms which are already exemplars of CSR. Even more ironically, if CSR really does equate with enhanced business performance, then its promotion through procurement is eroding the competitive advantage of those firms that have already seen the light. But the unconvincing nature of arguments in support of CSR based on the 'business case' has already been demonstrated. The promotion of CSR undoubtedly makes more sense in terms of the 'sustainable communities' agenda. In this respect, policies of the decentralisation of procurement to a local and regional level are surely an essential pre-requisite. Yet, recent trends towards the policy of decentralisation of procurement have a long way to go before they overcome decades of increased direct control by the Treasury. The accumulated effects of 'enterprise' initiatives such as compulsory competitive tendering (CCT), 'best value' and public–private partnerships (PPP) are not so easily eradicated. And few would feel nostalgic about the stifling stagflation of the 1970s. The essential dilemma is how to reconcile the quest for local-ised sustainable communities with the cathartic effects of globalisation. For firms rooted in localised contexts, CSR is the very essence of a sustainable community. But arguments in favour of localisation consistently struggle to survive the competitive rigours of the global market place.

Conclusions

The origins of CSR can be traced back to the Great Depression of the 1930s and the subsequent correctives of Keynesian economics. It was against this background that debates around the CSR concept grew and developed throughout the 1960s and the 1970s. Theoretical justifications were provided by systems thinking and stakeholder theory, both of which resonated with the idealism of the time. However, as the 1970s drew to a close, the political climate on both sides of the Atlantic became much less supportive of CSR. The 'enterprise culture' was born from a combination of a shift to the political right coupled with the rapid acceleration of globalisa-tion. The new political climate served to discredit the consideration of social externalities that lay beyond the narrow confines of economic efficiency. Milton Friedman famously condemned social responsibility as a threat to free society.

The advent of the enterprise culture in the United Kingdom had a signific-ant impact on the construction sector. Many changes resulted from the bitter aftermath of the national building strike of 1972. Government and industry employers colluded in the incentivisation of self-employment, labour activ-ists were blacklisted by the Economic League and DLOs were opened to private competition in the cause of economic efficiency. Such courses of action led to an industry where the majority of operatives was notionally self-employed, with sustained adverse implications for health, safety and training. The discourse of the enterprise culture also presented a challenge to long-standing notions of professionalism. The changing political climate initiated the reconstitution of CSR, resulting in the requirement that it must be judged by the business case. Herein lies the essential paradox at the heart of current debates: CSR allegedly goes beyond the business case, and yet must be judged by the business case. Furthermore, research motivated by the need to link CSR to improved profitability has been stubbornly inconclusive. In consequence, enlightened self-interest became the dominant justification. Even if CSR cannot be proven to contribute to short-term profit, it can still seemingly be justified in terms of the long-term economic interests of the firm.

There is currently little intention on the part of the government to regu-late firms in terms of compliance with CSR. Such an approach would be anathema to the enterprise culture. DTI policy has therefore settled for a regime of 'soft regulation', whereby firms are encouraged to conform to voluntary standards. An ever-present caveat is that CSR in any case is good for business. Whilst social welfare is best achieved through unregulated enterprise, government retains the right to advise business on how to act in its own interest. Herein lies a second essential paradox at the heart of CSR. Yet, industry leaders understandably accord government agencies with little legitimacy in terms of their commercial acumen. In contrast to CSR, *Rethinking Construction* was enthusiastically endorsed by industry because it conformed to the established industry recipe. The message promoted instrumental improvement measures whilst reinforcing the industry's obsession with narrowly defined efficiency. In many ways, *Rethinking Construction* is the epitome of the enterprise culture. Any consideration of social externalities became marginalised from the discourse of industry improvement. The follow-up report, *Accelerating Change*, displayed a slightly more sympathetic view on some issues, but the subsequent *Respect for People* report was far too wedded to instrumental tool kits to offer any solace to CSR. The latter report agonised rather too much about the need for a proven business case. The overall message was that 'respect for people' is a good thing, provided that it contributes to profit.

But there are causes for optimism. There is undoubtedly a groundswell of interest within society at large concerning the need for sustainable development, and CSR fits well with the sustainability agenda. The recent focus on sustainable communities is to be welcomed, especially for the way in which it challenges the need for a narrowly constituted business case. The increasing use of procurement as a driver for CSR is also a positive development. This remains true even if at present there is little in the way of enforcement, once firms have completed the bureaucratic exercise of pre-qualification. Procurement mechanisms could further be geared towards localised CSR, if coupled with a greater commitment to the decentralisation of procurement policy. But such a trend would require a reversal of decades of increased centralised prescription by the Treasury.

As a final word, it is clear that the ascribed meaning of CSR has evolved over time and will continue to do so. Commitment to CSR will always be shaped and constrained by the prevailing political discourse. Furthermore, CSR will continue to be enacted differently in different contexts. Grand generalisations should therefore be treated with caution. But it is also necessary to recognise that the construction industry and society are not independent entities. Idealistic people who care deeply about broader society are no less likely to be found in construction firms than elsewhere. Such individuals are indeed capable of breaking out from the constraints of selfish vested interest to forge a more socially responsive collective future. In consequence, inspiration is more likely to be found in the unpublicised actions of real people embedded in localised contexts than in the grand exhortations of New Labour apparatchiks.

Postscript

On 28 June 2007, Gordon Brown finally replaced Tony Blair as Prime Minister of the United Kingdom. The Department of Trade and Industry (DTI) was immediately superseded by the Department for Business, Enterprise and Regulatory Reform (DBERR). The new department combined the functions of the DTI with those of the Better Regulation Executive (BRE), previously part of the cabinet office. The colocation of responsibilities for regulation and enterprise seems unlikely to reverse the established policy trends described in this chapter. The astute observer will notice that the Department for Business, Enterprise and Regulatory Reform sounds much more palatable than the Department for Business, Enterprise and Regulation.

Note

1. The expression 'Essex Man' is a political and social stereotype that came to prominence in the United Kingdom during the 1990s, in part, to explain the electoral successes of Margaret Thatcher. The term is indicative of the many working-class males in the South-East of England who deserted the Labour Party to vote Conservative. The expression is especially linked with the 'right-to-buy' policy whereby council house tenants became property owners. It is frequently used in a derogatory way to suggest somebody who is uncultured and motivated only by narrow economic self-interest.

References

Aupperle, K.E., Carroll, A.B. and Hatfield, J.D. (1985) An empirical investigation of the relationship between corporate social responsibility and profitability, *Academy of Management Journal*, 28: 446–463.

Balabanis, G., Phillips, H.C. and Lyall, J. (1998) Corporate social responsibility and economic performance in the top British companies: are they linked?, *European Business Review*, 98(1): 25–44.

Beck, M. and Woolfson, C. (2000) The regulation of health and safety in Britain: from old labour to new labour, *Industrial Relations Journal*, 31 (1): 35–49.

Berrien, F.K. (1976) A general systems approach to organization, in Dunnette, M.D. (ed.), *Handbook of Industrial and Organizational Psychology*, Chicago: Rand McNally College Publishing.

Bowen, H.R. (1953) *Social Responsibilities of the Businessman*, New York: Harper and Row.

Bowley, M.E.A. (1966) *The British Building Industry: Four Studies in Response and Resistance to Change*, Cambridge: Cambridge University Press.

Burns, T. and Stalker, G.M. (1961) *The Management of Innovation*, London: Tavistock Publications.

Business in the Community (2000) *Winning With Integrity*, Report of the Business Impact Project Team, London: Business in the Community.

Carroll, A.B. (1979) A three-dimensional conceptual model of corporate social performance, *Academy of Management Review*, 4: 497–505.

Carroll, A.B. (1999) Corporate social responsibility: evolution of a definitional construct, *Business and Society*, 38 (3): 268–295.

Clark, J.M. (1939) *Social Control of Business*, New York: McGraw-Hill.

Clarke, S. (2003) The contemporary workforce: implications for organisational safety culture, *Personnel Review*, 32 (1): 40–57.

Cochran, P.L. and Wood, R.A. (1984) Corporate social responsibility and financial performance, *Academy of Management Journal*, 27: 42–56.

Construction Skills Certification Scheme (2007) Scheme Booklet: Get Qualified, Get on, http://www.cscs.uk.com/upload_folder/downloadmaterials/scheme-booklet-may-07.pdf (accessed 4 June 2008).

Davis, K. (1960) Can business afford to ignore social responsibilities? *California Management Review*, 2: 70–76.

Davis, K. and Blomstrom, R.L. (1966) *Business and its Environment*, New York: McGraw-Hill.

Dawson, S., Willman, P., Bamford, M. and Clinton, A. (1988) *Safety at Work: The Limits of Self Regulation*, Cambridge: Cambridge University Press.

DETR (1998) *Rethinking Construction*, London: Department of the Environment, Transport and the Regions.

DEFRA (2006) *Procuring the Future*, Sustainable procurement national action plan: recommendations from the sustainable procurement task force, London: Department for Environment, Food and Rural Affairs.

Department of the Environment (1985) *Lifting the Burden*, Cmnd 9571, London: HMSO.

Department of Employment (1986) *Building Businesses... Not Barriers*, Cmnd. 9794, London: HMSO.

Druker, P.F. (1984) The new meaning of corporate social responsibility, *California Management Review*, 26: 53–63.

DTI (2001) *Business and Society: Developing Corporate Social Responsibility in the UK*, London: Department of Trade and Industry.

DTI (2002) *Business and Society: Corporate Social Responsibility Report 2002*, London: Department of Trade and Industry.

DTI (2004) *Corporate Social Responsibility: A Government Update*, London: Department of Trade and Industry.

du Gay, P. and Salaman, G. (1992) The cult(ure) of the customer, *Journal of Management Studies*, 29: 615–633.

Exworthy, M. and Halford, S. (eds) (2002) *Professionals and the New Managerialism in the Public Sector*, Buckingham: Open University Press.

Flynn, R. (2002) Managerialism, professionalism and quasi-markets, in Exworthy, M. and Halford, S. (eds), *Professionals and the New Managerialism in the Public Sector*, Buckingham: Open University Press, 18–36.

Fournier, V. (2000) Boundary work and the (un)making of the professions, in Mallin, N. (ed.), *Professionalism, Boundaries and the Workplace*, London: Routledge.

Freeman, R.E. (1984) *Strategic Management: A Stakeholder Approach*, Boston: Pitman.

Freeman, R.E. and Reed, D.L. (1983) Stockholders and stakeholders: a new perspective on corporate governance, *California Management Review*, 25: 88–106.

Friedman, M. (1962) *Capitalism and Freedom*, University of Chicago Press, Chicago.

GCCP (2000) *Achieving Sustainability in Construction Procurement*, London: Government Construction Clients' Panel.

Gray, J. (1998) *False Dawn: The Delusions of Global Capitalism*, London: Granta.

Griffin, J.J. and Mahon, J.F. (1997) The corporate social performance and corporate financial performance debate: twenty-five years of incomparable research, *Business and Society*, 36: 5–31.

Gyi, D.E., Gibb, A.G.F. and Haslam, R.A. (1999) The quality of accident and health data in the construction industry: interviews with senior managers, *Construction Management and Economics*, 17: 197–204.

Harvey, D. (1989) *The Condition of Postmodernity: An Inquiry into the Origins of Cultural Change*, Oxford: Blackwell.

Harvey, M. (2001) *Undermining Construction*, London: Institute of Employment Rights.

Harvey, M. (2003) Privatization, fragmentation and inflexible flexibilization in the UK construction industry, in Bosch, G. and Philips, P. (eds), *Building Chaos: An International Comparison of Deregulation in the Construction Industry*, London: Routledge, 188–209.

Haslam, R.A., Hide, S.A., Gibb, A.G.F., Gyi, D.E., Pavitt, T., Atkinson, S. and Duff, A.R. (2005) Contributing factors in construction accidents, *Applied Ergonomics*, 36: 401–415.

Health and Safety Commission (HSC) (2003) *Health and Safety Statistics Highlights 2002/03*, Sudbury, Suffolk: HSE Books.

Hencke, D. (2000) Left blacklist man joins euro fight. *The Guardian*, 9 September.

Hollingsworth, M. and Tremayne, C. (1989) *The Economic League: The Silent McCarthyism*, London: National Council for Civil Liberties.

IIP (2004) Unlock your organization's potential: an overview of the standard framework, http://www.investorsinpeople.co.uk/Documents/IIP_StandardOverview1.pdf (accessed 8 June 2008).

ILO (2001) *The Construction Industry in the Twenty-First Century: Its Image, Employment Prospects and Skill Requirements*, Geneva: International Labour Office.

Johnson, H.L. (1971) *Business in Contemporary Society: Framework and Issues*, Belmont, CA: Wadsworth.

Kast, F.E. and Rosenzweig, J.E. (1970, 1985) *Organization and Management: A Systems and Contingency Approach*, 4th edn, New York: McGraw-Hill.

Keat, R. and Abercombie, N. (1991) *Enterprise Culture*, London: Routledge.

Kirkham, L.M. and Loft, A. (2000) Accounting and the governance of the public sector: the case of local authority direct labour organisations in the UK in the 1970s, paper presented at *IPA Conference*, London.

Kreps, T.J. (1940) *Measurement of the Social Performance of Business*. US Temporary National Economic Committee, Investigation of Concentration of Economic Power (Monograph No. 7.), U.S. Government, Washington, DC.

Langford, D.A. (1982) *Direct Labour Organizations in the Construction Industry*. Aldershot: Gower.

Lawrence, P.R. and Lorsh, J.W. (1967) *Organization and Environment*, Cambridge, MA: Harvard Press.

Legge, K. (1995) *Human Resource Management: Rhetorics and Realities*, London: Macmillan.

McGuire, J.W. (1963) *Business and Society*, New York: McGraw-Hill.

Maclagan, P. (1995) Ethical thinking in organizations: implications for management education, *Management Learning*, 26 (2): 159–177.

McWilliams, A. and Siegal, D. (2001) Corporate social responsibility: a theory of the firm perspective, *Academy of Management Review*, 26 (1): 117–127.

Murray, M. (2003) Rethinking construction: the Egan report (1998), in Murray, M. and Langford, D. (eds), *Construction Reports 1944–89*, Oxford: Blackwell. 178–195.

NAO (2005) *Sustainable Procurement in Central Government*, London: National Audit Office.

ODPM (2003) *Sustainable Communities: Building for the Future*, London: Office of the Deputy Prime Minister.

ODPM (2004) *The Egan Review: Skills for Sustainable Communities*, London: Office of the Deputy Prime Minister.

Reeves, R. (2003) Reality bites, *Management Today*, April: 33.

Rethinking Construction (2002) *Respect for People: A Framework for Action*. The Report of the Rethinking Construction's *Respect for People Working Group*, London: Rethinking Construction.

Rice, A.K (1963) *The Enterprise and Its Environment*, London: Tavistock Publications.

Rosthnor, J. (2000) Business ethics auditing – more than a stakeholder's toy, *Journal of Business Ethics*, 27: 9–19.

Ruf, B.M., Muralidhar, K. and Paul, K. (1998) The development of a systematic, aggregate measure of corporate social performance, *Journal of Management*, 24 (1): 119–133.

Smith, A. (1937) *An Enquiry into the Nature and Causes of the Wealth of Nations*, Cannan, E. (ed.), New York: Modern Library. [first published 1776].

Snell, R.S. (1999) Obedience to Authority and Ethical Dilemmas in Hong Kong Companies, *Business Ethics Quarterly*, 9 (3): 507–526.

Strategic Forum (2002) *Accelerating Change*, London: Rethinking Construction.

van Marrewijk, M. and Hardjono, T.W. (2003) European corporate sustainability framework for managing complexity and corporate transformation, *Journal of Business Ethics*, 44: 121–132.

Vollmer, H.M. and Mills, D. (eds) (1966) *Professionalization*, Englewood Cliffs, NJ: Prentice-Hall.

Winch, G. (1998) The growth of self-employment in British construction, *Construction Management and Economics*, 16: 531–542.

Chapter 3

A business case for developing a corporate social responsibility policy

Louise Randles and Andrew Price

Introduction

The rationale of this chapter is to place the broader context of corporate social responsibility (CSR) within the business strategy of construction companies, thereby offering the relevance and efficacy of adopting CSR within the industry. The chapter highlights where CSR has been adopted in businesses outside of construction as well as within construction and what the successes and failures of this approach have been. The chapter concludes with an approach to adopting CSR within the construction industry.

The 1970s saw Friedman (Porter and Kramer 2003) argue the case for organisations to increase their profits rather than engage in social responsibility, and many took the same view. This view is not so widely held in business today, partly to avoid expensive court cases and protect brand image, but, perhaps more importantly, owing to increased organisational and individual social conscience. CSR relates to the role of business in society with the recognition that its impact is far wider than purely financial returns. The reason why a business adopts a socially responsible attitude is just as important as how it is achieved. It can be argued that some businesses have adopted CSR as a defensive measure; for instance, there are companies that have adopted CSR after their supply chain practices, such as unfair trading practice or the use of child labour, have been exposed to the public. There is no doubt that many companies have learned that labour rights abuses can cost them both financially and in terms of their reputation (Hilton and Gibbons 2002). Although this reactive approach may be the initial motivation, if the CSR ideals are fully incorporated into the business, the end result could well be the same as that for a business operating in a more proactive manner as in the case of The Body Shop (Hopkins 2003). Construction has traditionally been a rather conservative industry and slow to change; however, the world in which it currently operates is rapidly changing, and many construction organisations will have to change the way they operate, if they are to keep up with stakeholders' expectations. The embryonic beginnings of this change can be seen in the adoption of sustainable construction designs and methods, for example, The Eco Centre in South Tyneside (BSRIA 2000). The inclusion of waste management practices on larger sites and improved health and safety procedures are now prevalent throughout the industry.

The principles of sustainable construction have been well voiced within the construction industry with many companies already operating sustainable policies, however, much more needs to be done. The emphasis has so far been on environmental and economic considerations, which, although important, are only two-thirds of the 'triple bottom line'. The next logical step is to understand and fulfil the requirements of the social dimension of sustainability. Embracing CSR, which also includes economic and environmental considerations, needs strong emphasis on stakeholder interaction, as emphasised by Hopkins who provided the following definition of CSR:

> CSR is concerned with treating the stakeholders of the firm ethically or in a responsible manner.
>
> (Hopkins 2003: 1)

This requires consideration of internal and external stakeholders, taking into account impacts on society and the environment, a concept often lost in the frenetic business world. A more rounded definition of CSR is the one given by the World Business Council for Sustainable Development:

> Corporate Social Responsibility is the continuing commitment by business to behave ethically and contribute to economic development while improving the quality of life of the workforce and their families as well as of the local community and society at large.
>
> (World Business Council for Sustainable Development 2008)

Balabanis *et al.* (1998) concluded that economic performance is linked to CSR performance and disclosure, lending some weight to the business case for adopting CSR policies. Idowu and Towler (2004) noted that many companies have recognised the enormous benefits that can emanate from making their CSR policies and activities known. There are benefits to be reaped by adopting CSR policies within any business irrespective of whether they can be quantified in monetary terms or not. The remaining sections of this chapter describe how and why a business should develop a social conscience and what relevance this can have to the construction industry. The following section addresses corporate transparency and the thorny issue of what is important, taking into account the requirements of stakeholders and shareholders. This leads on to marketing, an area that construction is not usually associated with. Finally, employee and human rights are discussed, as this is increasingly being placed at the heart of good business practice. The chapter draws to a conclusion by summarising the business case for adopting CSR within the construction industry.

Developing a social conscience: business ethics

The phrase 'developing a social conscience' is probably not one readily associated with construction companies. However, the terms 'social responsibilities' and 'stakeholder awareness' are infiltrating the industry, and it is worth remembering that business ethics, whether they are perceived to be 'righteous' or not, apply to all businesses. The belief that businesses should not only behave according to certain moral standards, but also demonstrate a level of social responsibility and accountability to a range of stakeholders that stretches from employees and customers to suppliers and the wider community has risen up in the management agenda (Moon and Bonny 2001). But why has this happened? The increasingly litigious society is one reason, combined with a genuine desire to improve living and working conditions. The recent decision by international giant GAP to withdraw all 'made in India' clothes as the result of a sting operation that highlighted child labour within its supply chain (Kalra 2007) is an example of the prevailing attitude towards child labour in the wealthier countries. However, merely pulling out of a country does not ensure social responsibility. What is to become of those children? Is there a more palatable alternative being offered to ensure that they and their families can survive? Disassociation is not enough. Shareholder interest in a business's ethics and values is increasing, and more investment institutions are using ethical indices to support their decision-making processes. One incentive for adopting CSR policies is the growing evidence that CSR can help to reduce staff turnover, increase loyalty and raise brand awareness whilst minimising exposure to risks, legal and otherwise. This is achievable only if CSR is embodied wholly within the business and not merely employed as a risk minimisation technique.

So how does ethics correspond to real-life complexities in construction organisations? Having a code of ethics is not a panacea for developing an ethical company. McNamara (2008) noted that many ethicists believe it is the *developing* and *continuing dialogue* around the code's values that is most important. The code should be a living document regularly updated using reflection and dialogue, thereby creating ethical sensitivity and consensus. This may seem obvious, but many fail to recognise that it is the journey rather than the destination that often provides the greatest learning opportunity.

When addressing complex issues in what can be a stressful site environment, codes of practice, policy documents or value statements are often good starting points. However, the initial stages in the development of a suitable 'code of ethics' for construction companies are not so simple. In an industry that is mainly client-led (tender requirements/trends often dictate the pace of change within the industry), many organisations tend to wait for the client to demand change before taking the appropriate action. In CSR, this reactive approach can create problems with materiality/context, as the client will have a different set of stakeholders and hence a different set of priorities. The more forward-thinking companies have already begun to address the issue of CSR and are making significant progress in the development of a relevant but achievable policy. Companies such as HBG that have striven to incorporate CSR within and in line with their core business are noted within the industry; HBG won the newly created *Construction News* quality awards in 2005, and judges at the 2006 awards, where again it was a finalist, commented that 'HBG has a thorough and systemic approach to CSR that few companies in their [sic] sector can match'. In 2006, HBG upgraded its policy statement by publishing and distributing it (HBG 2006). It is this systemic approach and a strong recognition by the company's directors that the company has social as well as financial accountabilities that have led to the development of CSR within the business. Companies that have not heard of CSR or that wish to improve their understanding of it should look to those that have begun to integrate CSR within their core business and also refer to the DTI-funded CSR Competency Framework (DTI 2005) which sets out five attainment levels: awareness, understanding, application, integration, and leadership. Most construction companies are currently approaching the awareness phase, although there are a few who have exceeded this; however, in comparison with other industries there are no strong CSR leaders in the sector yet. This leaves plenty of opportunity to be in at the start for any company looking to lead the way. An alternative to the DTI framework is indices proposed by Business in the Community (BITC), which has launched a learning network for CSR practitioners in December 2007, that aims to help practitioners learn how to measure and report on their responsible business practices (BITC 2008). There are four key impact areas: responsible business strategy, workplace, environment and community. Both the DTI and BITC aim to encourage businesses to look outside of their core business and reflect on the interactions with other stakeholders and then incorporate improvements into their business strategy to improve performance and outcomes for all.

Corporate transparency and materiality

The complex issue of what to report is one being looked at by the UK Government with a desire to legislate for CSR, evident in the Private Member's Bill first introduced by Linda Perham MP. The Bill, intended to force larger companies (those with a turnover in excess of £50 million) to be open about their social, environmental and economic impact on the communities around them, was debated in the House of Commons; but like so many Bills it ran out of time (Armstrong 2004). The Bill was strongly opposed by David Varney, Chairman of mm02 and Business in the Community (an organisation of 700 member companies), who stated in his report *Motivate Don't Legislate*:

> We believe that the adoption of legislation as to what companies should do would be a retrograde step. To make it compulsory, for example in the

format which Linda Perham's Bill advocates, would require the redefinition of corporate responsibility to something many of our 700 member companies might fail to recognise. Compulsory corporate responsibility runs the risk, as with many regulatory outputs, of enforcing the lowest common denominator rather than promoting the best possible practice.

(Varney 2003)

This is an understandable viewpoint and one that many construction companies may share owing to the potential additional bureaucracy that is usually associated with new legislation. So what is the UK Government doing to promote CSR? There is a clear desire to *encourage* CSR, as demonstrated by the introduction of CSR ministers and the CSR Academy. The real change came when the UK government published draft regulations in the Operating and Financial Review (Ernst and Young 2005) for consultation. The proposals required publicly quoted companies to provide a narrative report setting out the company's business objectives, its strategy for achieving them, and the risk and uncertainties that might affect their achievement. It required companies to report on other matters where these are necessary for an understanding of the business, such as employee, environment, social and community issues (see HM Government 2008). The construction industry has an interest in all these areas: e.g. the building of the country's infrastructure, environmental degradation, waste minimisation and employee welfare. The responsibility of deciding what to disclose in the report will fall upon the directors, an incentive, if needed, for all directors to consider a CSR policy to address this. The report will need to place more emphasis on the outcomes and not just the 'measurements for assessment' which a lot of the rather lengthy corporate responsibility reports do at the moment. Yener (2002) stressed that the corporate reporting must be of a good quality, focusing on material matters for that particular industry and its stakeholders. The issue of 'better market information' has been taken up by the Institute of Chartered Accountants of New Zealand (2003) which advocated that good market information will promote the efficient functioning of markets through being:

- relevant (including timely)
- understandable
- reliable
- comparable (between entities and over time).

The concept of 'better market information' is essentially what the government is trying to achieve within the OFR, thus allowing stakeholders to make a more informed decision about whom they do business with. Crowther and Rayman-Bacchus (2004) view this lucidity as important, particularly to the external users of such information, as these users lack the background details and knowledge available to internal users, thereby allowing non-specialist investors greater transparency. In essence, reporting on social matters is about getting your business in order, to enable a business to earn and keep its license to operate. This is achieved by establishing who the key stakeholders in your business are, examining what is material to them and providing that information in a transparent manner.

The incentive to report on social matters for the moment primarily lies in becoming a better corporate citizen, which arguably could affect the bottom line for the better. Environmental reporting in the construction industry is much more common now than it was; however, the option of providing *regulatory relief* for companies that operate Environmental Management Systems (EMS) is one which is unlikely to be taken up after studies sponsored by the European Union International Institute for Sustainable Development (2007). Also, the Environment Agency have established that sites that have a certified EMS are not in any way more likely to comply with legal requirements than those that do not. The trend to adopt EMS and reporting is growing in spite of this, which gives some encouragement to take up social reporting voluntarily.

Profit and business ethics

The idea of ethics in business is certainly not new, but more and more businesses are facing the dilemma of how to achieve an ethical yet profitable business. There are many names frequently associated with the term 'ethical business', such as The Body Shop, Ben and Jerry's, B&Q and Innocence. These companies have taken ethical trading into the heart of their business, but mistakes have been made along the way that have affected profits. For example, Roddick (2004) in an interview with The New Strait Times openly admitted that The Body Shop had entered into a venture that involved establishing a greetings-card operation with a community that needed investment. However, the cards failed and did not sell in the primarily cosmetic-based shops. This created the dilemma of a community that had begun to rely on this trade and a product that was not selling. The lesson to be learnt from this is that ethics has a place in business, but should not put profits at risk. The best way to incorporate ethics is to look at the business's core operations and determine how to perform them in an ethical manner. Ethics should thus be a part of every decision made. Businesses need to be profitable for long-term survival in a competitive marketplace. However, if ethical trading is costing money, and if it is done voluntarily, a business should determine the value that it is adding and market that value. The simple equation presented below illustrates the cost differential (if there was one) that would need to be marketed.

$$\text{Added ethical value} = \text{Ethical product cost} - \text{Cost of equivalent non-ethical product}$$

It is this *added ethical value* that needs to be sold to the consumer as better quality or simply the feel-good factor of ethical purchasing with the added bonus of security of brand when the product falls within a supply chain.

This is a business practice that has worked well for the companies mentioned above with the exception of B&Q (2008) which has gone about it in a different way. B&Q has not gone down the road of highly publicising its environmentally and socially acceptable products, but has taken the long-term view of investing in communities while still making a profit. The full picture of B&Q and its social responsibility can be found at the B&Q website. In essence, B&Q began to look at the ethics of its business when someone asked the CEO whether B&Q bought its timber from sustainable sources. This spurred him on to learn more. The concept of sustainable-timber sourcing and investment in the communities surrounding that timber thus began to emerge. The fact that B&Q did not initially set out to become a socially responsible business makes it an interesting case. The CEO saw it as a good business practice that at the same time maintained a profitable business. The success of the B&Q brand and its success as a CSR pioneer highlight the fact that being socially responsible can go hand in hand with success and profits even for a mass-market operator. The construction industry is often remembered for its mistakes as much as for its successes: bridges that make pedestrians sick when they walk on them; and a massive cost overrun on the Eurotunnel rather than the many wonderful feats of engineering that were achieved. Moon and Hagan (2001) stated that it is easy to make the case that ethics has a powerful, practical and immediate impact on profitability, which creates a very different perspective compared with the sentiments of Friedman (Goodpaster and Matthews 2003) that businesses should seek to increase profits rather than engage in social responsibility. There are many opportunities to increase profits while engaging in social responsibility. This is done to the best effect by aligning a business's core values with its brand. This is also true for construction, which is part of the very fabric of society with many stakeholders. Its core values therefore have to be acceptable to a diverse audience, and the brand it portrays should reflect these values; for example, a utilities company that provides essential services such as clean water and wastewater management in the UK could have a core value of bringing clean water to countries in less developed countries. A brand image that reflects this core value would be a powerful marketing tool in all the countries where it operates. The prominence of 'ethical' companies through everyday products and services is growing, thus indicating the general public's desire for accountability within the marketplace, and construction is no exception to this.

An excellent example of a company that sees this added ethical value is Marks and Spencer (M&S). The 100 point 'Plan A' announced in January 2007 by M&S as part of its £200 million eco-plan is an example of a client leading the change towards a wholly ethical business that can provide traceability throughout its supply chain. The plan is set to aid M&S to become carbon neutral by 2012 by addressing issues under the five headings of climate change, raw materials, waste, fair partner and healthy eating (Greenbiz 2007), thus extending its influence beyond its stores and covering all its stakeholders including employees.

Clients increasingly expect suppliers to sign up to the CSR principles they have adopted as a prerequisite to winning or retaining a contract (Freshfields Bruckhaus Deringer 2006). Clients are looking for compliance and alignment with their CSR policies when assessing tender returns, and construction suppliers will have to show the same when preparing their tender returns.

The construction industry will have to take note of the need to prove traceability within its own supply chain as well as the responsibility towards its employees if it wants to do business with M&S and other socially conscious businesses, of which there are many.

Shareholders and stakeholders

Corporate governance is increasingly being used as a yardstick for company performance. Indices, such as the Dow Jones Sustainability Index (DJSI) or the FTSE4Good, help to make an assessment of a company's social and environmental policies to enable shareholders and stakeholders to make an informed decision when they consider their investments. In Spain, listed companies are required to disclose on the Internet information relating to their corporate governance, thereby reaching a wide stakeholder base. This method of disclosure is becoming increasingly popular in many countries and across a wide range of businesses. The need to address what are quite often conflicting requirements of shareholder and stakeholder groups has led to the development of several methods of measurement and indicators. One of these is the Global Reporting Initiative (GRI 2008) (see website http://www.globalreporting.org/Home). The generic indicators used within the GRI are a start, but can they really be all things to all people? Arguably, sets of industry sector-specific indicators are required. The new measures to evaluate performance have to be set within the context of a changing external environment (Crowther and Caliyurt 2004) whilst addressing the needs of the internal environment. The indicators used must be flexible enough to fit in with existing project management structures yet be meaningful to a diverse range of stakeholders. There are many CSR reports that read like public relations polemic, and construction must avoid this mistake, if it is to be taken seriously by its stakeholders.

Traditionally, construction has based its responsibilities and reporting procedures on operational and financial parameters. However, during the previous decade there has been a move towards a widening of these responsibilities with a growing consensus that society can collectively create more value if corporations invest in the broader development agenda (Shaughnessy 2004). The fact that corporations have become larger and more powerful has resulted in the role of the firm in society being reassessed, none more so than in construction. Margolis and Walsh (2001) consider that it is not simply a matter of having the role of the firm reassessed under the weight of rising social concerns, but rather, an emergence of an ever-shifting role from the reality of taking on new responsibilities. The globalisation of business means that firms may find themselves operating in countries with fewer welfare rights than within their own countries. It is becoming increasingly difficult to overlook this inequality and double standards when considering the overall ethical standards and international reputation of their business. Construction organisations need to assess their supply chains for compliance with the standards they report upon, while reassessing the supply chain as a value chain, and market their stakeholders appropriately.

CSR is not an optional 'add-on' to business core activities; it must be part of the way the business is managed for it to be meaningful. Integration within the company's ethos is essential to the success of any CSR policy as perceived by shareholders, directors, employees and often the wider community it serves. Kanter (2003) believes that investment by both parties (corporate and community) builds mutuality ensuring that the community partner will sustain the CSR activities when contributions from the business taper off. This is especially true for construction when it is involved in Public–Private Partnerships and social housing.

Marketing your stakeholders' CSR rhetoric?

Mention the word marketing, and it can create an image of the people who sit around a table creating ads to convince us that we cannot live without their products. So how do businesses convince stakeholders without sounding sycophantic? A business must be well prepared if it is considering becoming more socially responsible and wants to try something new and innovative without becoming the target of sceptics and journalists just waiting to report any misdemeanours. It is crucial to build trust by ensuring that the business has the ideals of CSR embedded as part of its business strategy and believes in the social value of what it is reporting. It must be honest in its achievements and not over-report them. There is nothing stakeholders dislike more than being misled and nothing competitors or reporters like more than exposing hypocrisy. Failure is imminent if the business is simply paying lip-service to CSR and has the outdated notion of CSR being a 'fluffy' concept for 'tree-huggers' that the business now has to put its name to, in order to comply with the OFR or because their competitors are doing so. If the board is not convinced that CSR is good for business, it should not practise it, because if it is practised half-heartedly more damage than good could be done to the business. However, if those in high places are serious about becoming better corporate citizens, they need to take a fresh look at the business and see where CSR could do more, whilst adding value to the business.

A long-term-investment view needs to be taken of CSR. There are risks, but they can be minimised by investing in a business's core strengths. In the construction industry the idea of growing a company's own workforce had been diminishing for a long time. The problem of taking this short-term view is clear, and the shortage of skilled trades throughout the industry is causing considerable concern. There are some in the industry striving to address this within their CSR policies and then marketing what they are doing to increase interest from both prospective recruits and clients.

Construction is a client-led industry; it is the clients who commission and initiate construction projects. The new Construction Design and Management (CDM) regulations have recognised this and put clients' responsibilities at their heart when influencing projects (see HSE 2007). Construction cannot profess to be the innovator of CSR; this title does not belong to any one sector; however, the construction sector can strive to be at the forefront of developing CSR alongside its clients. British Land has been working with others in the property and construction sectors to develop a set of procurement and CSR guidelines towards establishing a consistent CSR approach in the construction sector. This highlights the increasing shift towards the clients' interest in how a project is delivered rather than merely what is delivered, and it is this shift that will dictate what the industry follows. The thought that the industry can lead in the arena of CSR is one which the industry should strive for. However, the speed at which the industry adopts it will more likely be client-led.

The need to promote urbane marketing is becoming evident with both clients and recruits becoming more sophisticated in their expectations. The reputation of a company can be spread far and wide at the touch of a button; therefore, the marketing of CSR is increasingly important. An example of a successful idea is the 'Make Poverty History' wristbands becoming a universal success. They became the 'must have' accessory across a wide age and social range. However, there were problems within the supply chain, and some bracelets had allegedly been manufactured in Far Eastern sweatshops (McLean 2005). So what was a clever successful marketing idea to raise awareness of poverty has possibly turned into an embarrassing mistake! The need to go about marketing in a thorough and logical manner, with a rigorous investigation into one's own business methods and those of the essential global supply chain, is evident.

Employees and human rights

Employees create organisational performance and so should be one of the most important stakeholder groups. Yet few employees would consider this to be true, or is it just a lack of self-esteem? Doubtful. It is more likely that the adage 'customer is always right' has been so well used over the last couple of decades that it is now ingrained. This too is changing; the people who invest their intellectual property in a company also think of it as 'their' company to varying degrees, and it is this investment that will drive the change in employees' feeling from that of a company cost to one of a company asset.

So, can construction become a caring, sharing industry? The Human Rights Act and current skills shortage have prompted the managers and directors to remember who delivers their products, buildings and infrastructure. The identified requirements (needs) and the unidentified requirements (expectations) of stakeholders will have to be incorporated from the outset in every project. In a recent set of interviews undertaken within the civil engineering sector, it was highlighted that while remuneration was important, it was not what solely motivates employees (Randles *et al.* 2005). Moreover, there were many factors considered when an individual decided to stay with a company – including flexibility of hours and the degree to which his/her opinions were listened to or not. In this increasingly technological world, employees are replacing the traditional 'hard assets' of a company, as demonstrated by the number of construction companies that have turned into management providers rather than remaining as builders. The need to increase human capital and hold on to it to create social capital has never been greater.

Human capital refers to the properties of individuals and social capital to connections among individuals – social networks and the norms of reciprocity and trustworthiness that arise from them (Putnam 2000). Human capital can be invested in via education, training and medical provision, but to increase social capital the company must hold on to the human capital and create the type of networks where that human capital can flourish. Handy (2003) asks a fundamental question: whom and what is a business for? The terms of business have changed. Ownership has been replaced by investment, and company assets are increasingly found in its people, not in buildings and machinery. It is this trend that will drive the change towards more socially responsible business methods including the treatment of employees. Harnessing diversity and promoting a feeling of being valued within the workplace is good for business and implicit with regard to the human rights of employees. The modern professional employees have high expectations of their employer, and if these expectations are not met they simply move on.

The UK Government's White Paper *Fairness for All* focuses on building an inclusive society highlighting the fact that globalisation and migration are established features of modern Britain. By 2006, for the first time, there were more people aged 55–64 than in the age-group 16–24. Alongside this, 22 per cent of adults are covered by the Disability Discrimination Act, with many experiencing difficulty with the built environment (DTI 2005). These social and demographic changes mean that construction will have to adapt its recruitment and design priorities to take these into account, if it is to become part of an inclusive society.

Attracting new graduates

The decline in students' entering the construction industry in the past was hardly surprising given its image. *Bob the Builder* and the image of women in construction with sagging trousers and hard hats as shown in previous advertising campaigns did nothing to reverse the trend. Students have a vast choice of career paths and are generally way ahead of the industry in their views on social and environmental issues. They increasingly do not want to work in polluting industries or industries that have poor health and safety records. The construction industry is also one of the largest users of natural resources with a poor record of workforce diversity, and it needs to address these issues if it wants to attract new graduates into the industry. The demise of the maintenance of the rail infrastructure, widely publicised in the press, has done nothing for the brand image of several large construction firms, compounding the recruitment problems. Again, the damage done to companies like Nike was enormous when the non-governmental organisations (NGOs) publicised the sweatshop conditions in parts of their supply chain (Hilton and Gibbons 2002). The future is emerging of fair trade and balanced living in tune with the environment; if the construction industry cannot align itself with such concepts, the best students will find an industry that will. The traditional long hours and site conditions that the construction industry is trying to move away from in certain circles are still prevalent in others. The sophisticated student wants a work–life balance rather than having the 'earn all you can' mentality of the Thatcher years. The process of developing a CSR policy could help the industry discover a new marketing opportunity to attract new graduates. It could help reassure potential recruits that the industry is evolving into a credible, responsible career path for the brightest graduates to enter.

Projects and good business

Brooks and Thomas (2005) advocate that strategy and strategic management can be summed up as the pursuit of sustainable competitive advantage for organisations, and this is evident within the concept of CSR. The aim of CSR and strategic management to highlight the mission of the business and reflect upon its interaction with its stakeholders (advantages and disadvantages) is one aspect that makes CSR a strategic management issue and vice versa.

The propensity of organisations to produce mission statements and CSR policies is one step which can, if taken to the heart of the business strategy, provide a sustainable competitive advantage as well as improve working conditions for employees and improve projects for other stakeholders. An example of how the two interact is that most major construction organisations have health and safety as one of their major strategic management policies with the aim of reducing accidents at work and providing a healthier working environment. The fact that health and safety also features in many of their CSR policies is no surprise. The construction industry can effectively measure health and safety, and constructors H&S performance is a requirement of many tenders; therefore, by actively promoting health and safety through policies and then filtering this down operationally, CSR is actively affecting working practices. This is evident in the strict health and safety rules now in force on all major construction sites. Organisations can then use the fact that they have a good measurable record in this area, making it a competitive advantage within the market place. The wider context of projects and good business is one area where construction can use its experience to provide infrastructure in the public realm to improve the surroundings of the community. The involvement of that community in the design and build process will go some way in ensuring the success of that project. A recent example of this successful inclusive involvement is the Grove Village regeneration project in Manchester where the first housing PFI contract was won by a consortium that advocated local involvement alongside employment as a major part of their bid. The Grove Village project has proved successful, and this is in no small part due to the local involvement in the decision-making process. Grove Village won the Manchester Evening News property awards in 2006 for the best regeneration project. In a statement from Grove Village they attributed this success to:

> Putting the needs of the local community at the heart of the process has been a key driver for Grove Village. From the start tenants and residents have been involved in the new design of the estate through planning exercises and open days. Tenants continue to be consulted in the ongoing development of the estate.
>
> (Grove Village 2006)

It is this inclusivity that is at the heart of CSR and good business, and a living example of how strategic management and CSR can effectively influence operational outcomes.

Conclusion

To return to the beginning, Hopkins' (2003) definition of CSR, being concerned with treating the stakeholders ethically or in a responsible manner, gives some idea of what CSR means, but not how important CSR is to a modern business. The increasing sophistication of clients and future employees, not to mention shareholders and all those other stakeholders, makes it essential for businesses to discover what their social responsibilities are and live up to them. The success of B&Q and The Body Shop is evident; they have a clearly integrated CSR mission that has enabled them to become the sustainable businesses that they are today. They have, to some extent, protected themselves against negative publicity by making CSR an integral part of how they operate their business and have been transparent about their failings as well as their successes, giving them credibility. By ensuring that their CSR standards are continued throughout their supply chain, they minimise risks and maximise marketing opportunities.

The question is shifting away from 'does ethics or CSR affect the bottom line?' to 'will my bottom line be affected by not incorporating ethics or CSR?' To put a monetary value on CSR is difficult; how are ethical considerations measured? A high turnover of staff can be measured in loss of human capital with the subsequent loss of tacit knowledge; the same is true for customer retention. Currently, it is difficult to recruit and retain qualified/competent staff at both craft and professional levels (Dainty and Ison 2003). The minimisation of risks, whether they are social, environmental or simply loss of brand reputation, can all have an economic impact. Nobody expects construction to suddenly become philanthropic. There is, however, an obvious need to change the image of construction companies in the United Kingdom, if they are to attract and retain the best staff and customers. CSR could be a good place to start this sea change in operational culture and consequently in image (Randles, Price and Carrillo 2003). Whether it is motivated by the need of directors to comply with the OFR, the need to increase staff and customer retention or a genuine interest in becoming a better corporate citizen, CSR is here to stay. Awareness is growing, and leadership looms for those in the construction industry who adopt this very modern way of working.

The particular problems with the construction industry that may affect the pace of the adoption of CSR and subsequent integration are centered on the transient nature of the workforce and projects undertaken. The very fact that every project is different – whether that is due to the construction technique, design or simply the location – brings its own specific challenges. This can be mitigated somewhat by building long-term partnerships with suppliers and subcontractors. The use of frameworks is becoming popular with clients and contractors alike, as this affords the opportunity for the client to influence its partners while building long-term relationships. The contractor benefits from continuity of work, which helps to stabilise the company giving it the ability to forward-plan and incorporate long-term challenges such as the training of the next generation of employees. Although the popularity of procurement routes such as PFI/PPP and frameworks has increased in recent years, and this clearly has given the continuity required to implement long-term relationships and hence CSR opportunities, there will always be the one-off clients and projects. It is this sector of construction that has the greatest challenge when it comes to integrating CSR within its core business. The dilemma facing firms in this sector is not how they incorporate CSR within their business, but how they ensure they are not compromised by factors outside of their control, such as their fragmented supply chain. The need to constantly review the requirements of their stakeholders (who will change frequently) and ensure compliance with their CSR credentials may require a dedicated resource that could be seen as an ongoing cost that is not relevant to their business. Clearly, there are obstacles to overcome before CSR can become mainstream within the construction industry, and what works for one company may not necessarily work for another. Is doing nothing an option? Possibly. But the fact is that embracing CSR requires the detailed appraisal of the business; it may highlight areas of opportunity that can be exploited, and can itself be the reward for the expenditure or effort.

So how does a business become a better corporate citizen while increasing the sustainability of its business? Taking the attainment levels given in the CSR Competency Framework – awareness, understanding, application, integration and leadership – it is clear there is a lot of work to be done, if construction is to move through these levels towards leadership. The following steps should help to include CSR within the corporate mission and integrate it throughout the business.

- First, assess what level the company is at with regard to its internal and external CSR.
- Appoint a champion who is high enough up the ladder to make the necessary decisions and is passionate about how CSR can improve the business.
- Look at the stakeholder group and assess who the key stakeholders are, what information they require and how the business affects them.
- Canvass the opinions of local governments. Align the business objectives with this information.
- Provide the CSR information in a transparent manner that is accessible to all stakeholders and shareholders. (When this is right, it is time to reap the rewards and market the information in such a way as to add value to the business.)
- Do not forget to admit where improvements can be made and to show progression and intention of becoming a better corporate citizen.

The need to address internal issues first is one point often overlooked and needed to gain credibility before tackling the supply chain. This does not preclude working with the supply chain, as ongoing measures for improvement, while investigating what industry-specific indicators are needed.

The business case for adopting CSR is one of 'sustained competitiveness'. By integrating CSR throughout the business, it will become part of how a business conducts itself within the communities it serves, whether that is locally, nationally or globally. This will have a positive effect on staff recruitment and retention, increased loyalty from stakeholders and shareholders and the minimisation of risks from negative working practices whilst ensuring the business stays ahead of future legislation and competition.

References

Armstrong, M. (2004) The business of ethics. *Guardian Unlimited.* July 5, www.guardian.co.uk/business/2004/jul/05/ethicalbusiness.money (accessed 8 June 2008).

B&Q (2008) www.diy.com (accessed 16 June 2008).

Balabanis, G. *et al.* (1998) Corporate social responsibility and economic performance in the top British companies: are they linked? *European Business Review*, 98: 25–44.

BITC (2008) CR Index. http://www.bitc.org.uk/what_we_do/cr_index (accessed 6 June 2008).

Brooks, S. and Thomas, P. (2005) *CSR and Strategic Management: The Prospects for Converging Discourse,* Cambridge: Critical Management Studies Conference.

BSRIA (2000) *Sustainable Construction: The United Kingdom Viewpoint.* Report 13, UK: 41.

Chartered Accountants of New Zealand (2003) Improving corporate reporting: a shared responsibility chartered accountants of New Zealand, report for the minister of commerce http://www.nzica.com/AM/Template.cfm?Section=News_Files &Template=/CM/ContentDisplay.cfm&ContentID=7487 (accessed 6 June 2008).

Cowther, D. and Raymann-Baccus, L. (2004) Perspectives on corporate social responsibility. Aldershot, England: Ashgate.

Crowther, D. and Caliyurt, K. (2004) *Stakeholders & Social Responsibility,* Malaysia: Anstead Service Centre, 9.

Dainty, A. and Ison, S. (2003) *Constructing Nottinghamshire,* UK: CITB, 2.

DTI (2004) Fairness for all: a new commission for equality and human rights, DTI.

DTI (2005) The CSR Competency Framework, www.csracademy.org.uk/competency. htm (accessed 10 October 2007).

EMAS. http://www.bsdglobal.com/tools/systems_emas.asp (accessed 6 June 2008).

Ernst and Young (2005) Changes in corporate governance: key issues http://www.ey.com/global/download.nsf/International/ChangesInCorpGov.pdf/$ file/ChangesInCorpGov.pdf (accessed 6 June 2008).

Freshfields Bruckhaus Deringer (2006) *The Development and Impact of CSR in the Construction Industry,* Briefing Note.

Goodpaster, K. and Matthews, J. (2003) Can a corporation have a conscience? *Harvard Business Review on Corporate Social Responsibility*, USA: HBS Press, 141.

Greenbiz (2007) Marks & Spencer Launches 200 Million Pound Eco Plan, 16 January 2007. Online. Available at: http://www.greenbiz.com/news/news_third.cfm? NewsID=34442 (accessed 22 November 2007).

GRI (2008) The global reporting initiative http://www.globalreporting.org/Home (accessed 6 June 2008). International institute for sustainable development (2007). European eco-management and audit scheme.

Grove Village (2006) Grove village, Press release, 16th October 2008. http://www.grovevillage.co.uk/newsitem04.htm (accessed 16 June 2008).

Handy, C. (2003) What's a business for? *Harvard Business Review on Corporate Social Responsibility*, USA: HBS Press, 71–72.

HBG (2006) *Corporate Social Responsibility.* Online. Available at: http://www. hbgc.co.uk/documents/document_112.pdf (accessed 22 November 2007).

Health and Safety Executive (2007) Want construction work done safety, A quick guide for clients on the construction (design and management) regulations 2007, http://www.hse.gov.uk/pubns/indg411.pdf (accessed 8 June 2008).

Hilton, S. and Gibbons, G. (2002) *Good Business*, UK: Texere.

Hopkins, M. (2003) *The Planetary Bargain: Corporate Social Responsibility Matters*. London: Earthscan Publications Ltd.

Idowu, S. and Towler, B. (2004) A comparative study of the contents of corporate social responsibility reports of UK companies, *Management of Environmental Quality: An International Journal*, 15: 420–437.

Kalra, R. (2007) Counterview: child labour is only a lesser evil, *The Times of India*, 1 November 2007. Online. Available at: http://timesofindia.indiatimes.com/ Opinion/ Editorial/ Counterview_Child_labour_is_only_a_lesser_evil/articleshow/ 2506589. cms (accessed 22 November 2007).

Kanter, R. (2003) From spare change to real change: the social sector as beta site for business innovation, *Harvard Business Review on Corporate Social Responsibility*, USA: HBS Press, 200.

McLean, C. (2005) Where's your wristband?, timesonline, 19 June. www.timesonline. co.uk/tol/life_and_style/women/style/article5.31988.ece (accessed 8 June 2008).

McNamara, C. (2008). Complete guide to Ethics Management: an ethics toolkit for managers. http://www.managementhelp.org/ethics/ethxgde.htm (accessed 6 June 2008).

Margolis, J. and Walsh, J. (2001) *People and Profits?* USA: Lawrence Erlbaum Associates, Inc.

Moon, C. and Bonny, C. (2001) *Business Ethics*. London: The Economist Books.

Moon, C. and Hagan, J. (2001) *Business Ethics*. London: The Economist Books.

Porter, M. and Kramer, M. (2003) The competitive advantage of corporate philanthropy, *Harvard Business Review on Corporate Responsibility*, USA: HBS Press, 29.

Putnam, R. (2000) *Bowling Alone: The Collapse and Revival of American Community*, New York: Simon and Schuster.

Randles, L., Price, A. and Carrillo, P. (2003) The contribution of CSR to sustainable construction, *Social Responsibility*, Malaysia; Anstead Publications.

Randles, R., Price, A.D.F., and Carrillo, P. (2008) Towards a corporate social responsible construction sector. Internal Report, Leicestershire, UK: Loughborough.

Roddick, A. (2004) CSR Interview, Malaysia. *The New Straits Times*, 26 November.

Shaughnessy, H. (2004) Implementing CSR communications for business results. *Conference Report*, 24–25 February 2004, London: CSR Datanetworks.

UK Government (2008). The UK government gateway to corporate social responsibility. http://www.csr.gov.uk/ (accessed 6 June 2008).

Varney, D. (2003) Motivate don't legislate. BITC. Available from the world wide web: http://www.bitc.org.uk/resources/viewpoint/motivate.html (accessed 23 January 2003).

World Business Council for Sustainable Development (2008) Corporate social responsibility. http://commdev.org/section/tools/csr (accessed 6 June 2008).

Yener, D. (2002). Developing effective corporate governance practices: a brief overview of policy and practice. http://www.cipe.org/pdf/whatsnew/events/budaconf/ developing.pdf. (accessed 06 June 2008).

Part II

Impact of construction on communities

The role of construction and infrastructure development in the mitigation of poverty and disaster vulnerability in developing countries

Girma Zawdie and Mike Murray

Introduction

There has been a sea change in corporate business behaviour in recent years following the growing concern with such contemporary issues as climate change, world poverty and sustainable development. Global responsibility has now emerged as an objective at the heart of corporate investment decisions. In this respect, the significance of the role that construction activities play in the development process through the supply of sustainable infrastructure systems capable of responding to prevailing social, economic and environmental pressures is widely acknowledged.

Investment in construction constitutes about 50 per cent of all investments in capital goods in many countries; and in many developing countries, the construction industry constitutes about 4 per cent as against the 7 per cent average for developed countries (World Bank 1994). Whether growth in the share of the value of construction activities in total GDP (gross domestic product) is a cause or a consequence of economic growth is not clearly established. However, the underlying pattern of the relationship between GDP per capita and the share of construction in total GDP – with the latter first increasing at lower levels of development, then levelling off and finally declining at higher levels of development (see, for example, Bon and Crosthwaite 2000) – appears to suggest that investment for capacity expansion in construction is relatively more important for developing countries than it is for the developed ones. This is perhaps not surprising in view of the fact that growth in developed economies is largely based on knowledge-driven post-industrial activities, whereas most of the developing economies are largely agrarian and rural, and growth of these economies is subject to acute infrastructure constraint.

A major challenge for construction activities in developing countries – or any other industrial activities for that matter – is one of adjusting business behaviour to changing local and global circumstances. The problem is that construction activities in developing countries have often been organised and managed primarily with the aim to respond to short-term socio-economic and political exigencies. Consequently, there has been little or nothing by way of investment in sustainable physical infrastructure to provide the basis for long-term development. This has been exacerbated by the prevalence of capacity or knowledge deficit and the absence of policies and strategies for capacity-building through knowledge transfer, knowledge exchange and knowledge generation. The upshot of this has been to limit the scope of the construction mission in developing countries largely to short-term pre-occupations.

There are two reasons for the short-term orientation of investment in construction and infrastructure in developing countries. First, because of the labour-intensiveness of construction activities, investment in the construction industry has traditionally been considered as a convenient mechanism for coping with the growing problem of urban unemployment. The merit of employment generation through investment in construction and infrastructure projects has been widely orchestrated by national governments and multilateral institutions like the International Labour Organization (ILO). Though employment generation is important for developing countries, it is not, however, necessarily consistent with the much broader objective of sustainable development. Moreover, if investment in construction is to be appraised chiefly on grounds of its short-term employment effect, it would restrict the construction sector from developing the capabilities that developing countries would badly need to cope with emerging socio-economic and environmental pressures as they evolve.

Second, investment in construction activities has traditionally been invoked for setting in motion rapid economic growth through the so-called 'Keynesian multiplier'. The 'multiplier' argument, however attractive to developing countries, falls short of addressing the complex nature of the challenges arising in the development process. The merit of the 'Keynesian multiplier' as a mechanism for generating self-sustaining economic growth in developing countries is somewhat misplaced, as it attributes undue credit to investment in such countries where, in fact, the potential 'multiplier effect' is highly constrained by the prevalence of structural bottlenecks and often 'leaks out', causing a balance of payments pressure on the macro-economic system.

The focus of policy on the employment and growth effects of investment in construction, largely driven by political and commercial concerns, has rendered the construction industry in developing countries falling short of providing the basis for sustainable infrastructure. This is broadly reflected by the fragility of the infrastructure system already in place in these countries. It is, for instance, observed that in many parts of sub-Saharan Africa, infrastructure projects, once delivered, are often laid to waste. According to the World Bank (1994), about one-third of the roads built in sub-Saharan Africa during the 1970s were out of use either for want of upkeep or else for failure of appropriate project choice in the first place. Ofori (2007a,b) argues that in most developing countries, the construction industry has failed to fulfil its expected role of providing a basis for socio-economic improvements, in general, and improvement of living conditions, in particular.

It may be argued that businesses operating in developing countries are too limited to think global and to espouse ethical issues. But then, policies in these countries have not been known to be so comprehensive as to enable business behaviour to evolve with global and ethical senses of responsibility. The globalisation of the construction industry itself and the coming into the picture, in recent years, of a range of international conventions pertaining to the global agenda of sustainable development have, nonetheless, seen business behaviour change, albeit to a limited extent, in favour of corporate social responsibility (CSR). This is, however, more apparent in developed than in developing countries. Some engineering and construction companies operating in developing countries do have strategic objectives that explicitly align their business practices with social goals through, for example, direct hands-on contributions to poverty reduction and post-conflict and disaster relief efforts. A case in point is the experience of Engineers Against Poverty (EAP 2004), who have set out to address the detrimental impact of globalisation on some of the world's poorest countries and to mobilise the private sector to help meet the Millennium Development Goals.

The remainder of this chapter is in five sections. The part following this introduction sets the context by examining current concerns about poverty and disaster reconstruction works and how construction activities can be organised and managed with the view to achieve the Millennium Development Goals. This is followed by a discussion of the dynamics underlying the state of construction and engineering capabilities in developing countries. The third section outlines and discusses the factors that account for the fragility and inadequacy of infrastructure provision in developing countries. This is followed by a discussion of the strategic issues that policy and planning would need to focus on to promote infrastructure development. The fifth part draws some conclusions.

The global agenda for poverty alleviation and the role of construction

Planning and policy in developing countries have to contend with both short- and long-term problems in such a way that meeting short-term needs through, for instance, disaster reconstruction works, does not compromise long-term commitments to poverty alleviation programmes. The two are not mutually exclusive, since neglect of short-term problems would make long-term problems more intractable, and losing sight of the long-term in favour of short-term pre-occupations would result in the deepening and broadening of the poverty problem, as is the case in many developing countries today, where the majority of the populations subsist on less than US$1 a day (World Bank Group 2007). Much of the problem of world poverty today is attributable to the failure of policy and planning to reconcile the apparent conflict between short-term and long-term interests in a comprehensive framework.

The challenge of world poverty has to date exercised a wide range of initiatives of which the UN Millennium Development Declaration (alternatively referred to as the Millennium Development Goals or MDG) is the most comprehensive, aiming to halve the number of people who live on less than US$1 a day by 2015. In seeking to achieve this, the MDG focuses on the objective of poverty reduction through job creation and access to safe water and affordable and healthy shelter without, however, compromising the global agenda for sustainable environment as noted below:

> Development, at its most basic level, begins simply with access to clean water, sanitation, and housing. Meeting these fundamental needs unlocks human potential, restores dignity, and enables people to go about their day-to-day activities, to generate social networks and business opportunities, and to progressively create the social, physical and economic infrastructure that in turn fuels further development and brokers the relationship between people and their environment.
>
> (Silva 2007: 15)

Ofori (2007b) calls for construction firms to be MDG-compliant in their operations, so that the provision of infrastructure in developing countries is underpinned by poverty reduction objectives. This is particularly important since, according to projections of current trends, in 15–20 years' time, about 80 per cent of the world's new infrastructure will be constructed in developing countries (Watermeyer 2006).

Table 4.1 summarises the contribution which construction can make towards the achievement of the eight development goals. It is apparent that engagement with global poverty is by no means limited to the realm of philanthropy. Indeed, the opportunities for firms in the engineering and construction sector to contribute to poverty reduction through their core business activities are significantly greater than what they have often been perceived to be. This is because of the significant interface of the value chains of such firms with the local economies in the course of creating and maintaining infrastructure (Lynch and Matthews 2007). The initiative of business enterprises to be actively engaged in poverty reduction schemes is leveraged by the availability of multilateral and bilateral funding support for infrastructure development in poor countries (Mansfield and Zawdie 1993).

In macro-economic terms, investment in infrastructure projects would be expected to impact on the local economy by relaxing the infrastructure constraint on growth. The output of construction activities is also expected to expand during periods of economic growth, causing further economic growth. In the long run, investment in construction projects is presumed to have a capacity-creating effect, providing a basis for technology transfer and technological learning, thereby leading to managerial and operative

Table 4.1 Millennium development goals and role of construction

Millennium development goals	Contribution of construction
Goal 1: Eradicate extreme poverty and hunger	Effective and efficient production of buildings and infrastructure
	Maximum linkages of construction to other sectors of national economy to create stimulus for development Generation of employment opportunities through appropriate choice of technology and procurement
	Continuous development of industry
Goal 2: Achieve universal primary education	Design and construction of suitable school buildings (in local economic, climatic contexts)
	Contribution to economic growth and national development to create jobs for graduates
Goal 3: Promote gender equality and empower women	Creation of job opportunities for women and youth (Goal 8) at all levels in construction, with close attention to working conditions on sites, remuneration and career progression
Goal 4: Reduce child mortality	Construction of hospitals and infrastructure Provision of job opportunities to generate income
Goal 5: Improve maternal health	Effective site management to avoid health hazards
Goal 6: Combat HIV/AIDS, malaria and other diseases	Initiatives to avoid spread of HIV/AIDS by construction workers (through education)
Goal 7: Ensure environmental sustainability	Sustainable construction – cradle-to-grave considerations of all aspects of construction
	Effective management of completed buildings and infrastructure
Goal 8: Develop a global partnership for development	Construction as a partner for development – study the role of construction in development in order to enhance it
	Construction as a creator of wealth and less of a burden in imported inputs
	Research on, and develop, construction industries in developing countries to enable them to play a role in globalising economies
	Effective logistics of construction in landlocked and small island developing states
	Effective construction technology transfer inconstruction – from research to practice; from industrialised to developing countries
	Partnership among industry, government and researchers Global networks of researchers to study matters on construction and MDGs

Source: Adapted from Ofori (2007b).

skill development. This may not happen to be the case, however, if the infrastructure projects are not appropriate to the socio-economic and cultural circumstances of developing countries, since the weak economic base in these countries cannot readily absorb such inappropriately designed projects.

Technically speaking, projects would be considered to be appropriate to developing country conditions if they have low 'incremental capital–output ratio' (ICOR) – in other words, low investment requirements per unit of expected output. Projects with high ICOR represent high-cost and risky ventures. Where the mechanism of technology transfer fails to deliver appropriate project designs and/or effective technological learning, high-cost projects with high ICOR values are likely to abound, aggravating rather than alleviating poverty conditions. Projects with low ICOR values are invariably based on simple technologies, and their operation would not require highly sophisticated managerial and technical skills, which are anyway in short supply in developing countries. Characteristically, low ICOR projects are labour-intensive and high ICOR projects, capital-intensive.

The task of poverty reduction in regions like sub-Saharan Africa would, therefore, necessarily focus on infrastructure development based on low–cost 'poverty projects', which by virtue of their low ICOR, are geared to facilitate access of the rural and urban poor to basic-needs services. However, infrastructure investments in sub-Saharan countries have for the most part of the last four decades concentrated not on 'pro-poor'and rural-oriented projects but rather on grandiose urban-oriented ones with unfavourable ICOR values. In these circumstances, it is perhaps not surprising that the massive investment channelled into infrastructure development in many sub-Saharan countries has not made much dent into the poverty problem they have had to contend with for so long.

In theory, since economic growth is determined as a proportion of investment rate, given the capital–output ratio, and in view of the resource circumstances of developing countries, in general, there is good reason to believe that investments in pro-poor infrastructure projects are most likely to be growth-promoting and poverty reducing, whereas anti-poor infrastructure projects with high capital–output ratios are likely to be growth-constraining and poverty perpetuating. It is in this context that the significance of the contribution of construction activities to the achievement of MDGs will have to be considered.

Construction capability and challenges of development and reconstruction

There is evidence to show that local construction firms are emerging in developing countries to cater for the various categories of the construction market, and that in their bid for increasing market share, these local firms are supported by national governments who would make it increasingly difficult for international construction firms to win contracts without conditions that would require them to broaden and deepen their local involvement (Zawdie and Langford 2002). Domestic construction firms may have limited capabilities and would, in most cases, hardly be expected to stand on their own to deliver on competitive grounds, responding to the infrastructure demands of poverty reduction and disaster-management programmes. It would, nonetheless, be rash to write off the credibility of these firms, as they are yet to evolve along their respective learning curves. The knowledge and experience these firms acquire while implementing projects in joint venture initiatives are crucial for capacity-building in construction. However, prospects for capacity-building for any firm would ultimately depend on the rate and effectiveness of learning from experience. South Korea's experience of learning that enabled it to evolve from being an importer of turnkey projects to being a competitive exporter of such projects in a space of two decades is an instructive case in point (Dahlman *et al.* 1987; Kim 1988; Lall 1988).

Local construction firms in developing countries would be expected to build their capabilities through learning from technology transfer and also from the challenges of local problems through the identification, design and development of projects that would meet the infrastructure needs of local communities vulnerable to poverty and disaster conditions.

Capacity-building through learning from technology transfer

Working as partners in joint venture projects or as subcontractors on the supply chain, local construction firms in developing countries have the opportunity to learn from the experiences of international construction firms who play an important role as vehicles of technology transfer to developing countries. Through their activities in developing countries, they can impart technical and managerial skills from which local counterparts can benefit greatly. The extent of the learning benefit to be derived by local firms would, however, depend on the attitude of the local construction firms towards their expatriate counterparts.

The exposure of local firms to the business culture of international firms is generally known to elicit three types of responses, which qualify them as *encapsulators, absconders cosmopolitans* (Hall and Hall 1990). The 'encapsulator' would totally withdraw from the local culture to hang out with expatriate firms. Such firms are uncritical in their learning from the experience and expertise of expatriate firms; and to the extent they fail to weave expatriate culture into local culture, their contribution to the capacity-building exercise would be marginal. The 'absconder' is characteristically averse to alien culture and would totally reject the idea of learning from expatriate firms. Where such firms dominate, there would be little or no scope for capacity-building. On the other hand, the 'cosmopolitan' keeps a foot in both camps. Such firms would be critically selective when learning from their expatriate counterparts and applying what they have learned to local conditions. As such, they could make a difference to local capability development. Given these qualifications for learning and capacity-building, policy in developing countries would be expected to aim at providing the conditions that would enable 'encapsulators' and 'absconders' to become 'cosmopolitans'. Local firms would also be attracted to the 'cosmopolitan' view, if policies were supportive of competition, innovation and development of the enterprise culture.

The fact, however, remains that most of the local construction firms in developing countries are small and relatively inexperienced without the managerial and technical capability to handle major construction projects that governments in these countries would like to have to be able to cope with the long-term challenge of poverty alleviation and the short-term challenge of disaster-management. On the other hand, dependence on international construction firms for the construction of infrastructure projects presents a two-edged sword. On the positive side, international construction firms would be expected to impart management and technical skills to their local counterparts in return for the profit they earn and the experience they acquire in project management while operating in developing countries. Moreover, donors and international funding agencies like the World Bank favour international firms because, with experienced staff and a proven track record regarding performance overseas, they are reckoned to be more credible than domestic construction firms. It is also noted that the internationalisation of construction activities in developing countries has the added advantage of widening the scope for competition and efficiency of project delivery, much in the interest of client governments.

However, there is also a downside to the operation of international construction firms in developing countries. Langford and Rowland (1995) argued that the complex, multi-organisational and cross-cultural setting envisaged in the internationalisation process is likely to create formidable challenges for construction management in general. In developing countries, the problem with construction firms is that many would prefer to work alone despite the requirement by national governments that they enter into joint venture arrangements with local firms (Miles 1995; Miles and Neale 1991). This would make the commitment of international firms to joint venture initiatives in developing coutries somewhat lukewarm. It is, therefore, argued that international firms sometimes enter joint venture arrangements merely 'to hide behind a political association' and widen their influence to keep the costs of bidding down to a minimum (Mansfield and Zawdie 1993). Joint venture of convenience between international firms and their local counterparts based on the 'political motive' of the former are, however, rarely likely to be functionally well integrated. In these circumstances, dependence on international construction firms can be costly and disadvantageous for the host countries. The problem would be further aggravated if the international firms assumed monopolistic tendencies in their business behaviour effectively precluding local firms from participating in the construction of infrastructure projects. Furthermore, corruption within international construction projects (see Chapter 7) will also reduce the overall effectiveness of technology transfer by reducing the optimum level of funding available for assistance and development.

It is, therefore, apparent that the State in developing countries has a key role to play in promoting the transfer of technologies that are capable of creating learning opportunities for local construction firms. But for lack of policies geared to this end, the knock-on effect of technology transfer in terms of local capability development has been marginal. Local firms lack the technical and managerial capabilities for surveying, designing, developing, operating and implementing infrastructure projects. No wonder, therefore, that cases involving participation of local construction firms in bidding to undertake major infrastructure development projects and winning such contracts have been few and far between.

There has, however, been increasing concern in recent years as to how to maximise the learning benefits to local firms, deriving from the activities of international construction firms in developing countries. The World Bank and other international funding agencies have been strongly advocating that domestic firms should seek to 'cut their teeth' on the market for maintenance and labour-based projects, these being areas of comparative advantage (Beusch 1994). Local firms, learning on the back of association with international firms as subcontractors and joint-venture partners, could also gain knowledge and experience that would ultimately enable them to compete with the international construction firms in bidding for complex civil engineering and building projects (Kim 1988). Two decades on, the emerging global economies of India and China refelect the evolution of a developing country to a global economic player, facilitated through the growth in construction GDP, albeit, it is uncertain how much of this growth is aligned with their indigenous industries' acquisition of new technology. Moreover, attention to CSR issues (i.e. employment of child labour, low wages, dangerous working environments) during such transformation offers 'cultural dilemmas' for international firms working in such countries.

It can be concluded from the discussion above that while moving up the learning curve is often a daunting challenge for local construction firms in developing countries, the problem could be mitigated if international firms operating in developing countries were sensitive to the implications of their business behaviour for the achievement of poverty alleviation. This would require international firms operating in developing countries to assume corporate social – and, indeed, global – responsibility, and national governments to ensure through policy provisions that the mechanisms for effective learning from joint venture initiatives and for the development of enterprise culture are put in place.

Capacity-building through learning from challenges of poverty and disaster

Lessons could also be learned by local construction firms in developing countries from the challenges encountered in the identification, design and development of projects that would meet the infrastructure needs of local communities. Developing countries in general suffer from the secular problem of poverty and the occasional occurrences of disasters, natural or man-made. The two problems can impact on one another in a vicious circle, with vulnerability to disasters increasing as a direct result of the acuteness of poverty, and failure to cope with disasters leading to increased vulnerability of communities to the poverty trap. Policy intervention to break this vicious circle would call for the provision of infrastructure systems that would facilitate implementation of mitigation strategies.

Challenges of poverty alleviation programmes

The task of poverty alleviation requires, *inter alia*, the provision of more and better housing; roads; facilities for communication and water and power supply; and facilities that would enable local communities to have ready access to health, sanitation and education services. Responding to the challenges arising from the provision of these facilities would offer local construction firms a learning ground to develop engineering and management capabilities. Perhaps the most profound effect of infrastructure development is perceived to be on the attitudes and values of the rural poor, who constitute the majority of the population in most developing countries, since improved transport and communication systems enhance their mobility and access to information and new ideas by facilitating wider interaction with the world beyond the rural horizon (Zawdie and Langford 2002). Of particular significance in this respect is investment in rural infrastructure projects which constitute a major component of poverty reduction and sustainable-development programmes (Ahmed and Donovan 1991; DFID 1997; Galenson 1994; Pouliquen 1999; Riverson *et al.* 1991). Such programmes often seek to use rural infrastructure investments as a tool for creating local employment through the promotion of labour-based and foreign-exchange-saving construction methods, and for supporting local initiatives by involving grassroots organisations and communities in the planning and management of projects. However, the development experience of many poor countries, such as those in sub-Saharan Africa, shows that the economic and social significance of rural infrastructure has not been matched by policy and planning commitments to channel investment funds into rural infrastructure development projects. This is in large measure a result of the urban bias of development strategy, which many of these countries opted for in the first place.

Edmonds and De-Veen (1992) note that it would make economic sense to use labour-based techniques to develop a rural road network, and they find it paradoxical that most infrastructure building and maintenance programmes in developing countries are heavily capital- and equipment-intensive. Such practices, however, go against the grain of CSR in business behaviour, allowing the adoption of infrastructure projects that are inappropriate to the socio-economic circumstances of developing countries and militating against the achievement of poverty alleviation objectives. Jowitt *et al.* (2004) argue that the focus of development should be not on engineering or technology but on people, and that it is crucial that the knowledge and aspirations of local people are mobilised if the challenge of poverty is to be faced head-on, and if the engineering profession is to 'avoid the imposition of inappropriate development and replace it with development that has a deep, lasting and sustainable impact on poverty'.

Not all international construction firms operating in developing countries, however, fail on grounds of CSR. For instance, in its operation in Kenya, Gibb Africa, an engineering consultancy firm, uses labour-intensive construction and local materials and trains local people to maintain roads. Apparently, more profitable projects are often sacrificed so as to offer local stakeholders the ability to prioritise the building programme. On another Gibb Africa project, tender documents required contractors to educate labourers and the local communities about the dangers of HIV/AIDS and the company reimbursed the contractor to do this (Karekezi 2005). This is particulary important given that, as Lerer and Scudder (1999) note, the in-migration of construction workers to construction sites may result in an increase in the spread of sexually transmitted diseases and human immunodeficiency virus (HIV).

The consultation with local communities noted above appears to align itself with what Lynch and Matthews (2007) term a 'social licence to operate'. They argue that planners, financers and contractors should seek informed consent and support from local stakeholders to construct and operate a project in the latter's local environment. This is, of course, considered de rigueur for CSR-oriented constructors in developed countries, and Ofori (2001) has suggested that this should be tracked as part of a tool kit for measuring construction industry development in developing countries. More recently, Engineering without Frontiers (EWF) has recommended that environmental and social objectives should receive more prominence in procurement. They call for contract documents to set out requirements and responsibilities for achieving social key performance indicators (KPIs) and insist that these should be included in conditions of contract with details and defined benchmarks in specifications (ICE 2006). The tangible benefits of this practice are noted below:

> The emergence of this second generation of CSR activities – in which business competencies are aligned with social needs for commercial advantage – is an important development, the implications of which should not be underestimated. When CSR is effectively incorporated

into core business activities, when it becomes fused with business development strategies and aligned with the development priorities of low and middle income countries, then it is likely to gain momentum that could significantly impact on poverty reduction.

(Matthews 2004: 24)

The role of non-governmental organisations (NGOs) in combating poverty, through mobilising local communities into a wide range of development activities, has also been noted by Zawdie and Langford (2002). Funded by donor agencies, NGOs play a major role in providing support to local initiatives aimed at meeting the infrastructure needs of local communities. NGO-supported community projects generate income through employment creation, create learning opportunities, and empower people as individuals and as communities by enabling them to participate in the design, development, financing and management of the projects. One such NGO – Engineers Without Borders-UK (EWB-UK) – provides opportunities for partnership between engineering students, industry, and local communities. The mission of EWB-UK, a student-run charity etsablished in 2001, is to facilitate human development through engineering contributions to international development. Newby (2007), a founding trustee, notes that many young engineers find inspiration through the work placements undertaken overseas and that this demonstrates that there is more to engineering than number crunching and more to be achieved than making something cheaper, bigger, or faster.

One major area of infrastructure development for which support to community initiatives has particular relevance is the housing problem, which has given rise to policy concern about the adequacy of its supply, and its affordability and sustainability. Governments in developing countries have sought to address the problem in response to the United Nations Declaration of 1974, which encouraged developing countries to expand low-cost housing on a self-help basis through the establishment of cooperatives, utilising, as much as possible, local raw materials and labour (MacPherson 1982). More recently the concern has been orchestrated through the Johannesburg Plan of Implementation that came out at the 2002 World Summit on Sustainable Development and the Millennium Development Goals. This concern has not, however, yet been matched by achievement. In part, this is a result of demand outstripping supply in the face of rapid population growth and high urbanisation rates. But more important is the absence of institutional and policy frameworks that would enable communities to design, finance and implement housing schemes compatible with the local social, economic and environmental conditions (Ramsay 1993; Wilkinson 1993).

Learning from the challenge of meeting local development needs calls for community-based initiatives; such initiatives, however, tend to thrive where communities have access to support-systems provided by governments or NGOs. Experience, however, shows that the scope for such learning has been marginal at best because of the absence of institutional mechanisms and support systems for promoting grass-roots participation in decision-making processes and for the development of skills through the accumulation of knowledge and experience. As discussed above, NGOs have in recent years sought to address this problem, but their impact has been limited, when viewed in the light of the size of the problem. What is more, they would not be expected to make much progress while goverment policy remains restrictive or there is no policy at all.

Challenges of disaster management

The number of natural disasters hitting the world each year is estimated to be in the range of 300–500, resulting in an average death toll of 60,000 with 95 per cent of these occurring in developing countries (Campher 2005). These disaster events are set into two categories: the low-frequency, high-impact hazard type, such as the Indian Ocean tsunami disaster; and the high-frequency type with greater cumulative effects such as tropical cyclones, storm surges and floods. Bangladesh has been suffering from floods and storm events for over 35 years and in the Sahel Region of sub-Saharan Africa, drought has become a chronic problem, resulting in famines, death, massive rural-urban migration and destitution. In addition to these natural disasters are man-induced disasters, a good part of which is precipitated by wars and conflicts that cause the loss of hundreds of thousands of lives and extensive damage to properties and infrastructure, and also by the absence of CSR in business behaviour.

One of the principal reasons for the high disaster vulnerability of communities in developing countries is the fragile and inappropriate nature of the infrastructure system in these countries. Many developing countries are in this respect ill-prepared to withstand the shocks of disaster, so that even minor disasters could result in human tragedies of inordinate proportions. In such instances, the impact effects of natural disasters are amplified by inadequacies in the supply of relevant infrastructure. Thus, for example, the great Ethiopian famine of 1974, which claimed the lives of many people in northern Ethiopia and led to social and political instability in the country that had a lasting effect, could have been avoided had the road network in the country been developed enough to allow the smooth flow of food grains from the food-surplus pockets in the south and west to the food-deficit regions in the north (Sen 1981). In Bangladesh, the provision of robust flood defence systems could have had a mitigating effect on the extent of disaster caused by the recurrence of floods and storms. Many developing countries have also had their share of man-induced disasters, resulting from the collapse of constructed facilities built on or near high-risk areas, as, for example, geologically unstable areas prone to earthquakes or landslides. The total collapse of the Highlands Tower Condominium in Kuala Lumpur, Malaysia, in December 1993, which claimed the lives of more than 40 people and caused widespread panic, is a case in point, demonstrating failures in design and engineering works; shoddy construction; negligence of the local authority in approving the project without consulting experts on soil and the risk of hill slope development; and failure of the construction procurement system governing the planning, building and maintenance of such constructed facilities.

Rapid urbanisation often results in 'fast-track' construction with high risk implications for people's lives and wellbeing owing to unsuited land use, inadequate protection of urban infrastructure, ineffective building code enforcement, poor construction practices and limited opportunities to transfer or spread risk (UN 2005). Carrion (2005) notes that in Quito, as in many other cities in South America, a significant proportion of low-income families have established settlements (*barrios*) on the steep slopes surrounding the city. The environmental impacts of these settlements include deforestation, the weakening of soils due to inappropriate construction techniques and the associated risks of landslides and floods. Carrion further notes that 60 per cent of the 1600 homes in these *barrios*, which were constructed by 'pirate developers' without permission from the local authorities, do not comply with anti-seismic standards. Shoddy and inappropriate construction is almost a common feature of the infrastructure system in developing countries. It thrives on the back of rapid urbanisation and the pressure to grow the economy fast to meet the demands of a growing population; the lack of corporate responsibility on the side of those who deliver the infrastructure; and the neglect of policy on the side of governments.

Disasters are best avoided, or else mitigated. The last three decades have seen disaster-relief activities in developing countries being a major preoccupation of NGOs and donor agencies. Anecdotal evidence suggests a shift in corporate response to disasters occurring in developing countries whereby disaster relief initiatives extend beyond the conventional financial donations to include direct involvement and participation in relief works. Post-disaster rehabilitation and reconstruction works involve high costs; and this offers good business opportunities for construction firms. For instance, reconstruction and relief works in tsunami-affected areas in Sri Lanka were estimated to cost US\$2.5 billion (UN 2005).

Disasters occurring in developing countries attract donation of funds from governments and the general public in developed countries to facilitate relief works through NGOs. The problem, however, is that the response to the business potential created by these funds is not often complemented with CSR. However, NGOs such as RedR Engineers (see case study below) and the Disaster Resource Network (DRN) are excellent examples of individuals and corporations within the engineering–construction industry, humanitarian response organisations and government agencies that engage in international disaster response whilst simultaneously engaging in CSR-related actions. DRN has launched the New Engineering and Construction Initiative 2006–2010 with the aim to expand its role in disaster response and mitigation.

Case study: a need for short-term and long-term assistance

RedR UK

The mission of RedR, originally called Registered Engineers for Disaster Relief, is 'to relieve suffering caused by disasters by selecting, training and providing competent and committed personnel to humanitarian programmes worldwide'. The organisation was founded to provide a register of qualified engineers and technical personnel to work for up to three months with front-line agencies in disasters, and hence meet the needs of disaster victims. In 1979 Peter Guthrie was working as an engineer with Scott Wilson Kirkpatrick when a huge exodus of Vietnamese boat people caused a refugee crisis in Malaysia. Peter Guthrie was recruited through Oxfam to work in the refugee camps for three months. He was the only engineer there:

> When I returned I saw the pressing need for engineers to help in this sort of work and compiled a register of engineers who could be called upon at short notice to work with frontline relief agencies.
>
> (RedR UK 2007)

International engineering consultancy Arup is a patron of RedR since its formation, with support staff who are RedR members, and Silva (2007) describes how its employees assist in the humanitarian sector through relief and assistance in many post-disaster or post-conflict situations. After the Indian Ocean tsunami of 26 December 2004, six Arup staff gave a total of 33 man-months' expertise in water, shelter, infrastructure, and project management to several NGOs and UN agencies in Sri Lanka, Aceh, and India. This was in addition to £200,000 donated directly through NGOs. These activities are clearly related to the core values of the organisation that date back to Ove Arup's Key Speech (1970) document, a review of which by Hamil (1993) notes the desire for Arup to be a humane work organisation that fulfils a wider social purpose.

While disaster situations offer opportunities for business initiatives, there are concerns that the focus on short-term disaster relief activities would pre-empt resource commitments for long-term development projects. However, in the context of poor countries, the presumption of conflict between the short-term and the long-term is misplaced insofar as failure to cope with the consequences of natural disasters is in large measure conditioned by under-lying weaknesses in long-term development trends – i.e. weaknesses which are further exacerbated by failures to resolve problems in the short-term period. Indeed, the task of poverty alleviation calls for both short-term and long-term policy engagement; and most NGOs are now increasingly integrating relief work with development work. Thus, the argument that NGOs should be limited to relief activities and that development projects should be the preserve of governments is hardly tenable, particularly since NGOs are proving effective in development and networking activities at grass-roots level. Not surprisingly, NGOs have become increasingly attractive to donors as conduits for channelling aid resources to developing countries.

There is, however, a potential 'moral hazard' deriving from the participation of NGOs and construction firms in disaster-relief activities and poverty reduction programmes through the provision of infrastructure, where, particularly, local politics hamper transparency in the award of contracts and terms of references. Where there is no adequate mechanism for monitoring CSR, as is the case in many developing countries, infrastructure providers like NGOs and construction and engineering companies could be part of the problem rather than part of the solution. In such situations, developing countries would be left with inappropriately designed and poorly built infrastructure, while the net benefit offered by the aid funds is reaped by NGOs and construction and engineering companies.

Factors affecting the state of infrastructure in developing countries

The socio-economic significance of infrastructure projects for developing countries can hardly be over-emphasised. Why then, it may be asked, has the infrastructure system in developing countries remained weak and fragile, making the population vulnerable to perpetual poverty syndrome and the risk of disaster events, such as earthquakes, floods, droughts and conflicts.

The absence of robust infrastructure systems in developing countries can be attributed to at least six factors (Zawdie and Langford 2002). First, project financing poses a major constraint on infrastructure development in poor countries in general. The investment requirement of infrastructure projects aimed to meet the needs of national economies with a growing population is generally substantial, but the poverty of the economies means that the scope for mobilising domestic resources for infrastructure development is narrowly limited. Consequently, foreign aid obtained in the form of loans and grants from bilateral and multilateral sources has been a major component of investment finance for infrastructure development (Mansfield and Zawdie 1993). However, such sources of investment finance cannot be expected to be open indefinitely. Moreover, they are subject to political conditionalities that may not be favourable to developing countries.

Second, poor investment decisions with respect to the choice of investment projects have the effect of amplifying the problem of infrastructure development financing which many developing countries have to contend with. The problem is particularly apparent when the bulk of investment resources is committed for capital-intensive projects with high ICORs without careful consideration of the appropriateness of infrastructure projects to the objectives of sustainable development. Investment decisions of this type broadly reflect the 'urban-bias' and 'top-down' orientation of plans and policies in developing countries.

Third, the dominance of state participation in the delivery of infrastructure projects in many developing countries is considered to be a major factor militating against prospects for the development of sustainable infrastructure in these countries. Most of the infrastructure projects are publicly funded and managed initiatives, which are known to have experienced severe financial difficulties at various times mainly because their operation is inefficient as a result of being largely based on non-commercial principles (Donaldson *et al.* 1997; Kessides 1995; World Bank 1994).

Fourth, financing of infrastructure investments in developing countries has almost invariably depended upon the provision of public funds and foreign aid. The role of private finance, in general, and foreign direct investment (FDI), in particular, has been negligible; and prospects for these to replace government funding and foreign aid as a strategy for financing infrastructure development do not appear promising as most private investors do not consider poor countries attractive for investment, partly for reasons of political instability, and because of the low average incomes and the small size of markets for infrastructure services. Thus, for example, of the 1170 privately financed infrastructure projects worldwide between 1985 and 1995, only 5.5 per cent were in the sub-Saharan region (Donaldson *et al.* 1997).

Fifth, in many developing countries, project implementation failures are known to have occurred as a result of contractors finding it difficult to obtain foreign exchange, manpower and machines in time (Wells 1986). Another commonly observed problem in these countries, particularly where the clients are government organisations, is failure to make payments in time, thereby adversely affecting the cash flows of all contracting agents on the supply chain. Small contractors are most hard hit by this and are often the first to be forced out of business because of their relatively weak financial positions.

Sixth, the weakness of the local construction industry in terms of engineering and management capability is a major factor accounting for the poor record of infrastructure development in poor countries. Although local construction firms have emerged in these countries to exploit the commercial benefits offered by investment plans for infrastructure development, there are questions about the adequacy of their professional experience in the conduct of competitive business, the level of their engineering expertise, and their ability to deliver projects according to schedule as defined by the terms of contract. Shortfalls in engineering, managerial and organisational capability on the side of construction firms also cast doubt on the sustainability of the projects delivered. This situation is best demonstrated by an experience from Ethiopia (Zawdie and Langford 2002) described below.

In 1993, the Ethiopian Road Authority awarded turnkey contracts for 15 rural road construction projects to different local construction firms. The contracts involved the survey and design of roads as well as their construction, all to be done in a package for delivery in 3–5 years' time. However, five years after the award of the contracts, only one road had been completed.

Another was not even started. Project implementation performance was, on the whole, rated to be 25 per cent. In the event, those contractors who seemed to have no chance of completing the road projects were forced to terminate their contracts, whereas those judged to be capable of successfully completing their projects under increased supervision and strict follow-up schedules were allowed to continue. For the domestic firms to meet their contractual obligations, it was imperative that they set up survey and design offices in their respective establishments. This, however, was a tall order for the domestic firms in view of the shortage of the relevant managerial and technical skills, knowledge and experience. The domestic contractors decided to undertake the task without any prior knowledge of the alignment of the roads, the quantities of work involved, the terrain classifications and the volume and hardness of the rock excavations involved.

Strategic issues for construction and infrastructure development

It is apparent from the discussion above that the development of sustainable infrastructure in developing countries is generally constrained by factors that could be broadly categorised into three sets of issues – namely, shortfalls in the provision of finance; deficiencies in the engineering and management capabilities of construction firms; and failures in the CSR of the infrastructure providing agencies.

Financing infrastructure development

Traditional sources of finance for infrastructure development have included public funds often supplemented by loans and grants from bilateral and multilateral sources. These have, however, given rise to the predominance of state-driven strategies for infrastructure development, and have, in many cases, resulted in investment projects invariably aimed at social goals – including, *inter alia*, the creation of employment and the subsidisation of prices to consumers (World Bank 1994) – and the commercial objectives of competitiveness, efficiency and profitability. In sub-Saharan countries, publicly funded infrastructure systems have, for the most part, been inefficiently managed, and, not surprisingly, they have seldom generated the resources required for 'greenfield' investments and for the repair and maintenance of existing facilities.

There is now increasing recognition of the significance of private sector finance as the way forward for mitigating the problem of inadequate and unsustainable infrastructure. This is based on the understanding that private sector investments in infrastructure development are likely to be more efficient than publicly funded investments, because they are commercially motivated. Private involvement in the provision of public goods can be achieved through the entry of new competitors into markets monopolised by state-owned enterprises and/or through the privatisation of these enterprises. This strategy represents a shift from the traditional state-driven, top-down system of planning to the use of markets in infrastructure provision and from dependence on foreign aid to the recognition of the growing significance of FDI and other private finance initiatives (PFI) or even private–public partnerships (PPP) for infrastructure provision.

The problem, however, is that most investors would consider developing countries, in general, and the poorer ones like those in sub-Saharan Africa, in particular, too risky for business. This is not surprising considering the smallness of the market; the persistence of political and social instability; the high transaction costs and uncertainties that investors face while dealing with national governments; the lack of transparency in the legal and policy frameworks for negotiating agreements and contracts and making investment decisions; and the weaknesses of domestic financial institutions in these countries. If, therefore, private finance is to be a viable option for infrastructure development, governments in developing countries would need to address the factors influencing investors' perceptions of risk with the view to create the policy environment that would enable commercial initiatives to thrive. This task constitutes a major aspect of structural adjustment programmes which many developing countries have had to pursue, albeit reluctantly, upon the advice of the World Bank and the International Monetary Fund (IMF).

Corporate social responsibility and construction

The inappropriateness of many construction projects in developing countries reflects the absence of CSR from the activities of construction and engineering companies operating in these countries. 'Inappropriate construction' is taken here to encompass both design and workmanship defects in quality and the construction of infrastructure and buildings on plots of land that are susceptible to natural and man-made risks. When clients focus on quick delivery and low cost of project or the delivery of 'prestigious projects' irrespective of cost – as has often been the case with governments in developing countries that have no democratic accountability – there is no legal constraint binding construction and engineering companies to operate within the moral framework of CSR. This has invariably resulted in the 'moral hazard' of landing tax payers with shoddy, unsustainable and inappropriate infrastructure.

Concern with poverty reduction and sustainable development has created increasing awareness that failure to address such issues as transparency, accountability and overall corporate governance could have serious implications not only for the industry but also for the wider community and the environment in terms of cost, delays and quality of constructed facilities. There is, indeed, a general understanding that construction and engineering companies that integrate CSR into their business strategies are most likely to derive the benefits of enhanced reputation, competitive edge in contract bidding and better community relations.

Government will also have to engage in the CSR agenda by providing the legal and institutional framework that would enable the construction industry to grow and to protect the interests of clients who use the constructed facilities, thus ensuring that quality and safety are not compromised for short-term corporate gains. This would require, among other things, construction and engineering companies operating in developing countries to formulate and implement safety training programmes at project level, in every project. A growing number of companies originating from developed countries are adopting CSR-related codes of business principles, and many are known to sponsor charitable foundations operating in developing countries. However, most companies would argue that their prime social responsibility would be to conduct their core activities efficiently and to the highest standards of integrity, and also in ways that are agreeable and appropriate to local social, cultural and resource circumstances in developing countries.

Donor agencies working on development programmes have also a crucial role to play in ensuring that the CSR agenda is implemented, so that development funds can be effectively channelled, benefiting communities in developing countries. This means that donor agencies should also work in many other capacities: as policy advocates within their own governments and within multilateral institutions and as mediators, facilitators, technical advisors and disseminators of good practice and learning (World Bank 2004).

Training, education and research for sustainable construction

The deficiency in the supply of infrastructure underlying the chronic problem of poverty is broadly reflective of the prevalence of the skill and knowledge gap in developing countries. One aspect of CSR, as noted above, is facilitating the provision of relevant skills and knowledge that would be instrumental in construction capability development. We have also seen that technology transfer through the cross-border operation of construction and engineering companies and NGOs and charities can serve as a mechanism for bridging the skill and knowledge gap that has been a major factor constraining the development of sustainable infrastructure in developing countries.

The need for more and better training, education and research to respond to the prevailing skill and knowledge gap is further orchestrated by CIB, the International Council for Research and Innovation in Buildings and Construction, which established a CIB task group (CIB TG29) to examine the construction industry in developing countries. In 2002, the task group evolved into a fully fledged working commission (CIB W107) with a strategic research agenda for poverty alleviation through construction and post-disaster reconstruction schemes. Another task group, launched in 2005, explores labour issues such as 'construction employment and poverty reduction' and 'appropriate technologies and employment generation'. Indeed, Singleton (2003) has called for sound engineering solutions to poverty alleviation and makes specific recommendations for 'appropriate engineering', that is, the application of labour-based construction, as differentiated from labour-intensive construction. The latter substitutes human labour for machines whereas labour-based solutions challenge designers to change the technology involved to what is appropriate for manual labour.

Ofori (2007a) argues that researchers in construction have a duty to contribute to the wider national and international effort to lift millions of people in developing countries out of poverty. The publication of research findings is one way in which the academic community can assist. Through publications such as *The Journal of Construction in Developing Countries* (published jointly by CIB-W107 and the Akademi Sains Malaysia) and guest editorials (e.g. Ngowi 2002, *Building Research and Information: Challenges Facing Construction Industries in Developing Countries*) researchers have the power to influence policy. However, Ofori has concluded that much of the contemporary research into construction in developing countries has been wanting, with little of it classified as profound. He challenges the research community to engage in research areas that would build upon (rather than repeat) the existing body of knowledge and contribute to the formation of policies and initiatives that lead to real improvements in the lives of the citizens in developing countries. He identifies areas of research priority for construction in developing countries to include the following:

- Private sector involvement in the provison of infrastructure and other major construction projects.
- Integration of construction in the advent of globilisation and liberalisation of economic régimes.
- Formation of regional economic blocks and common markets, including among developing countries.
- Global consensus on the need to fight poverty.
- International concerns with sustainable debvelopment, especially environmental issues.
- Threats of pandemics such as HIV/AIDS and avian influenza.
- Cultural issues on sizeable projects.

It is hoped that research along these areas could possibly help fill the knowledge gap that hitherto constrained the development of construction capability in developing countries.

Conclusion

From the discussion in the foregoing sections of this chapter, it is apparent that developing countries will need to address construction and infrastructure development as a key factor in strategies for mitigating poverty and disaster vulnerability. Many developing countries, particularly those in sub-Saharan Africa, find themselves locked in a vicious circle of poverty with infrastructure that is too fragile and inadequate to enable them to cope with disaster situations and to build capabilities for sustainable development or to achieve the Millennium Development Goal of halving the number of people who live on less than US$1 a day by 2015.

It is important, therefore, to note that while provision of sustainable infrastructure is crucial for the mitigation of poverty and disaster vulnerability, the systemic nature of the problem means that it can best be resolved if considered in relation to – and not in isolation from – the respective social, cultural, economic, political and environmental contexts. In this chapter, we have attempted to explain the persistence of the vicious circle of poverty and disaster vulnerability partly by shortfalls of investment finance for infrastructure development; partly by deficiencies in construction and engineering capabilities; and partly by failures in the corporate social responsibilities of the agencies involved in the provision of infrastructure. In all these, the state has a major role to play as the provider of the legal and policy frameworks that would enable NGOs, grass-roots communities, engineering and construction companies and private investors to participate in the business of designing, developing, financing and implementing infrastructure projects. State policies should also promote capacity-building in construction and infrastructure provision through learning from technology transfer via international construction firms. In addition, capturing and disseminating locally acquired experiences, gained from responses to challenges precipitated by the chronic problem of poverty, is of considerable value.

It is also important for governments in developing countries to provide the institutional basis for implementing CSR in the provision of built infrastructure facilities. One way of doing this is by ensuring that the terms and conditions of contracts for construction projects are carefully drawn up in the light of the socio-economic and environmental circumstances of their respective countries; that the tasks of contract award and contract management are conducted with transparency and accountability; and that contract award is conditional on sharing of knowledge and good practice with local communities. This would reduce, if not remove, the scope for irregular practices that have hitherto constrained the development of sustainable infrastructure. Under the regime of CSR, governments in developing countries would promote construction and infrastructure investments geared towards the development of pro-poor projects that are characteristically capital- and foreign-exchange-saving, local resource using, employment generating and income redistributing. It is possible to reach the majority of the population in developing countries with such projects, rather than with the capital-intensive, high-ICOR infrastructure projects that have traditionally been favoured, often for no good reason but that they appear modern and prestigious. In fact, such high-cost projects have in many instances been known to constrain rather than facilitate economic growth and exacerbate the poverty and disaster vulnerability of the poor in developing countries.

References

Ahmed, R. and Donovan, C. (1991) *Issues of Infastructure Development: A Synthesis of the Literature*, Washington, DC: International Food Policy Research Institute.

Arup, O. (1970) *The Key Speech*. Online. Available at: www.arup.com/_assets/_download/download5.pdf (accessed 10 June 2007).

Beusch, A. (1994) Training for labour-based road construction and maintenance, *Science, Technology & Development*, 12 (2–3): 247–259.

Bon, R. and Crosthwaite, D. (2000) *The Future of International Construction*, London: Thomas Telford.

Campher, H. (2005) Disaster management and planning: an IBLF framework for business response. London, UK: International Business Leaders Forum.

Carrion, D. (2005) Dealing with illicit construction in Quito, cluster 1, session 1.10 addressing the root causes of vulnerability of human settlements in mega-cities, *Proceedings of World Conference on Disaster Reduction*, 18–22 January Available at: http: //www.unisdr.org/wcdr/thematic-sessions/presentations/session 1-10/quito-mr-carrion.pdf (accessed 9 June 08).

Dahlman, C.J., Ross-Larson, B. and Westphal, L.E. (1987) Managing technological development: lessons from newly industrialising countries, *World Development*, 15 (6): 759–775.

Department for International Development (1997) *Eliminating World Poverty: A Challenge for the 21st Century*, London: HMSO.

Donaldson, D.J., Sader, F. and Wagle, D.M. (1997) Foreign direct investment in infrastructure: the challenge of southern and eastern Africa, world bank and the international finance corporation, Washington DC.

Edmonds, G.A. and De-Veen, J.J. (1992) A labour-based approach to roads and rural transport in developing countries, *International Labour Review*, 131 (1): 95–110.

Engineers Against Poverty (2004) Corporate social responsibility as a strategy for poverty reduction: fact or fiction? *Engineers Against Poverty Conference Report 13*, October. Online. Available at: www.engineersagainstpoverty.org/docs/EAP %20conference%202004.pdf (accessed 2 April 2007).

Galenson, A. (1994) *Provision of Rural Infrastructure in Sub-Saharan Africa*, Washington, DC: World Bank.

Hall, E. and Hall, F. (1990) *Understanding Cultural Differences*, Somerville, USA: International Press.

Hamil, S. (1993) Building on ethics: Ove Arup and corporate social responsibility, in good business? *Case Studies in Corporate Social Responsibility*, Bristol: SAUS Publications.

Institution of Civil Engineers (2006) *Enhancing Procurement Procedures for Infrastructure Works and Services to Deliver Social Objectives and Strengthen*

Operations and Asset Performance, Council Paper. Online. Available at: www. ice.org.uk/downloads//EAP-EWF%20council%20paper%20final.pdf (accessed 10 August 2007).

Jowitt, P., Beckmann, K., Matthews, P. and Cameron, P. (2004) *Engineering the World Out of Poverty*. Online. Available at: www.engineersagainstpoverty.org/ docs/SustDevIntnl_EwF%20article.pdf (accessed 2 April 2007).

Karekezi, P. (2005) Working lives, honest brokers, *New Civil Engineer*, 17 March, p. 28.

Kessides, C. (1995) *The Contributions of Infrastructure to Economic Development: A Review of Experience and Policy Implications*, Washington, DC: World Bank, Discussion Papers.

Kim, S. (1988) The Korean construction industry as exporter of services, *World Bank Economic Review*, 2 (2): 225–238.

Lall, S. (1982) *Developing Countries as Exporters of Technology*, London: Macmillan.

Langford, D.A. and Rowland, V.R. (1995) *Managing Overseas Construction Contracting*, London: Thomas Telford.

Lerer, L.B. and Scudder, T. (1999) Health impacts of large dams, *Environmental Impact Assessment*, 19: 113–123.

Lynch, M. and Matthews, P. (2007) An economic and social performance framework for engineering and construction firms operating in developing countries, *Proceedings of (CME 25) Construction Management Economics: Past, Present and Future*, 16–18 July, University of Reading (CD copy only).

MacPherson, S. (1982) *Social Policy in the Third World: The Social Dilemmas of Underdevelopment*, London: Wheatsheaf.

Mansfield, N. and Zawdie, G. (1993) The role of donor agencies and consulting firms in technology transfer to developing countries: the case of consulting projects, *Science, Technology & Development*, 11 (2): 99–112.

Matthews, P. (2004) Corporate social responsibility as a strategy for poverty reduction: fact or fiction? *Engineers Against Poverty (2004) Conference Report 13*, October. Online. Available at: www.engineersagainstpoverty.org/docs/EAP %20conference%202004.pdf (accessed 2 April 2007).

Miles, D. (1995) *Constructive Change: Managing Technology Transfer*, Geneva: ILO.

Miles, D. and Neale, R. (1991) *Building for Tomorrow: International Experience in Construction Industry Development*, Geneva: ILO.

Newby, T. (2007) Engineers without borders, Royal Academy of Engineering, *Ingenia*, 32: 30–35.

Ngowi, A. (2002) Editorial: challenges facing construction industries in developing countries, *Building Research and Information*, 30 (3): 149–151.

Ofori, G. (2001) Indicators for measuring construction industry development in developing countries, *Building Research and Information*, 29 (1): 40–50.

Ofori, G. (2007a) Guest editorial: construction in developing countries, *Construction Management and Economics*, 25: 1–6.

Ofori, G. (2007b) Millennium development goals and construction: a research agenda, *Proceedings of (CME 25) Construction Management Economics: Past, Present and Future*, 16–18 July, University of Reading, p. 1 (abstract).

Pouliquen, L.Y. (1999) *Rural Development from a World Bank Perspective: A Knowledge Management Framework*, Washington, DC: World Bank.

Ramsay, G. (1993) Informal housing, affordability and building standards, in Dewar, D. (ed.), *The De Loor Task Group Report on Housing Policy for South Africa: Some Perspectives*, Cape Town: Working Paper No. 1, Urbanity and Housing Network, University of Cape Town.

RedR UK (2007) *About Us*. Online. Available at: www.redr.org.uk/en/About_Us/History.cfm (accessed 20 December 2007).

Riverson, J., Gaviria, J. and Thriscutt, S. (1991) *Rural Roads in Sub-Saharan Africa: Lessons from World Bank Experience*, Washington, DC: World Bank.

Sen, A.K. (1981) *Poverty and Famines: An Essay on Entitlements and Deprivation*, Oxford: Clarendon Press.

Silva, J.D. (2007) From relief to long-term development, *The Arup Journal*, 2: 14–15.

Singleton, D. (2003) Poverty alleviation: the role of the engineer, *The Arup Journal*, 1. Online. Available at: www.arup.com/_assets/_download/download67.pdf (accessed 2 April 2007).

United Nations (2005) *Proceedings of the Conference Building the Resilience of Nations and Communities to Disasters*, World Conference on Disaster Reduction, 18–22 January, Kobe, Hyogo, Japan, Geneva: United Nations.

Watermeyer, R. (2006) Poverty reduction responses to the millennium development goals, *The Structural Engineer*, 2 May.

Wells, J. (1986) *The Construction Industry in Developing Countries: Alternative Strategies for Development*, London: Croom Helm.

Wilkinson, P. (1993) Housing need and housing policy in South Africa: a critical approach to the strategy approach adopted, in Dewar, D. (ed.), *The De Loor Task Group Report on Housing Policy for South Africa: Some Perspectives*, Cape Town: Working Paper No. 1, Urbanity and Housing Network, University of Cape Town.

World Bank (1994), *World Development Report 1994: Infrastructure for Development*, Washington, DC: Oxford University Press.

World Bank Group (2007) *Poverty Net, What is Poverty*. Online. Available at: http://web.worldbank.org/WBSITE/EXTERNAL/TOPICS/EXTPOVERTY/0,,contentMDK:20153855~menuPK:373757~pagePK:148956~piPK:216618~theSitePK:336992,00.html (accessed 14 July 2007).

Zawdie, G. and Langford, D.A. (2002) Influence of construction-based infrastructure on the development process in sub-Saharan Africa, *Building Research & Information*, 30 (3): 160–170.

Chapter 5

Community interaction in the construction industry

Krisen Moodley and Chris Preece

Introduction

Since the later period of the twentieth century, society has taken a progressively deeper interest in the behaviour of organisations. The socio-political lobby has become more influential and sophisticated and community politics is gaining prominence in shaping local decision making. Furthermore, business and development activities have the power to shape and alter the way society evolves through their actions and utilisation of resources. This has a far-reaching influence over social order. Consequently, greater accountability and community involvement are demanded by society.

The construction industry has a major impact on society. It has the power to shape and alter the way communities develop. Unlike many other industries, it has the ability to have a more profound impact through its products. The products of other industries are consumed and do not have a long-term presence. The impact of construction can be seen all around us. Construction is also a major employer and is used as a vehicle for economic and social development. The activities of construction companies impact on every community they work in on a local, national and international level.

Business and society

Over the past three decades, the study of business and society has been central to the development of the modern organisation. The development of the business–society paradigm has it roots in the initial work of Preston (1975: 434–453). This work focused on the management of social issues by organisations. It provided guidance as to what was a social issue and on whether the issue needed action. This eventually led to the understanding that the firm had to determine those issues for which it was socially responsible. Carroll (1979: 497–505; 1991: 38–48) built on this with his corporate social performance model that encapsulated corporate social responsibility (principles) and corporate social responsiveness (strategies). Wartick and Cochran

(1985: 758–769) defined dimensions of social responsibility as economic, legal, ethical and discretionary. This follows on from Caroll's original model. Economic and legal principles are easy to define but ethical and discretionary are more difficult. There are no ethical principles that can be cited or enforced. It is difficult to determine whether something is being implemented or is merely window dressing. The Wartick and Cochran approach is limited, as it has to fit into the categories considered. It was an external construct imposed on business that did not reflect the way in which business operated. An approach that reflected business relationships was seen to be more appropriate.

Freeman's (1984) landmark work provided an integrated business and society model through the development of stakeholder theory. The stakeholder model provided an integrated approach to the question of the interaction of business with society. Research in Canada, cited in Clarkson (1995: 92–117), distinguished between stakeholder issues and social issues. The key point was that firms had their relationships with their stakeholders, and not with society. Under other corporate-responsibility and social-performance models, it became difficult to determine what were social issues and their relevance to the firm. The stakeholder approach put the firm at the nexus of relationships it has with its stakeholders.

Freeman (1984) suggested that 'A stakeholder in an organisation is (by definition) any group or individual who can affect or is affected by achievement of the organisations' objectives'. Identifying and analysing stakeholders is a simple way to acknowledge the existence of multiple constituents of the firm. By adopting an open systems perspective, we can advocate that stakeholders have some sense of participation in the affairs of the firm.

Donaldson and Preston (1995: 65–91) indicate that there are descriptive, normative, instrumental and managerial aspects to the stakeholder model. The descriptive approach is a mode of describing what the organisation is, that is, a network of competitive and cooperative interests having intrinsic value. The instrumental approach is the framework for examining connections between the practice of stakeholder management and the achievement of corporate goals. The normative approach is based on the belief that stakeholders have legitimate interests that are of intrinsic value. This supports the belief that stakeholder goals are ends in themselves, rather than means to add efficiency and profitability. The intrinsic-worth approach primarily sees stakeholding as a matter of social conscience and what is morally good. The managerial approach is concerned with attitudes, structures, practices and policies that are needed to deal with the practical problems of stakeholder management rather than analysis. Stakeholder theory offers different directions and possible paths.

The stakeholder model offers ethicists an approach to the business and society problem. This model allows the identification of groups that have an interest in the firm and therefore have intrinsic value to the firm. The concept that the firm should have a social conscience provides a starting point for ethical approaches. This intrinsic-value approach, from a narrow sense, draws a deontological view of the moral imperative of duty. The moral-rights view indicates that stakeholders have rights that include consultation and involvement in decision making. The concept of inclusiveness is central to stakeholding and ethical thinking. In support of 'virtue ethics', supporters draw on the work of Aristotle and claim that a virtuous organisation would have inclusiveness in the community as part of its virtues.

Stoner and Winstanley (2001) indicate that the 'duty of care principle' requires stakeholder needs to be taken into account and possibly be consulted with. The Kantian approach tends to see the moral imperative that people are seen as ends in themselves rather than as means to an end. Wheeler and Sillanpaa (1997) indicated that an organisation that aspires to be a stakeholder corporation would almost certainly aspire to be described as ethical. They talk of 'shared values' rather than policies, strategies and documentation. Corporate citizenship also requires the creation of internal structure perspectives before dealing with external relationships. Good corporate citizenship is a part of every facet of the organisation.

The adoption of a more philanthropic outlook for the organisation focuses on the adoption of a total stakeholder approach to strategy. Freeman and Liedtka (1997) and Porter and van der Linde (1995) focus on the need for organisations to take cognisance of all the stakeholders of the company. By adopting this approach, greater attention can be given to external environmental influences and to people, communities and societies that are present throughout the life cycle of a company's development and its products. A shift in emphasis means that issues outside the market also gain support, thereby allowing the company to prosper in the total environment. The adoption of the stakeholder approach provides the base to justify the application of a new philanthropic paradigm.

Construction and the community

The construction industry is inextricably linked to the communities they work within. The industry changes the physical landscape and leaves products that last a lifetime within a community. There are growing concerns over the protection of the environment on which development is likely to encroach. This often leads to confrontations that are by no means isolated and extend to all types of construction and development projects. Can a new focused approach that takes cognisance of the concerns of people be more successful than the current approaches that often ignore concerns of society in general?

To function more effectively, the industry needs to consider different strategies when interacting with the external environment. The construction industry is linked to the communities within which they operate through the planning processes, labour, resources, production processes and, finally, the finished products. As communities become more sophisticated, they acquire the means to disrupt, delay or ultimately stop projects from achieving completion or conversely help expedite them. In the 1990s effective direct-action campaigns influenced a number of transportation infrastructure projects in the United Kingdom: Newbury, second runway at the Manchester airport, etc. It would therefore be logical for construction companies to seek the support of local communities to ensure projects are completed satisfactorily.

Increasingly, clients with a social remit or customer base require greater social input from their contractors and consultants. A structured approach to social policy is required to meet this need and provide viable deliverables. A more focused community strategy to develop closer relationships with target communities would therefore seem ideal to help in the implementation of construction activity. Clients do not want to alienate their potential customers by starting with negative perceptions about their business as they acquire new facilities.

Community perceptions of the construction sector

In 1996, the Construction Industry Board report, *Constructing a Better Image*, highlighted a range of problems with the image of the construction industry in the United Kingdom. This report was part of the broader review of the industry following the publication of *Constructing the Team* (Latham 1994). *Constructing a Better Image* indicated that the industry was perceived as promoting adversarial relationships, environmental insensitivity, poor working practices, and had a reputation for underperformance and cost overruns. The report recommended a sustained campaign to improve the image of the industry, including the setting up of the Considerate Constructors Scheme and the observance of the National Construction Week.

In 2002 the Building Research Establishment (BRE) published a report on community impacts, and according to it, many of the problems highlighted in 1994 still existed (Mindy 2002). The report, based on a series of case studies, highlights several problems during construction. These are as follows:

- noise, pollution and dust;
- parking, road closure, traffic, access and pavement obstruction;
- security, health and safety;
- general disorder, litter and workmen's negative attitudes.

Although the BRE report was based on a relatively limited number of projects, these projects do highlight concerns as they were carried out by reputable construction companies working for major procuring clients. Whilst these problems terminate once the projects are finished, construction activity, however, leaves a more permanent reminder of the impact of construction projects. The BRE report indicates that construction projects are not stand-alone activities and that the planning system needs to recognise this. They influence the whole environment that surrounds them in that they change views, influence wind patterns, impact on transport and can have far-reaching social and economic impact. The view that construction can disrupt communities was supported in the *Engage* CIRIA report (Sommer *et al.* 2004). The perception lingers that the construction industry does not engage adequately with the rest of society.

There is a perception that the industry responds more slowly at a social-responsibility level than most other industries. The KPMG 2005 survey of the use of sustainability reports indicates that only 28 per cent of construction companies produce such documents. This lags behind sectors such as mining, oil and gas, utilities, chemicals and synthetics, and forestry products. However, this is a considerable improvement over 2002, when only nine per cent of the construction industry produced sustainability reports. The impact of the industry is enormous, yet the level of concerns of organisations working in the industry would seem not to reflect this.

Community interaction

The development of a community strategy is concerned with the orientation of the company to key stakeholder influences. Corporate involvement in the community is not only geographic and project-specific, but requires a long-term outlook and a philosophy. This would seem to be contrary to the short-term approach of traditional construction projects. Therefore, the strategy needs to operate at two levels: a project-specific level and an industry/environmental level.

The industry/environmental strategy: This strategy is aimed at issues that influence the firm irrespective of the project or location. The community policy aims to support activities that are of strategic interest to the firm, as well as its norms and values. The other influential factors are, in general, social, economic, political and environmental issues. This part of the emerging strategy is concerned with the long-term approach needed to make community policy meaningful. The industry/environmental part of the strategy is easily implemented because it relates more directly to the corporate ethos of the firm. The values a firm adopts will provide the springboard to community-related activity. A company committed to education and training of its workforce would be able to translate these activities into a community policy. A literacy initiative, skills development programme, etc. may follow as a community strategy. A great deal of this activity will emerge from the broader CSR strategy.

The project strategy: This part of the strategy is aimed at local and project concerns. Ideally, local policies fit into the industry-wide strategy, but sufficient flexibility has to be built in to allow implementation at project level to be meaningful. Local economics, local values, local politics and local infrastructure play an important part in community strategy at project level.

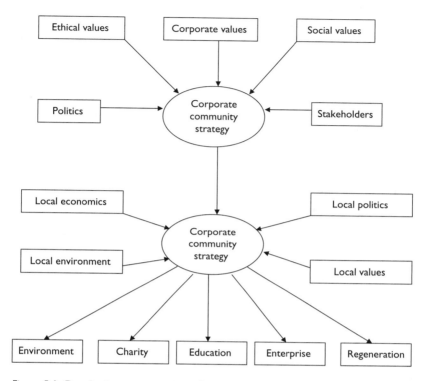

Figure 5.1 Developing a community policy.
Source: Adapted from Preece *et al.* (1998).

In the construction industry the project-level activity is often the driver for community policy. The community policy implemented locally acts as the easiest driver for CSR goals (see Figure 5.1).

Implementing the community policy

The starting point for the community policy is vested in the corporate goals and values of the company. These are specifically the value system of the organisation and its CSR goals. The strategic objectives of the company should include a clear altruistic commitment. All subsequent action can then be evaluated against these aims. Without a value-based system, the actions that a company takes will be seen as nothing more than public relations. The policy should not only be community-based, but be part of a broader corporate responsibility ethos. This removes the charges of image building rather than having concerted responsibility values.

Once the commitment to a philanthropic policy has been established, the next stage is to decide on the possible courses of action open to the firm.

The activities of the firm establish those that will most benefit from having a community policy associated with them. The current social issues and needs provide the areas from which the policy can proceed. The aim is to support a community issue that would be beneficial to the firm from more than a purely publicity context. The activities would contribute to the business interests of the firm.

An essential part of the community strategy is the commitment of resources to the policy. This may take the form of finance, staff or equipment. The commitment of resources for what are often seen as intangible returns is often the most difficult decision to make. Inadequate resources and the lack of genuine commitment to the policy will lead to the failure of the scheme. If the policy is tied in with the corporate goals, then the schemes are more likely to have adequate resources allocated to them.

The construction industry is noted for its instability. In a climate of change, it is difficult to convince the employees that a community policy is important or that the firm is concerned about social welfare. The case for the policy has to be made, if there is to be commitment to its implementation. The managers and staff who are required to implement these activities need to be aware of the importance of their actions, that there is a firm commitment to the policy and that it will not be axed during the next cost-cutting exercise.

Once the policy has been sold internally, it has to be tested in the environment in which it will operate. Involvement in costly schemes that do not satisfy community needs is futile and often counter-productive to the aims of the policy ideals. Consultation with interest groups on the actions and activities to be undertaken is important. The practicalities of the implementation, the problems and the possible outcomes are more likely to be identified during this process. The principal danger is that there is often a wide variety of possible interest groups that may be operating to their own agendas, therefore, care must be taken to ensure the programme is not hijacked. A useful dialogue will, however, ensure that the policy can be refined before being introduced. The use of third-party organisations such as Business in the Community, United Way or other charities can help shape and frame community policy. The approach undertaken by the organisation must fit into its overall values.

The policy should be fairly clear at this stage. Its aims and mechanisms should have been identified after the internal and external consultation stages. It is unlikely within the construction industry context that the execution of these activities will be taken by people exclusively involved in community policy. It is likely that construction project managers and regional managers will more likely be the group that is involved in many of the day-to-day decisions on these activities. These individuals will have to be made aware of the objectives of the community policy and how they help in developing the profile of the company. The business case for the community policy must be made. A clear and easy-to-use management interface has to be developed between those responsible for the community policy and those involved directly in its implementation. The support system for those staff who are not familiar with the policy should be introduced with a view to assist in decision making and action plans. Community activity should not be seen as outside the normal routine activity of the firm. If community activity is part of the culture of the organisation, it is more likely to be successful.

Corporate-level involvement: charity and charitable foundations

Community relations can extend to non-construction activity. The traditional form of community interaction has always been charitable giving or work in a local community. The UK housebuilder Persimmon has had its corporate headquarters in the city of York for 30 years. It has made a charitable pledge to the York Minster development campaign for ten years.

York Minster is the symbol of the historic city and offers Persimmon the opportunity to be involved with a project that is at the heart of civic pride. The firm also extends its charitable activity towards training schemes for apprentices for the Minster Project. New apprentices are required for the traditional skills that are needed for the repair and conservation of the Minster. Skills development will help the upkeep of the Minster in the long term. However, this does not exempt the company from facing rigorous scrutiny when it operates within the city. It still faces the criticisms and planning objections that are the normal part of a critical review in any community.

Charitable giving is the traditional way of engaging in the community. In situations such as this, the company may not benefit from adopting a stakeholder approach, as it is unable to influence business and politics. Companies in such situations are likely to lean towards an instrumental approach to stakeholding that can be used to increase efficiency and economic performance. Campbell (1997) suggests that adopting a stakeholder approach is the key to a better understanding of how to make money. However, the social contract approach can also be considered as part of the instrumental approach. In the case mentioned above, the organisation can be seen to be binding itself to its host environment. This rationale is an economic one.

CRASH

Corporate-level community involvement can be implemented either directly or through a third party. Companies that do not have the experience to implement community programmes may tap into the expertise of other charitable or community organisations. The company can then use the expertise of the charity to meet its community involvement goals. A good example of such an organisation is CRASH.

In 1991, Tony Denison founded CRASH in response to the problem of homelessness. Denison, whose background was in construction, soon enlisted the help of the construction industry to convert empty office spaces into temporary shelters. In the same year, the Rough Sleepers Initiative was started by the government, which soon led to a partnership between the government, the corporate sector and voluntary agencies. CRASH aims to help homelessness charities to improve their premises and facilities and the accommodation provided to homeless people. CRASH brings together expertise and help from professionals and companies in the UK construction and property sector to help in these projects. Charities that work with the homeless can call upon the pool of expertise from CRASH partner organisations to help produce better-quality projects. CRASH gets suppliers, property owners, architects, surveyors and engineers to provide advice and services on a non-profit basis. In certain cases the advice is provided free of charge.

CRASH describes itself as a second-tier charity and does not work directly with the homeless. The organisation supports those who work directly with homeless people. CRASH does not take on new-build or public-sector projects and does not award grants. This suggests CRASH is primarily a facilitation charity. The very nature of CRASH makes it an ideal charity to be supported by companies in the construction industry. The skills of these companies are easily deployed and fit into an overall philanthropic strategy. CRASH offers an opportunity for organisations to make contributions to philanthropic activities in a related environment. Involvement with CRASH or similar organisations gives the companies an opportunity to work with communities without concerns about the effectiveness of their actions.

Involvement with CRASH is a reaffirmation of the charity principle. The idea of those who are wealthier members of society being charitable to those less fortunate is a very old notion. By participating in a charity, the notion of assuming some responsibility for the plight of others is engendered. The cynical view might be that those that support charity are helping to counteract critics who claim that construction organisations are uncaring and interested only in profits. All organisations are driven by senior management, and their ethical values ultimately define the organisation's attitude to charity. Respect, integrity and empathy are characteristics that are central in developing individual ethical virtues (Carter 1985; Robinson *et al.* 2007). Mostakova-Possardt (2004) refers to these individual characteristics as the critical moral consciousness. In charitable giving, the individual morality of the leaders of construction organisations is important.

Charitable foundations

Another means of achieving community responsibility is through the use of charitable foundations. A charitable trust or foundation is a legal organisation that can be set up by anyone who has decided that he/she wants to set aside some of his/her assets or income for charitable causes. Each trust can define its remit and the manner in which it intends to spend its money. The use of a charitable trust allows the organisation to control the manner in which it wants to carry out its activities. It allows the organisation to control its philanthropic activities either directly or through third parties.

The Ove Arup Foundation was endowed by the Arup Group with an overall mission of the advancement of education directed towards the promotion, furtherance and dissemination of knowledge of matters associated with the built environment. These objectives are in acknowledgement of Sir Ove Arup's contribution over many decades to engineering, 'with emphasis on the multi-disciplinary nature of design in engineering and architecture'. The Ove Arup Foundation has at its heart a mission to stimulate new activities that will change the way people think and help them to acquire and apply the knowledge needed to improve their natural and built environment. The work of the foundation is directed towards the development of new courses, research and grants to promote learning in engineering and the built environment. Its activities cover the entire spectrum of education from school to higher education and research. This charitable trust focuses its efforts on areas that are closely related to the company that set up the trust. This approach allows the Ove Arup Foundation to perpetuate the ethos and ideas of its founding father. The Arup Group works alongside the Foundation as an integral part of its social responsibility programme. The Group also has a more extensive programme of socially responsible activities that are committed to projects. The Arup Group and the Ove Arup Foundation work in tandem to achieve social and community goals. The Arup approach heads towards enlightened self-interest, as it is a company that is socially aware without giving up its economic interests. The benefit of this approach is that the organisation gains economically, the customers gain value and the employees grow by behaving in a responsible manner. Understanding profound change and the ability to retain its values allows the organisation to survive.

The Wates Foundation was founded by the Wates family which is associated with the Wates Construction Group. The Foundation believes that the quality of life in society can be improved by informed and independent intervention to alleviate distress, deprivation and social exclusion. The Foundation aims to alleviate distress and improve the quality of life by promoting a broad range of social priorities with a core area of special focus. There is emphasis on the physical, mental and spiritual welfare of the young and disadvantaged aged 8–25. Racial equality is stressed throughout. Grants are concentrated on projects in the Greater London area with a preference for south London. The Foundation focuses on activities that are related to social and community improvement.

The activities of the Foundation do not relate to the activities of Wates Construction but are part of corporate responsibility activities of the commercial organisation. The Wates Foundation works towards helping and developing specific communities. Its actions are designed to help improve a particular location. In this case, the activities of the charitable foundation are not related to the construction activity. The Wates construction company has its own corporate responsibility programme, but recognises the role played by the charitable trust in helping communities. The Wates approach does show a recognition of the broader problems faced by communities and the need to tackle them. The philosophy is that a better society will benefit everyone. The purpose of the Wates Foundation is purely philanthropic. The basis under which it operates is to make social change rather than promote self-interest. This approach may be seen as the intrinsic-worth approach to stakeholders where the organisation's actions are seen as a matter of moral conscience. This view is supported by Freeman (1984) and Handy (1993, 1997) in that good in itself is morally justifiable. The Wates approach has no self-interest and follows a Kantian philosophy, that is, the moral imperative that people are seen as ends in themselves rather than as means to an end.

Project-level involvement: Considerate Constructors Scheme (CCS)

It is easier to implement community policies during projects. Resources are already present; however, the industry has not always been good neighbours. In1994, following Latham's review *Constructing the Team* the Construction Industry Council formed the Latham Review Implementation Forum (RIF). Working groups were set up, one of which being responsible for recommendations on 'The Image of the Industry'. This group identified improvements that could be made to the image of construction through better management and presentation of its sites. Two ongoing schemes were used as a template for a National Considerate Contractors Scheme. Between 1994 and 1997 the concept was developed in conjunction with the Chartered Institute of Building (CIOB).

The scheme evolved to become the Considerate Constructors Scheme so as to include all parts of the industry. In 1997 the big impetus for the scheme came with its adoption by a number of local authorities and it has now become an accepted part of good construction practice.

What are the principles behind the Considerate Constructors Scheme? The following are identified as key areas under the programme:

- Considerate – the work is carried out with consideration to the needs of traders, businesses, site staff, visitors and the general public. Attention should be given to the needs of people with visual or mobility concerns.
- Environment – noise from construction operations and other activities to be kept to a minimum. The use of local resources is to be encouraged. Attention should be paid to waste, recycling and pollution avoidance.

- Cleanliness – the working site is to be kept clean and in good order at all times. Barriers, lights, and warning signs are to be kept clean and well maintained.
- Good neighbour – general information regarding the scheme should be provided to neighbours affected by the work. Good communication with neighbours should be maintained.
- Respectful – respectable and safe standards of dress should be maintained. Lewd behaviour and derogatory language should not be tolerated. Pride in the management and appearance of the site should be achieved.
- Safe – operations should be carried out with care and consideration for site personnel and the public.
- Responsible – everyone associated with the construction should be aware of and comply with the code.
- Accountable – the scheme poster should be clearly visible with the details of the individuals responsible clearly displayed.

A more detailed account of the scheme can be found at www.considerate constructorsscheme.org.uk.

Complaints that cannot be dealt with directly can be referred to the CCS. The complaint is logged and referred to the site manager for a remedy. If the response is not satisfactory, the problem can be raised through the organisation to director level. If the problem persists, then the CCS can remove the company from the scheme.

The CCS is a good starting point for community involvement. It took the industry until the mid-1990s to realise that they were not good neighbours. The CCS represents a small start in the right direction. The growth of the scheme is largely based on the attitudes of individual companies and local authorities. In areas where the scheme is not mandatory, performance can fall unless individual companies adopt the scheme.

Critics of the CCS suggest that the approach does not go far enough and does not work through all levels within the industry. Without widespread enforcement for all types and sizes of projects, the improvements that CCS brings will be tarnished by the non-compliers. The construction industry is notorious for its slow change, and the CCS is a step in the right direction. It is not the 'Holy Grail' of community relations, but represents a first step that companies can take towards developing a better community profile. It is a sober reminder that it took until the last decade of the twentieth century for the construction industry to decide to be good neighbours and implement a scheme such as CCS.

Environment and the community

The construction industry does not have a great environmental record. An environmental policy is almost a pre-requisite for every major construction company. Current legislative trends suggest that all construction companies will be forced down the path of becoming more environmentally conscious. A company that is likely to be involved in an activity that is environmentally sensitive should focus its activity on community schemes that will bring it into closer contact with the interest groups. It reinforces the commitment of the company to the environment and provides a useful mechanism to gain feedback and advice on how to deal with future environmental problems.

A proactive response to environmental problems is a good way of setting the tone for a community initiative. Areas that are focused on during the development of environmental policy are renewable resources, pollution, energy consumption, fossil fuels, transport and pro-environmental initiatives. These activities have to be promoted as part of an ongoing commitment to improve social conditions. These elements remain a starting point, and more consideration should be given to design, sourcing and transportation of materials and components. Many of the more proactive construction companies are now publishing audited environmental reports that evaluate the organisation's environmental performance.

A leading example of a proactive environmental policy is that of Carillion, which has a long-standing record of publishing its environmental performance. It is possible to track the organisation's environmental performance over a period of time. At the heart of this policy is how the company performs at a community project level. The activities at project level support the broader environmental goals of the firm. Reporting environmental performance is being adopted by more companies which, along with third-party auditing, should help to improve the overall environmental performance within the industry.

The environmental strategy of a construction company can never be confined to the narrow remit of its corporate operations. Central to any scheme will be activity at local project level as well. Issues such as noise, dust, pollution and graffiti should become part of the environmental response to any project. Ultimately, performance at this level determines the public image of the firm in terms of its environmental activity. This is an area where the complete adoption of the CCS for all projects would be useful. The environmental appraisal of construction projects should also look at measures such as landscape rehabilitation, removal of waste and decontamination. Environmental or corporate responsibility reports are dependent on project-level performance. To meet the requirements of these reports, companies undertake project-level environmental schemes. They also become involved in environmental projects and schemes close to their projects. This may involve relatively simple activities such as removing litter and graffiti, planting trees, clearing watercourses, establishing natural habitat or creating gardens and green spaces. Though this does not establish the environmental credibility of the company, it does indicate a willingness to improve the environment in which the company is operating. As global environmental concerns increase, the environmental impact on local communities will become a bigger concern.

Enterprise, education and training

Enterprise, education and training need not be construction-specific to help society. Community involvement means attention can also be given to areas outside construction. Companies may help develop literacy programmes, offer childcare facilities, distribute sports equipment or provide advisory services for the local community to develop its policies.

Regeneration, suppliers and employment

Construction is a key driver for regeneration, particularly in urban areas. This offers another opportunity to develop a community policy and link it to the aims of the firm. The traditional view on regeneration was to develop the infrastructure and then allow the communities to fit into the schemes. This approach was identified as being flawed, and the goal of new regeneration schemes is a partnership between the local community, the constructors and the government/funding agencies. Regeneration schemes allow construction organisations a unique opportunity to build a long-term presence in the community.

The construction industry's status as a major employer within communities can also be brought into the remit of community policy. In particular, the extent of local labour used is allied to any additional training provided. Improved skills make people more employable and are a value-added benefit to society. In the United Kingdom the erosion of the construction skills base is a problem, and, therefore, the development of a skills base is important. Regeneration projects, often in deprived areas with high unemployment, offer opportunities for skills development. The number of new jobs created and the number of locals undertaking these jobs are good indicators of commitment to the community's economic regeneration. There is a counter-argument that stresses the mobility of labour. Using local labour helps to improve the economic multiplier effect and helps communities in the long term.

Materials, plant and equipment for the construction industry are sourced from the most efficient and cost-effective sources. It is possible to source materials and components in the global marketplace. Though it is often more cost-effective to source materials globally, there are possible negative impacts. Environmental costs associated with transport, ethical behaviour of suppliers, possible environmental damage and poor working conditions may all be part of the global sourcing issues. The cost–benefit analysis of sourcing locally should be carried out as part of a broader corporate responsibility remit. The benefits to local communities in terms of employment, economic regeneration, environmental impact and maintenance of skills bases have to be considered. The question arises: is construction a stand-alone activity or should it attempt to support and sustain economic growth within a community? The socio-economic impact of construction should not be underestimated.

Community relations and communication

Construction work will disturb the nature of a community no matter how hard the industry tries to minimise the impact. Following a sustained period of direct action in the 1990s and high-profile protests, the UK government decided to act on the way information is handled on public projects. A 'Code of Practice for the Dissemination of Information on Major Infrastructure Projects' was introduced in 1999 (ODPM 1999). The primary aim was to keep the communities, and public at large, informed of activities on projects, particularly during the planning stage. In order to minimise the impact, there is a need to inform the community of the various actions taking place to protect the reputation of the company and its clients.

Where the project is of a sensitive nature or of a substantial scale, it may be prudent to engage directly with the community by holding a public inform-ation meeting or a similar event. This event should have representatives of the client, the contractor and other relevant parties on hand to answer questions about the project. This provides an opportunity to remove any misconceptions and reassure the public about the project. Gambrill (2003) indicates that on the Channel Tunnel Rail Link (CTRL) project, an inde-pendent complaints commissioner was appointed. The commissioner dealt with complaints during the planning phase, compulsory purchase period and during the life of the project. This alternative dispute procedure was conceived during the passing of the Channel Tunnel Act, which met the requirements for having an independent complaints commissioner. This was a unique role in that the commissioner's office captured all the complaints irrespective of whom the complaint was lodged against. This allowed com-plaints to be coordinated at a single point. Complaints related to issues such as dust, flooding, settlement or traffic-related issues. The complaints com-missioner had to look into persistent, ill-founded and malicious complaints and deal with them accordingly.

Kennerley (2003) indicated that the success of the complaints commis-sioner was dependent on support staff and building good relationships with engineers and contractors. The success of the complaints commissioner helped reduce complaints and allow contractors and engineers to get on with the projects. The high-level forum which is part of the CTRL project also indicated that the complaints commissioner worked well on this scheme. The CTRL project is large and covers many communities, and by having a complaints commissioner there is an acknowledgement of the need to engage with communities.

Active communication is important from the very outset of a project irrespective of its size. A press release in both local and regional press outlining the project and providing details of programme and contacts is useful. However, this will depend on the size of the individual project, as small projects may not need this level of communication. Nevertheless, good practice is always to inform the community where possible. Communication with the community must continue through the life of the project. Traditionally, this was done in the form of leaflets or newsletters. Current technology allows for the development of a project website. Certain key information should be provided over the life of the project, irrespective of the technology used. Information on the project duration, progress, likely disruption and remediation, health and safety, and contact details should be made available on a continuous basis. The facility for the community to ask questions or make complaints should also be available throughout the life of the project. A critical area for construction–community interaction is communication. If misconception and rumours are disspelled, communities are more likely to accept construction. Communication with communities should not be seen as a chore and the facility to respond and interact should be an essential part of any construction project.

Case study: Costain Construction

Costain Construction is a good all-encompassing case study of a major firm in the sector that adopts a multifaceted approach. Most Costain employees at its London headquarters do not live in the area. However, they are keen to set an example. Through Business in the Community, they have developed relationships with a few carefully selected community projects. Costain works with a local primary school on a reading scheme at lunchtimes, and its Christmas cards are designed by the pupils and 'bought' by the company. At the local parish church, it provides accounting and public relations expertise for the church's 'Breakaway' project for homeless people, whilst the construction division reroofed the church 'at cost'. Costain's support for a local drop-in advice and community centre includes services ranging from painting and gardening to strategic and management consultancy. This is corporate involvement in a local community.

At a company and regional level Costain follows a similar approach. Education, environment and regeneration are at the heart of its approach. When it is possible, the company works with schools and universities, community groups and public, private and voluntary organisations to secure benefits in education, improve the environment and support the revamping of whole communities. Costain sees itself as a resource to be tapped by people in these areas. Involvement is governed by proximity to sites and an appropriate fit between community needs and the skills, experience and resources of Costain people.

(Continued)

The Group has created a 'Community Chest'. Instead of making gifts to needy causes, it now invites employees to bid for small sums to enable them to become genuinely involved in community projects. Only if employees can demonstrate their involvement, the commitment of their community colleagues and the clearly identified, targeted benefits to those involved in the project, will the 'Community Chest' pay out. This allows Costain to engage in community activity through all levels of the organisation as well as engaging its staff. The Costain model may not be the solution for all companies, but it does offer a template of how a company can engage with a community in the areas of education and training.

Conclusions

The construction sector has a major impact on the environment. Its products and processes are not static and can affect communities locally, nationally and internationally. The image of the sector is generally poor, despite notable examples of successful schemes and measurable improvements in performance. The challenge for construction firms is to combat negative and cynical public opinions. Corporate and project stakeholders are more demanding, and there is far more public and media scrutiny of the activities of businesses than in the past.

This chapter has shown that corporate involvement in the community is of benefit as it may help minimise delays during the planning phase and during the construction life. It can also help in preventing damage to the reputation of the contractors, consultants and clients. There is also the possibility of political and social support through the goodwill of communities. Engagement in communities also creates broader social benefits in improvement of skills, education and regeneration, which all help to improve society, as the construction industry starts to become involved in communities on a much larger scale.

The leading construction companies, such as those mentioned in this chapter, have often been committed to the communities in which they operate and are in the process of developing more comprehensive approaches to community relations. However, the vast majority of businesses in this sector, which tend to be at the smaller end of the market, are not. The poor reputation of construction is largely due not to the major organisations which most members of the community have never heard of, but to the smaller firms that carry out most of the work for the general public and end up being vilified on popular consumer rights television programmes. However, communities will not tolerate poor performance from large or small companies.

Approximately 30 per cent of construction activity is now registered with the Considerate Constructors Scheme, although not all parts of the United

Kingdom are compliant. Over the past ten years, its questions have evolved with the times to include whether contractors are monitoring their carbon footprints, whether they provide access for all and how many of their operatives speak English (Olcayto 2007). There is evidence that performance on project sites, and the way in which projects communicate with their neighbours have improved. However CCS is not the only programme, and there would seem to be a need for a single point of reference and a standardised benchmark. A new development in the scheme is to identify the 'ideal' site of the future.

The modern construction firm needs to understand its stakeholders at a corporate and project level. It needs to be able to communicate more effectively to ensure a positive business climate for its operations. Any type of work undertaken by a construction firm, whether it is renovation or new build, is, without doubt, going to have an impact on the environment and local communities. Despite the amazing achievements of the construction industry, contractors still have a poor reputation. Construction, no matter how large or small, can inconvenience local residents just as an inconsiderate neighbour can. Goggles and ear-defenders may assist workers; however, the public are still bombarded with clouds of dust, noise, assault courses for public walkways, vibration and traffic delays.

It takes construction firms many years to build up a reputation in society; this can evaporate within days. If any local residents are suffering as a result of an inconsiderate site, what they will see is a bad neighbour and a poorly run company – someone, or some business, that simply does not care. Respect for local communities is essential, as it is they who inevitably have a major input into whether a construction firm exists, or simply withers away.

The authors suggest a ten-point plan to help contractors address the challenges (Preece *et al.*, 2006). This set of guidelines will assist contractors in reducing the negative impacts on the environment and local communities, and help enhance mutual understanding and goodwill. A professional, effective approach to community relations begins right at the start, before the project has even been won. Included is advice throughout a project's timescale from putting the bid together, to when the work is finished. The ten points are as follows:

1. Form a community-relations plan as part of a bid for work.
2. Appoint an on-site liaison representative. He/she is to be responsible for the complaints procedure and hence dealing with the public.
3. Ensure that before site-work begins, the foundations for public contact are installed, i.e. Helpline, Webpage and Intranet.
4. Conduct residents' meetings and letter drops, both before and during the project.
5. Be a good neighbour – follow the site etiquette.
6. Conduct a thorough scheme assessment and make appropriate changes.

7. Be committed to conducted schemes throughout the project. Repair work and renovation should be accomplished.
8. Enforce close relations with public services, the media and, most importantly, the local residents – they are not as scary as they seem!
9. Join the Considerate Constructors Scheme (or similar).
10. Learn from mistakes – it is understandable to make a mistake once; take heed and do not repeat it again.

Given the demands of CSR in general, a movement has started in the construction sector which should bring about tangible improvements in performance and reputation over time and make a difference to communities. However, currently, the question remains, whether most of the construction businesses are doing anything other than developing a more sophisticated approach to public relations.

The real drivers for change in this industry come from government and clients. Given the greater emphasis on sustainability and corporate social responsibility across all business sectors, it is expected that the development of effective community relations practices will become increasingly important in the construction industry of the twenty-first century.

Websites

Business in the Community: www.bitc.org.uk
Carillion: www.carillionplc.com
Considerate Constructors Scheme: www.considerateconstructorsscheme.org.uk
Costain: www.costain.com
CRASH: www.crash.org.uk
Ove Arup Foundation: www.theovearupfoundation.org
United Way: www.unitedway.org
Wates Foundation: www.watesfoundation.org.uk

References

Campbell, A. (1997) Stakeholders: the case in favour, *Long Range Planning*, 30 (3): 446–449.
Carroll, A.B. (1979) A three-dimensional conceptual model of corporate performance, *Academy of Management Review*, 4: 497–505.
Carroll, A.B. (1991) The pyramid of corporate social responsibility: towards the moral management of the organisations stakeholders, *Business Horizons*, July/August: 38–48.
Carter, R. (1985) A taxonomy of objectives for professional education, *Studies in Higher Education*, 10 (20): 135–149.
Clarkson, M.B.E. (1995) A stakeholder framework for analyzing and evaluating corporate social performance, *Academy of Management Review*, 20 (1): 92–117.
Donaldson, P. and Preston, L.E. (1995) The stakeholder theory of the corporation: concepts, evidence, and implications, *Academy of Management Review*, 20 (1): 65–91.

Freeman, R.E. (1984) *Strategic Management: A Stakeholder Approach*, Boston: Pitman.

Freeman, R.E. and Liedtka, J. (1997) Stakeholder capitalism and the value chain, *European Journal of Management*, 15 (3): 286–296.

Gambrill, B. (2003) The Channel Tunnel Rail Link: community relations during implementation, *Proceedings of the ICE, Civil Engineering*, 156: 24–27.

Handy, C. (1993) What is a company for? *Corporate Governance an International Review*, 1 (1): 14–16.

Handy, C. (1997) *The Hungry Spirit*, London: Hutchinson.

Kennerley, J.A. (2003) The Channel Tunnel Rail Link: the complaints commissioner, *Proceedings of the ICE, Civil Engineering*, 156: 54–58.

KPMG (2005) *International Survey of Corporate Sustainability Reporting 2005*, University of Amsterdam and KPMG Global Sustainability Services, pp. 12–13. ISBN 90-5522-031-0.

Latham, M. (1994) *Constructing the Team: Final Report; Joint Review of Procurement and Contractual Arrangements in the United Kingdom Construction Industry*, London: HMSO.

Mindy, H. (2002) *Working with the Community*, Project Report Number 81169, Building Research Establishment.

Mostakova-Possardt, E. (2004) Education for critical moral consciousness, *Journal of Moral Education*, 33: 245–270.

Office of the Deputy Prime Minister (1999) *Code of Practice on the Dissemination of Information on Major Infrastructure Projects during Construction*, ODPM, October.

Olcayto, R. (2007) Taking care of business, *Construction Manager*, May: 15–17.

Porter, M.E. and van der Linde, C. (1995) Green and competitive: ending the stalemate, *Harvard Business Review*, 73 (5): 120–133.

Preece, C., Moodley, K. and Smith, A. (1998) *Corporate Communications in Construction*, Cambridge, UK: Blackwell Science Ltd.

Preece, C.N., Townsend, J. and Moodley, K. (2006) *Guidelines for Effective Community Relations Surrounding Construction Projects*, Construction Management Group, School of Civil Engineering, University of Leeds Report, pp. 1–29.

Preston, L.E. (1975) Corporation and society: the search for a paradigm, *Journal of Economic Literature*, 13: 434–453.

Robinson, S., Dixon, J.R., Preece, C.N. and Moodley, K. (2007) *Engineering, Business and Professional Ethics*, Oxford: Butterworth-Heinemann.

Sommer, F., Bootland, J., Hunt, M., Khurana, M., Reid, A. and Wilson, S. (2004) *ENGAGE How to Deliver Socially Responsible Construction: A Client's Guide*, Ciria C627, CIRIA.

Stoner, C. and Winstanley, D. (2001) Stakeholding: confusion or utopia. Mapping the conceptual theory, *Journal of Management Studies*, 38 (5): 603–626.

The Construction Industry Board (1996) *Constructing a Better Image. A report by Working Group 7 of the Construction Industry Board*, London: Thomas Telford.

Wartick, S.L. and Cochran, P.L. (1985) The evolution of the corporate social performance model, *Academy of Management Review*, 4: 758–769.

Wheeler, D. and Sillanpaa, M. (1997) *The Stakeholder Corporation*, London: Pitman.

Part III

Prevalence and nature of corrupt practices

Chapter 6

Corruption in the UK construction industry

Current and future effects

John Tookey and Dale Chalmers

Introduction

The artist and writer Ashleigh Brilliant once said, 'I either want less corruption, or more chance to participate in it'. The somewhat cynical and double-edged sentiment in his observation neatly summarises the attitude of Western society in general, and its construction industry in particular. The spectre of bribery and corruption has plagued society around the world for centuries in every walk of life. It represents, in the words of the song, 'the oldest, yet the latest thing'. Although often perceived as being a problem endemic in the developing world, in recent years even mature democracies such as the United Kingdom have experienced their fair share of corruption that has warranted criminal investigation. The now infamous 'Cash for Peerages Row' was an issue almost a century ago when David Lloyd George's Liberal Government of the day faced accusations of engaging in corrupt practice, which led to the Honours (Prevention of Abuses) Act 1925 being enacted.[1]

In construction, in particular, there has been a 'quasi-normality' in corruption and dishonest practices for many years. This has ranged from spurious claims and fraudulent invoices for services not delivered, down to the blatant cash-stuffed brown envelopes of the fully fledged bribe to get work (Anon. 2001: 12). In July 2005, Stansbury and Stansbury (2005) published an Anti-Corruption Code for individuals in the construction and engineering industry (England & Wales) on behalf of Transparency International (UK), which is an independent voluntary organisation, with the aim of improving ethical standards in business and government. The emphasis of this act was to attempt to make construction professionals aware that perceived 'industry norms' are actually criminal, despite their innocuous nomenclature. In developing countries, and those new to democracy, the UK system is often held up as a model to emulate for their fledgling systems. These countries, which are often exhorted to demonstrate transparency and good governance before aid funds are released, will be disconcerted to learn that the United Kingdom itself needs to rely upon a non-governmental organisation (NGO) such as Transparency International for a code to advise individuals and corporations to conduct their business affairs within the bounds of the Criminal Law in force in the country.

Corruption: developing or developed world issue?

How can we condemn corruption elsewhere in the world, whilst tacitly condoning it in the UK? It would appear that various UK governments, past and present, have forfeited the right to the moral high ground, since their acts, omissions or wilful blindness at home and abroad contribute to the perpetuation of corruption and bribery in the construction industry. The regularity of unfair, collusive and corrupt practices reported by the UK Office of Fair Trading would appear to demonstrate the widespread nature of the problem in the UK construction industry. For example, in early 2004, nine roofing contractors were fined for price-fixing in the West Midlands (OFT 2004a). As recently as 2005, the Office of Fair Trading (OFT 2005) reported that as many as 21 per cent of small and medium enterprises (SMEs) in the construction sector felt they could not compete freely and fairly for contracts put out to tender, as a result of collusion. This is further reinforced by a recent Chartered Institute of Building (CIOB) poll that indicated 41 per cent of its membership felt that corruption in the construction industry was 'widespread' – costing the industry around £3 billion per annum (CIOB 2006).

Furthermore, there unfortunately remains a pervasive air of criminality, and the potential for criminality, throughout the construction industry in the United Kingdom. At a lower level, there are significant opportunities for workers, managers and smaller companies to increase revenues by engaging in corrupt practices. These cover a spectrum of occurrences. At the 'ordinary-worker' level, corrupt practices include claiming government unemployment benefit whilst working in construction (Glackin 1998: 18), which is at endemic levels; forgery of *Construction Skills Certification Scheme* (CSCS) certificates, which purport to demonstrate that the holder has National Vocational Qualifications (NVQ) when they do not (Lynch 2004: 1) and illegal payment mechanisms – not to mention complete tax evasion – by self-employed workers on a huge variety of projects (d'Arcy 2001: 2). Frighteningly, even when illegal practices are reported to the UK authorities, there also seems to be a marked reluctance to engage in vigorous pursuit of those individuals whose corrupt behaviour appears quasi-officially classified in the 'no one got hurt so don't worry' category (Rogers 2005: 3).

Similarly, as we move up the 'corruption food chain', there are numerous further examples, such as labour-only subcontractors invoicing main contractors for workers who never existed being on site – the so-called 'dead men' scam. A recent notable instance was that of the £659 million dockyard upgrade at HMNB Devonport, which resulted in a substantial police investigation (Barry 2002: 5). In the same vein, a groundworks subcontractor at Heathrow Airport's new Terminal 5 project was removed from site for 'accounting irregularities' in the delivery of numerous wagonloads of lime for soil stabilisation works. In this particular corrupt practice, the subcontractor involved was invoicing the clients for thousands of tonnes of materials delivered to site that had, in fact, never been delivered, since the volume of trucks involved precluded the weighing of vehicles on entrance and exit from the site (Prior 2004: 6). Another variation on this particular scam is the rebadging of vehicles such as excavators (20-tonners became 24-tonners; 29-tonners became 34-tonners, etc.), so they appear capable of carrying heavier loads than they are actually capable of, and therefore permit corrupt managers to fraudulently acquire higher rental incomes and/or win contracts by deception (Anon. 2001: 12).

Inter-company corruption

However, it is at the high level of inter-company collusion and deception that most financial damage and wrongdoing can occur. Typically, this form of corruption takes place during tendering for contracts, where lucrative contracts are often 'gifted' to favoured subcontractors. A recent example that has been highlighted is the Hampden Park stadium project in Scotland (Staples 2002: 3) where investigations had to be launched into a contractor gaining substantial advantage in a tendering process and the management contractor receiving a significantly higher margin on the works undertaken as a direct result.

Staying in Scotland, another notable recent example has been the remarkable process of procuring the construction of the notorious Scottish Parliament building at Holyrood. The project for this landmark building, massively over budget and delayed years after scheduled completion, was embroiled in scandals throughout its gestation period. Oft-quoted examples of such scandals include the appointment of a management contractor who was selected in spite of submitting the most expensive tender; similarly, the final selection of the construction site came about in the most opaque of circumstances, involving, as it did, a chance meeting of key protagonists on a train from Glasgow to Edinburgh. The most highlighted aspect of alleged impropriety with regard to the project was the selection of Enric Miralles as lead architect, in spite of the fact that his company's initial submission was ranked 44th of 70 applications in the design competition (notwithstanding that the judges did not notice that his firm did not have sufficient personal indemnity insurance to work in the United Kingdom). The circumstances leading to the time and cost overruns in the Scottish Parliament project, along with some of the accompanying details of the management practices, not to mention the indifferent governance of the project, can be found in the highly illuminating *Fraser Report* (2004).

Championing the anti-corruption cause

Given the convoluted nature of contracts undertaken in the construction industry, corrupt practice at all levels and the inability of even governmental contracts to be free of corruption, an obvious question must be posed. If the UK Government (and, indeed, the Scottish legislative body) cannot be relied upon to take the lead to eliminate corruption in a major industry sector, who then should champion the cause? The potential candidates for this role include the Judiciary, Crown Prosecution Services in England and Wales (and *ipso facto* Procurator Fiscals in Scotland) and, indeed, the companies and construction professionals themselves including clients, contractors, subcontractors, consultants, employees and professional bodies. The alternative is for the construction industry to rely upon non-profit-making NGOs like Transparency International United Kingdom to highlight the intolerable presence of corruption in both public and private projects.

Although the overall utopian industry objective would be elimination of corruption, it must be accepted that this goal is unfeasible. People are ultimately people, and there will always be an obvious temptation to use the status and power of one's position in a professional role to lever personal advantage. This is the case, whether the incident considered is of a CEO of a large corporation receiving payments for preferring a particular bidder, or a cowboy builder exploiting the general public in undertaking unnecessary repairs on a house. Thus, before proposing a route towards improving the transparency and governance of UK construction, we must assess the current state of the industry's attitude to corruption. This will allow the assessment of whether the lessons that should have been learned from past experiences have been fully assimilated into current governance practice.

Defining corruption

Corruption itself is hard to define, since it has so much by way of subjective value attached to it depending on the country and culture in question. As such, it becomes necessary to fully define corruption before further analysis of the phenomenon can be undertaken. The *Chambers Dictionary* defines corruption as '[the quality of] bribery; dishonesty; rottenness; impurity'. Within the context of anti-corruption initiatives in the United Kingdom, the popular definition of corruption is 'the abuse of entrusted power for private gain' (Transparency International 2003).

Defining 'corruption' has proved extremely difficult for legislators when attempting to criminalise specific acts. The UK Government published a White Paper in June 2000 (HM Government 2000) for public consultation, setting out to codify existing laws into a single piece of legislation. After receiving criticisms and recommendations, a Draft Corruption Bill was subsequently published, which in turn was severely criticised by the

Joint Parliamentary Committee in its report, claiming particularly that the definitions of 'corruption' and 'corruptly' contained in the Bill were 'opaque'. The committee also argued that the Bill failed to cover corrupt conduct such as when the CEO of one firm bribes the CEO of another, or when an employer consents to the bribery of his agent. Undaunted, however, the response from the Government was a robust defence of the Bill's definition of corruption. When publishing its justifications in 2003, the Government rejected the Committee's recommendation that the essence of corruption could be better expressed in the following terms:

> A person acts corruptly if he gives, offers or agrees to give an improper advantage with the intention of influencing the recipient in the perform-ance of his duties. Alternatively a person acts corruptly if he receives, asks for or agrees to receive an improper advantage with the intention that it will influence him in the performance of his duties or functions.
> (HM Government 2003: recommendation 6)

Indeed, in construction, there are myriad possible mechanisms by which risk of fraud or corruption can be introduced into the process of infrastructure delivery (see Table 6.1). All of these mechanisms can be introduced in all types of economy, both in the developed world and in the developing world (Control Risks Group 2003).

The Government argued that '[t]he definition of corruption is based upon the principal construct which has existed in corruption law since 1906, and is expressed in two instances of case law as "a dishonest intention to weaken the loyalty of the servants to their master and to transfer that loyalty from the master to the giver" (Lindley 1957 case), and "dishonestly trying to wheedle an agent away from his loyalty to his employer" (Calland 1967 case)' (HM Government 2003: response 3). The Government went on to claim, '[There is] no evidence of activities of bribery outside of the agent/principal relation-ship which are morally reprehensible enough to be criminalized but which do not fall in the ambit of other statutes' (ibid.; response 14).

Much of the Government's defence against the various criticisms levelled at the Draft Corruption Bill stems from its need to cover extremely complex situations. It points out that the Committee's proposed definitions would not cover the situations where '[a] bribe is offered not to the person whose behaviour it is intended to influence. For example, giving a bribe to the wife of a public official in order to influence the official to award a contract would result in it being impossible to prosecute the wife who is the recipient of the bribe, nor the official who has kept his side of the bargain and awarded a contract' (ibid.; response 16).

Criticism from other quarters does not concern the legal problems of defin-ition. Rather it suggests that part of the reason why the Government has not yet enacted legislation under consideration for the majority of the cur-rent Government's term of office is 'the lack of political will within the

Table 6.1 Fraud and corruption within construction projects.

Project phase	Key risks
Design and construction planning	Budget overstated; project or project component not needed; project split to avoid tendering threshold
Designer selection	Improper relationship between project team and designer and/or contractor; abuse of tendering process for designer selection
Preparation of plans and specifications	Tailored to preferred contractor; vague and imprecise; cost element includes cost of corrupt payment made to secure contract
Invitation to tender	Improper relationship between project team and contractor; tailored to preferred contractor; restricted to preferred contractor; tenders from related parties; tenders from fictitious parties
Pre-qualifying review	Review designed to favour preferred contractor
Bidding	Bid information provided to preferred contractor; bids altered after opening; bidding cartel which decides in advance who will bid to win; bids not held securely by an independent party
Bid evaluation	Criteria ignored or distorted to ensure that preferred contractor wins; bids from non-preferred contractors are lost; bid evaluation process not documented
Construction process	Sub-standard work passed; theft of plant and materials; false claims for delay and disruption; variation claims for work not done; wrong overheads charged to project; costs charged from a different project; over-invoicing for materials and labour; order costs exaggerated and approved

Source: Adapted from Control Risks Group (2003).

Government' and that 'legislative reform is [not] high enough on the Government's agenda' (Macauley 2005). The Corruption Bill, when eventually enacted, will be an extremely important piece of legislation. Its impact will be felt domestically and internationally, and it is worthy at this early stage to address this Bill as if it is the final piece of legislation which, when enacted, ratifies the United Kingdom's signing of the UN Convention against Corruption in 2003. It is somewhat unfortunate, and wasteful in time and effort, that the Corruption Bill will not extend to Scotland. Instead, the Scottish Parliament will have to enact similar legislation separately from the rest of the United Kingdom before the UN Convention can be fully ratified.

One other important feature of the Draft Corruption Bill is that to satisfy the United Kingdom's international obligation, the Act will have an extra-territorial effect. Therefore, a UK based party committing an act of corruption on foreign soil will still be guilty of the offence of corruption if its actions can be brought within any of the prescribed offences. The fact that its definition of a corrupt act is ambiguous has led even the Government to concede, 'the approach we have adopted leads to a...[complex statute which]...may well not be immediately understood by the layman' (HM Government 2003; recommendation 6). Unfortunately, some corrupt acts not covered by the Act, falling domestically within other statutes, in an international sense may go unpunished since the corrupt act is committed abroad, and the applicable UK statute has no extra-territorial effect.

Internationally, defining corruption has proved equally troublesome, and most of the literature refers to criminality; e.g. in the United States 'corruption' is synonymous with the criminal label for extortion and bribery. As the New York State Organised Crime Task Force reported:

> In the construction industry only a thin line separates extortion and bribery. Illegal payments flow from contractors to officials. Sometimes money is paid to avoid an explicit or implicit threat; this is extortion. Sometimes money is paid to buy favours; this is bribery. Sometimes contractors claim not to know exactly why they pay; experience tells them that payoffs are necessary to assure that things will run smoothly.
>
> (Goldstock *et al.* 1989: 19)

Among the other crimes under the corruption banner, Goldstock *et al.* (1989) included collusive bidding, bid-rigging, theft, sabotage, fraud and racketeering. Similarly, in Japan, the construction industry was insulated from foreign competition and, as a result, an institutionalised system of bid-rigging known as *dangó* (agreement by consultation) flourished. *Dangó* is an independent arrangement among contractors to decide the successful bidder, or as Woodall (1996: 19) described, it is 'the contractors' practice of neatly dividing up and parcelling out their construction market by a gentleman's agreement'. Although the stifling of foreign competition allowed *dangó* to flourish, government officials benefited greatly not by anything so vulgar as receipt of cash from their favourite contractors, but by the practice of *amakudon* (descent from heaven). Career civil servants usually retired at the age of 50 but were able to follow second lucrative careers in senior management within the construction contractor's organisation whom they had favoured most when employed by the government (ibid.: 68–74).

Closer to home, the Irish Republic suffered greatly from corruption within construction. A particular concern was county councillors abusing the rezoning system initiated in the 1963 Planning Act, by the expedient of redesignating a piece of land for housing instead of agriculture:

> It did not matter that there were no roads or sewers to the land, or that there was plenty of building land elsewhere. The designation of land would be decided at periodic reviews but, crucially, county councillors were given the power to override the decisions taken by their officials.
>
> (Cullen 2002)

The corruption scandal that resulted, investigated by the 1997 Flood Inquiry, implicated local builders, county councillors, government ministers and even the former Taoiseach Charles Haughey. This body is still in existence sifting a quagmire of corruption. The vagaries of the definition of corrupt practices could, and does, lead to some companies engaging different standards of ethical behaviour depending on the country in which particular projects are undertaken. For these reasons, the highest standard should be thought most likely to reduce the construction industry's susceptibility to corruption in the United Kingdom.

Consequences of corruption

The primary reason for the construction industry meriting special treatment is that it would appear to be the most corrupt (BPI 2006). Construction industry insiders might have been aware of this fact for many years, but when Transparency International began publishing statistics on countries and sectors most prone to pay bribes, public works and construction outstripped all other sectors to emerge as the world leader. The European Commission (2003) estimated that the global cost of corruption amounts to approximately five per cent of the world economy. On a pro rata basis, in the UK construction industry this would equate to £4 billion of the total industry size of £80 billion (Construction Statistics Annual, DTI, 2005). Other authors (e.g. Stansbury and Stansbury 2005) assert that corruption in the UK industry is more widespread, possibly totalling up to ten per cent of total industry product. Why the construction industry is so prone to corruption is partly answered by these statistics and some of the tactics related in the first section of this chapter: essentially, being corrupt in construction is relatively easy. Thereafter, there is a circular argument that the industry is corrupt because it is easy to make money by engaging in corrupt practices, thereby making it more prone to corruption than other less lucrative sectors.

Another reason why the industry is prone to corruption is the complexity and number of disparate actors involved in any project, along with the contractual relationships arising *inter se* these participants. At each contractual interface the project is susceptible to corruption, and if corrupt practices are engaged in by any of the parties involved, the repercussions have a knock-on effect. Analysis of the complexity and vulnerability of a project to corruption at each of these contractual links has recently been published (Stansbury 2005: 36–50). The consequences of corruption in the construction industry, however, cannot simply be measured in monetary terms. The sad fact is that usually those engaged in corrupt practices are ambivalent about the consequences of their corruption. Although not directly translatable to the UK experience, some of the consequences of corruption are substantial, particularly in the developing world. Macauley (2005) observed:

> Corruption in [construction] plunders [and shapes] economies. Corrupt officials steer social and economic developments towards large capital-intensive infrastructure projects that provide fertile ground for corruption, and [neglect] health and educational programmes. The opportunity costs are tremendous, [hitting] the poor hardest. Were it not for corruption in construction, vastly more money could be spent on health and education and more developing countries would have a sustainable future supported by functioning market economy and the rule of law.

Example 1: Murder

In 2003 an Indian civil engineer, Satyendra Kumer Dubey, was murdered on his way home from work in a cycle rickshaw. Subsequently, the rickshaw driver disappeared, and two suspects who had been questioned by police were themselves murdered. The motive was that Mr Dubey had written to the Indian Prime Minister complaining of corruption in the construction of a 60-kilometre stretch of a new road. Despite a plea for confidentiality, the letter had been widely circulated among various government ministers (ibid.: 19–23).

Example 2: Death, destruction and homelessness

'Earthquakes don't kill people; collapsing buildings do. In the last 15 years 156,000 deaths, 584,000 injuries, nine million people homeless. Much of this devastation resulted from the collapse of buildings because concrete was diluted, steel bars were excised, or otherwise sub-standard building practices were employed' (ibid.: 19–23). Whilst earthquakes may not be prevented, it is possible to attenuate resultant disasters. Italian (Alexander 2005) and Turkish (Mitchell and Page 2005) disasters have been caused in part by the marriage of corrupt contractors and corrupt building inspectors and other public officials, resulting in ignored building codes – deadly in earthquake-zones.

Example 3: Monuments of corruption

Completed in the 1980s, the Bataan nuclear power plant in the Philippines was the largest investment project ever to be undertaken by the country. To date, it has never produced a single unit of electricity. The US giant Westinghouse admitted paying US$17 million as commission to a friend of the late Filipino dictator Ferdinand Marcos, who had personally intervened and overturned the original decision to award the contract to another company. Westinghouse maintains the payment was not a bribe. The reactor was sited on an active fault line that is part of the Pacific's 'rim of fire', creating a global hazard of nuclear contamination should the plant ever become operational (ibid.: 19–23).

These examples help illustrate that corruption in the construction industry has an insidious effect on ordinary people's lives. The harm that is wreaked by corrupt acts cannot be accurately measured, and the financial implication of corruption in the construction industry is but one disadvantageous consequence. A by-product of corruption is that it tends to self-perpetuate; in the United Kingdom, for example, it has been stated that the majority of individuals and organisations who engage in corrupt practices do so, not because they want to, but because they feel compelled to as a result of the way in which the industry and the political and business environment operate (Transparency International 2003: 5).

Perversely, however, corruption does not always have adverse consequences for the construction industry. In Japan the *dangó* system seemed to buck the trend, and even though the bidding is rigged and one might expect widespread negligence or venality, the contractors uphold high standards of accountability in meeting the exacting government specifications. Indeed, the 1995 Kobe earthquake inflicted the bulk of its devastation on older structures built to less stringent government specifications (Woodall 1996: 49).

Similarly, the Organised Crime Task Force (Goldstock *et al.* 1989: 78) in New York found evidence that the control of New York construction by the five Cosa Nostra families played a 'rationalising role' in the industry. The Mafia input to the construction marketplace served two essential and related functions. First, where there is fragmentation, its presence and power provide simplification and coordination; it can resolve labour disputes and assure that supplies and services are delivered on time. Second, it can ensure that extortion will not exceed predictable and 'reasonable' levels.

It is submitted that by eliminating corruption not only will the financial burden be eased to some extent on project costs, but also that impoverishment in the world will be lessened and human life will be held more sacrosanct. The result would be that the construction industry could return to its proper function and cast off its mantle as an industry where the real money is to be made by 'graft'. What then are the measures imposed in the United Kingdom to help achieve this objective?

UK measures

In comparison with those in other countries, UK measures, proposed or in existence, appear mundane and pedestrian. The Government's legislative output regarding corruption is generic across all industries, rather than specific to any particular sector. Neither do the more antiquated corruption statutes, still in force in England and Wales (and to a limited extent in Scotland[2]), address the problems of construction corruption. Various other statutes exist prohibiting corrupt practices, but are often breached in construction. These include the Competition Act 1998, Enterprise Act 2002 and Public Interest Disclosure Act 1998.

Competition Act 1998

Chapter 1, Prohibition of the Act, covers agreements and concerted practices that have the object or effect of preventing, restricting or distorting UK competition. Under the Act, companies engaged in anti-competitive behaviour can be fined up to ten per cent of turnover for each year of infringement (up to a maximum of three years).

Enterprise Act 2002

Companies engaged in anti-competitive activities may also be involved in hard-core cartel activity. In these instances the penalties include disqualification of directors, unlimited fines and imprisonment for up to five years.

The UK system has a slight twist in that under both the Enterprise Act 2002 and the Competition Act 1998, there is an amelioration of potential penalties if a company voluntarily provides relevant authorities with information about its involvement. This results in escaping criminal prosecution or receiving immunity or leniency.

Public Interest Disclosure Act 1998

This Act provides a system of protection for employees, whereby, in the event that an employee makes a protected disclosure to the proper authority and through the appropriate channels, the whistle-blower should not suffer detriment in his/her employment.

Government initiatives to reduce corruption in construction outside the legislative process have been sporadic, giving the impression of being uncoordinated and lacking a concerted effort to eradicate corrupt practices. Even publicising new initiatives is rarely timed successfully to coincide with the introduction of pertinent new pieces of construction legislation. Control Risks Group (2003) found that only 56 per cent of UK businesses had awareness of the OECD Convention; only 68 per cent of them were familiar with the United Kingdom's new laws implementing the Convention; and as few as 52 per cent of these companies had reviewed business practices in the light of the new legislation.

One of the few initiatives introduced in the United Kingdom by the Government was the Corporate Social Responsibility (CSR) Academy, established in June 2004, dealing with CSR generic issues and anti-corruption measures in addition. Businesses can obtain information, training and education, a competency framework and guidance on integrating CSR into their organisations (CSR Academy 2005). Other non-governmental organisations provide further detailed advice and help to construction professionals who join their associations. These include Society of Construction Law, Transparency International, and Public Concern at Work (PCAW).

Society of Construction Law

Seeking to harmonise the different professional ethics codes found in construction into a set of overarching principles, in 2003 this ethics group was established (Society of Construction Law 2006).

Transparency International

This was founded in 1993 as an independent, non-profit-making organisation with chapters in countries around the world. It incorporates various theme-based initiatives including the anti-corruption forum – an alliance of business associations, professional institutions and companies interested in domestic and international infrastructure construction and engineering sectors (Transparency International 2006).

Public Concern at Work (PCAW)

PCAW is an independent charity and legal advice centre, known as 'the whistle-blowing charity'. PCAW provides free confidential advice to people concerned about workplace wrongdoing, as well as training to employers, unions and professional bodies (Public Concern at Work 2006).

Currently, the UK approach is fragmented with no central controlling organisation in existence where advice can be sought on corrupt practices in the UK construction industry. Whereas elsewhere in the world major enquiries and task forces have been convened to deal specifically with corruption in a particular industry, the UK approach appears to favour more 'hands off' self-regulation. Unfortunately, professional bodies with specific ethical codes and standards for their members do not cover the majority of the actors in the industry.

Construction contracts

Whilst adopting a supportive posture for the stated aims of Transparency International (for example), the UK Government has generally adopted a position of virtual *laissez faire* in terms of new anti-corruption legislative provision. In many ways it can be argued that this position is justified since existing standard forms of contract should provide a structure that, to a significant extent, reduces or eliminates corrupt behaviour. Although there are numerous forms of contract currently in operation, this position is somewhat fantastic, to say the least, since corruption still manifestly occurs within the industry. It is therefore worthwhile to look at some of the provisions dealing with corruption within the standard forms of contract which the UK industry uses.

Joint Contracts Tribunal (JCT) provisions

Standard Building Contract with Quantities (SBC/Q/) 2005

Under the termination provisions, the contract addresses corruption in the following terms:

> Section 8: The Employer shall be entitled by notice to the Contractor to terminate the Contractor's employment under this or any other contract with the Employer if, in relation to this or any other such contract, the Contractor or any person employed by him or acting on his behalf shall have committed an offence under the Prevention of Corruption Acts 1889 to 1916, or, where the Employer is a Local Authority, shall have given any fee or reward the receipt of which is an offence under sub-section (2) of section 117 of the Local Government Act 1972.

Notably, only the employer has the right to terminate employment and has the discretion to terminate or otherwise, whereas the contractor is afforded no reciprocal right. The employer's right applies irrespective of whether contractor's employee (or agent) was involved in corruption with the contractor's knowledge, and applies even if the corrupt act was committed on a separate project. However, the right applies only if an offence under the legislation mentioned has been committed.

New Engineering Contract (NEC) provisions (edition 3)

This is a particularly important contractual form in the United Kingdom since the Olympic Delivery Authority (ODA) has chosen NEC3 contracts to procure all fixed assets and infrastructure (over £4 billion worth are estimated) for the 2012 Olympic Games to be held in London. However, the only element remotely dealing with corruption is contained within Core Clause 1, imposing an obligation on the parties to refrain from corrupt practices:

> Cl.10.1: The Employer, the Contractor, the Project Manager and the Supervisor shall act as stated in the contract and in a spirit of mutual trust and co-operation.

However, under Section 9, termination provisions, no specified right to terminate if corruption is discovered exists, unlike in the JCT contract. Under Cl.90.2 the employer may terminate for any reason, whereas the contractor can do so only in specified circumstances. The encapsulation in Core Clause 1, quoted above, of the parties' obligations *inter se* appears to satisfy the drafters, and perhaps it was felt that this simplistic statement, though lofty in its ambition, could be interpreted widely enough to combat corrupt practices.

Penalties for corrupt parties

Admittedly, if an actor within the industry is capable of committing and secreting an act of corruption, a mere breach of contract is not a deterrent. Consequently, the question must be raised as to what are the penalties for corruption in the industry and whether these are sufficient to deter. In the United Kingdom, the Office of Fair Trading (OFT) is charged with enforcing the Enterprise Act 2002 and the Competition Act 1998, and have it within its power to fine companies breaking the law by indulging in anti-competitive practices that distort the market or restrict competition. In 2004 the OFT issued a press release highlighting construction as one of five priority areas for work to align penalties for corrupt practices with offence severity, whilst explaining to businesses their rights and responsibilities (OFT 2004b). This has led to several notable cases in which parties to corrupt activities have been taken to task by OFT.

Case study 1

In February 2006, the OFT fined 13 roofing contractors £2.3 million in total (reduced to £1.6 million by leniency) after they collusively agreed to fix the prices of tenders for flat roofing and car park surfacing contracts in England and Scotland. The parties were found to have been involved in a series of separate arrangements in tenders for contracts in the South East, the Midlands, Doncaster, Edinburgh and Glasgow between 2000 and 2002 in breach of the Chapter 1 prohibition of the Competition Act 1998. OFT concluded that the parties' collusion in rigging prices was intended to distort competition and meant that buyers were unable to obtain competitive prices. In three instances, the successful contractor paid unsuccessful contractors compensation for backing off the contract or providing a cover bid that it knew would be higher than its own tenders.[3]

Case study 2

In June 2006, fines totalling over £1.38 million were imposed on three double-glazing contractors following an investigation into price-fixing in the aluminium spacer bar market used in the manufacture of double-glazing units. OFT concluded these companies participated in a cartel by agreeing to fix prices and share the market for glazing spacer bars, in breach of the Chapter 1 prohibition of the Competition Act 1998. The infringement occurred during the last part of 2002 and was brought to an early end as a result of OFT's intervention. In accordance with OFT's leniency policy, one company was granted 100 per cent leniency in recognition of its cooperation with OFT; similarly, a second company was granted 40 per cent penalty reduction under the leniency policy. Total fines imposed were thus reduced to £900,000. OFT's finding followed an earlier decision in November 2004, that price-fixing had occurred in another double-glazing raw material, desiccant, bringing total fines for double-glazing cartels to £3.7 million (reduced to £2.5 million by leniency).[4]

Lessons to be learned for the United Kingdom: which way forward?

It seems apparent that the United Kingdom needs to substantially up its 'anti-corruption game' in order for the widely held impression of propriety in governance to meet its actual performance in the arena. The anti-corruption and ethical-behaviour message needs to be inculcated at all levels in the industry through university courses, professional accreditation and trades-based day release programmes. Most importantly, this needs to happen soon and happen with a level of commitment that seems remarkably lacking in the current climate. As Hess and Dunfee (2003) note:

> Corporations that are serious in their attempts to reduce bribe payments should start with industry-wide initiatives to adopt the principles. An industry-based strategy will ensure that a corporation's competitors are playing by the same rules, as any firms attempting to free ride [should] be easily identifiable. Industry-based initiatives are [essential] because the publication aspect furthers the transfer of knowledge in fighting corruption.
>
> (Hess and Dunfee 2003)

Unfortunately, cynicism remains the greatest enemy of improving the ethical behaviour of the construction industry throughout the world, as eloquently observed by Goldstock *et al.* (1989):

> Cynicism is one of the greatest impediments to reforming the industry. If the industry participants believe that current Government initiatives are merely a political expedient and temporary response to the most recent flurry of exposes and like its predecessors, will soon dissipate, they will not commit themselves to a reform agenda.
>
> (Goldstock *et al.* 1989)

Furthermore, there is nothing to prevent employers from supplementing the implied terms within contracts of employment with the complete anti-corruption code of Transparency International UK. This could be included at a company induction at the same time as other procedures such as health and safety, and disciplinary grievance. Interestingly, this would bind both the employer and the employee by ethical behaviour standards, with a somewhat greater onus on the employer. The effect would be that an employer could be sued by any/all of its employees if it failed to comply with the standards of ethical behaviour laid down in the code. Thus, if an employee was aware that an order was unlawful, he/she could refuse to follow it and still have contractual protection. Similarly, an employee breaching the duty of non-disclosure of confidential information by rightly claiming that the disclosure was in the public interest by reason of illegality would be protected under the Public Interest Disclosure Act 1998.

Moreover, it remains essential to reassure employees that disclosure by following correct whistle-blowing procedures will not be professionally detrimental. Often whistle-blowers are victimised despite the protection afforded to them under the Public Interest Disclosure Act 1998. Indeed, PCAW (2006) reports that of employment tribunal cases convened to consider victimisation claims, 46 per cent of claimants were successful – implying that half of all whistle-blowers are victimised.

Come what may, it is essential that criminal prosecutions are pursued with vigour and determination. Arguably, corruption cases should not be dealt with in the lenient fashion previously described in the OFT case studies. Indeed, leniency in such cases is now a policy. The moral tone has been recently set in the international bribery case surrounding the Lesotho Highlands Water Project trials (Darroch 2005: 31–36) in which large international corporations and individuals were pursued for corruption irrespective of the size or type of bribes offered. Perpetrators of corruption need to grasp that their actions will be held to account without fear or favour, and that individuals who have perpetrated the corrupt act will be pursued in addition to the company concerned (where that company is vicariously liable) in both the criminal and civil courts. Furthermore, those found guilty of corruption should have the proceeds subjected to confiscatory legislation such as Proceeds of Crime Act, and sentencing should be such, to, in the words of Voltaire, 'encourager les autres' towards ethical behaviour.

There appears to be a case for placing companies and individuals found guilty of corruption on a tender blacklist for public sector works, since current punitive measures do not seem to have the necessary deterrent effect. However, during the introduction of EU Public Procurement Directives (European Commission 2003) into Scottish law, the anti-corruption measures within the directive proved controversial. Article 45 (ibid.) requires the exclusion of companies convicted of corruption, fraud, etc. to the benefit of law-abiding companies. However, it appears that this measure can be circumvented, since the Scottish Executive has in its draft regulation diluted this mandatory exclusion to provision applying only when the contracting authority has actual knowledge of conviction.[5]

A further refinement of the blacklist approach may be for any company tendering for a contract to provide a 'certificate of fitness to tender', including a statement to the effect that it has no current convictions or pending prosecutions for corrupt practices. This certificate could be obtained in a similar fashion to that adopted for financial standing, for example, Dunn and Bradstreet rating. The length of time for which a company may be blacklisted could be set in advance, and a pre-qualification routine to allow previous offenders back into the fold could be established, with a prerequisite that the company is independently audited (at its expense) to ensure that all anti-corruption procedures and safeguards are in place before reentering into the 'honest world' and bidding for work from the public sector.

Conclusion

The construction industry is one of the largest contributors to GDP and the largest employer in the United Kingdom, representing a huge proportion of the economy. The nature of the industry and complexity of the contractual and business operations within the industry leave it, as has been seen in the course of the discussion above, open to relatively easy exploitation by corrupt businesses and individuals. However, in spite of the obvious opportunities for corruption and cases of anti-competitive and unethical behaviour described, there appears to be no coherent and unified approach adopted to tackle the problem. On the contrary, there is a piecemeal approach to corruption and its elimination, primarily through the dubious effectiveness of self-regulation and contractual forms. Without doubt, this state of affairs demonstrates the need for organisations such as Transparency International and their ilk. Indeed, the fact of their existence, and the acceptance of Transparency International's anti-corruption code of conduct by the UK Government, are arguably a tacit acceptance that the problem is in many ways out of hand.

Unfortunately, the UK Government's willingness to eliminate corruption within the construction industry seems diluted, or at least attenuated, by its attempts to tackle social, corporate and governance impropriety elsewhere within society. These include the 'loans for knighthoods' scandal of early 2006 and the 'sex for visas' corruption scandal within the UK Immigration Service of mid-2006 which are now both subjects of ongoing police investigations at the time of writing. The apparent procrastination of the UK Government in introducing new corruption legislation with a unified Corruption Act, alongside the potential loopholes left when introducing the new Public Procurement Directive in Scotland, calls into question the existence of the political will necessary to enact genuine and effective anti-corruption legislation.

Transparency International itself, for example, whilst laudable in its aims, suffers by its anonymity. Its existence in the wider commercial world of construction is little known, which means that the stated aims of advancing anti-corrupt practice and ethical behaviour are inevitably dissipated. This would appear to be the ultimate fate of a not-for-profit organisation that is not tied to any particular government, or, indeed, to any particular business or labour interests. Transparency International could, it is contended, be seen as superfluous and therefore easy to ignore, since it does not have the capacity or strength to impose change on the industry. Yet, paradoxically, this weakness may also be a hidden strength, since its proposed changes can be seen by all stakeholders as not singling out any one participant in the construction process for undue attention by anti-corruption 'zealots'.

Future challenges to anti-corrupt practice in the UK construction industry

Without doubt, there is huge potential for the adoption of more wide-ranging legislation to tackle corruption within the UK construction industry. It will require moral ascendancy and political will to be able to force legislation relatively rapidly through the UK parliament(s) in the teeth of obvious objections that will be levelled by interest groups in the industry. However, the ultimate paradox created by the UK construction corruption debate over the next few years is that its genesis lies in two off cited UK governmental reports from the mid-1990s.

Latham's *Constructing the Team* report (1994) clearly exhorted contractors to 'partner' with suppliers and subcontractors – working closely and regularly with the same known faces in order to lock-in added value within the construction process for the benefit of the client, contractors and suppliers. Similarly, Egan's *Rethinking Construction* Report (1998) stated that best practice for the construction industry was to rationalise the supply chain. Egan stated clearly that one of the key problems within the construction industry was the 'Dutch auction' effect forcing down prices, and giving clients unrealistic expectations of cost and contractors, minuscule profits. He further emphasised the selection of 'prime' or 'preferred' contractors. Having a prime, or preferred, contractor codifies favouritism of certain key players in the industry. This in turn artificially limits both the total number of suppliers and sub-contractors that contractors and clients work with, as well as the diversity of the supply chain.

The challenge then is manifest in the apparent contradictions of current best practice as laid out by the government itself (Latham 1994; Egan 1998). How can construction professionals or contractors fit in with stated governmental ideas (Latham 1994) of best practice by 'partnering' with their suppliers on a project? By definition, the terms of the partnering arrangement are to regularly work with other companies or groups in a non-confrontational, collaborative manner. To a true believer in unadulterated transparent competition, this arrangement offers the likelihood of collusive and corrupt practices. Similarly, how can the same contractor rationalise its supply chain by using 'preferred contractors', ergo excluding tenders for contracts according to arbitrary measures of 'good' or 'bad' performance to increase margins and profits (Egan 1998)? Again, such practice can easily be alleged to be anti-competitive and potentially corrupt. These fundamental contradictions are likely to have the best business ethicists and anti-corruption campaigners scratching their heads at the mixed governmental messages.

Notes

1. Arthur Maundy Gregory, a corrupt adventurer and sometime spy, is the only person to be convicted under this Act. After successfully brokering peerages for Lloyd George, he finally received his come-uppance in 1933 after being caught

selling a knighthood. Sentenced to six months in jail, he was rewarded for his courtroom silence with a 'pension' from his ennobled clients – *Telegraph* 14 July, 2006.
2. Public Bodies Corrupt Practices Act 1889, The Prevention of Corruption Act 1906 and Prevention of Corruption Act 1916.
3. OFT Decision No. CA98/01/2006 Joined cases CE/3123-03 and CE/3645-03.
4. OFT Decision No. CA98/04/2006.
5. Draft Regulation No. 23 – Scottish Procurement Directorate Consultation Document; The Draft Scottish Regulations Implementation, Public Sector Procurement Directive 2004/18/EC and Utilities Directive 2004/17/E. August 2005.

References

Alexander, D. (2005) Italian study, *Global Corruption Report*, Transparency International UK, 26–29.

Anon. (2001) Letters page, *Construction News*, 28 June: 12.

Barry, S. (2002) Fraud probe at Navy Dockyard, *Construction News*, 1 August: 1.

Bribe Payer's Index (2006) *Transparency International UK*. Online. Available at: http://www.transparency.org/policy_research/surveys_indices/bpi (accessed 6 October 2006).

CIOB (2006) CIOB signs up to anti-corruption, *Press Release*, 8 June. Online. Available at: http://www.ciob.org.uk/news/view/1115 (accessed 10 October 2006).

Control Risks Group (2003) *Facing up to Corruption 2003 – Tackling the Hard Questions*, London, UK.

CSR Academy (2005). Online. Available at: www.csracademy.org.uk (accessed 10 October 2006).

Cullen, P. (2002) *With a Little Help from my Friends: Planning Corruption in Ireland*, Dublin: Gill and Macmillan.

d'Arcy, J. (2001) Jobs fraud 'rife' on channel link, *Contract Journal*, 19 September: 2.

Darroch, F. (2005) Case study: Lesotho puts international business in the dock! *Transparency International Global Corruption Report*, 31–36.

Department of Trade and Industry (2005) *Construction Statistics Report*, London: HMSO.

Egan, J. (1998) Rethinking construction. *Report of the Construction Industry Task Force*, DETR, London: HMSO.

European Commission (2003) *Comprehensive EU Policy against Corruption*, 28 May. Online. Available at: http://europa.eu/scadplus/leg/en/lvb/l33301.htm (accessed 10 October 2006).

Fraser of Carmyllie, Rt Hon Lord (2004) *Final Report of the Holyrood Enquiry*. Online. Available at: http://www.holyroodinquiry.org/FINAL_report/report.htm (accessed 10 October 2006).

Glackin, M. (1998) Benefit fraud in the firing line, *Building*, 17 July: 18–19.

Goldstock, R., Marcus, M., Thacher II, T.D. and Jacobs, J.B. (1989) Corruption and racketeering in the New York city construction industry. *The Final Report of the New York State Organised Crime Task Force*, New York: New York University Press.

Hess, D. and Dunfee, T. (2003) Taking responsibility for bribery, in Sullivan, R. (ed.), *Business and Human Rights, Dilemmas and Solutions*, London: Greenleaf.

HM Government (2000) *Raising Standards and Upholding Integrity: The Prevention of Corruption, White Paper*, London: HMSO.

HM Government (2003) *Reply to the Report from the Joint Committee on the Draft Corruption Bill (CM6086)*, London: HMSO.

Latham, M. (1994) *Constructing the Team*, Final report of the Government/industry review of procurement and contractual arrangements in the UK construction industry, London: HMSO.

Lynch, R. (2004) Cops probe CSCS fraud, *Construction News*, 24 June: 1.

Macauley, M.J. (2005) Corruption at the crossroads, *Global Corruption Report*, Transparency International UK, 112–115.

Mitchell, W.A. and Page, J. (2005) Turkish Study, *Global Corruption Report*, Transparency International UK 26–29.

Office of Fair Trading (2004a) OFT fines West Midlands roofing contractors for price-fixing, *Press Release 46/04*, 27 July.

Office of Fair Trading (2004b) Five priority areas identified in OFT annual plan, *Press Release 194/04*, 1 December.

Office of Fair Trading (2005) OFT urges SMEs to report anti-competitive practice, *Press Release 137/05*, 27 July.

Prior, G. (2004) BAA tightens security at T5, *Construction News*, 8 April: 6.

Public Concern at Work (2006), http://www.pcaw.co.uk/ (accessed 11 October 2006).

Rogers, D. (2005) Taxman drags heels over bogus workers, *Construction News*, 3 February: 3.

Society of Construction Law (2006) *SCL Ethics Group Statement of Ethical Principles*. Online. Available at: http://www.scl.org.uk/ethics.php (accessed 10 October 2006).

Stansbury, N. (2005) *Exposing the Foundations of Corruption in Construction*, Transparency International Global Corruption Report, 36–50.

Stansbury, C. and Stansbury, N. (2005) *Anti-Corruption Code for Individuals in the Construction and Engineering Industry (England & Wales)*, Transparency International UK.

Staples, J. (2002) Surveyor alleges underhand dealings on Hampden Park refit. *Scotsman Newspaper*, 3 December: 5.

Transparency International UK (2003) Anti-corruption initiative in the construction and engineering industry, *Report One: Introductory Report*, September.

Transparency International UK (2006) Anti-corruption forum, *Press Release 15/09/05*. Online. Available at: www.transparency.org/content/download/2238/13078 (accessed 10 October 2006).

Woodall, B. (1996) *Japan Under Construction: Corruption, Politics and Public Work*, Oxford: University of California Press.

Chapter 7

Corruption within international engineering-construction projects

Mike Murray and Mohamed Rafik Meghji

There are different ways to build an airport. You can give the work to the company who's going to do the best job at the best price, or you can give it to the company who gives you the biggest bribe, or get this, you can even decide to build an airport in the first place because you have received a bribe.

Corruption means that resources, often aid money, is being wasted on roads that fall apart before [they are] finished and railways that never make it to their destinations. But the OECD says that aid should only be cut when countries are corrupt from top to bottom. Corruption is a problem, the OECD says, that rich and poor need to solve together.

(Laurenson 2007)

Introduction

The construction industry is one of the most lucrative of industries; unfortunately, in 2002, Transparency International's (TI's) Bribe Payers Index survey found that it also has an unenviable stigma of being labelled the sector to be most heavily entangled in corruption worldwide (Transparency International 2002). TI confirms that corruption can be found in every phase of the construction process, from the preliminary planning stage, to the awarding of contracts, to the employment of subcontractors and to the operation and maintenance of projects:

The tender process may be corrupted by international pressure. Through offers of arms or aid the government of a developed country may influence a developing country to make sure that a company from the developed country is awarded a project, even if it is not the cheapest or best option.

(Transparency International 2002)

The extracts above taken from an OECD (Organisation for Economic Co-operation and Development) web-based video, suggest that the demand

and supply sides of construction, including financial institutions and government departments who support development, are both problem and solution regarding corruption in the construction industry. The OECD Anti-Bribery Convention of 1997 prohibits the bribery of foreign public officials in international commerce. The Convention binds the signatories to prosecute those engaging in corruption abroad.

The airport example is, indeed, topical with Lewis (2004) reporting on the establishment of the Trinidad and Tobago Transparency Institute (TTTI) and its actions to combat corruption in this Caribbean country. He cites King's (2002) report on corruption during the contract to build Trinidad and Tobago's Piarco airport. An investigation into the project revealed that the tendering system was corrupt, that money had been diverted into unauthorised accounts and that public funds were abused. In contrast, Transparency International (2000) note the new Hong Kong Airport at Chek Lap Kok was generally considered to have been exemplary in preventing corruption during its eight-year development.

Engagement in corrupt practices is not only illegal in worldwide jurisdictions; it also transgresses the ethics codes established by the construction industry's professional bodies. As such, the exposure, sanctions and eradication of corruption within the international engineering and construction industries fall under the Corporate Social Responsibility (CSR) umbrella. Indeed, the Chairman of Transparency International noted that:

> Transparency in public contracting is arguably the single most important factor in determining the success of donor support in sustainable development.
>
> Eigen (2005a)

Furthermore, the Major Projects Association (2006: 5) notes that the current emphasis on corporate responsibility has given rise to anti-corruption mechanisms and anti-corruption codes of conduct in many companies. The dilemma for engineering-construction companies in the West is the stringent application of such codes, particularly in overseas developing countries. Indeed, Corner House (2000) contends:

> If corruption is growing throughout the world, it is in large part fueled by policies and programmes that are being pushed by Western governments and which are further underwritten by poor governance and misdirected funds in the North.
>
> (OECD 2006)

The Organisation for Economic Co-operation and Development (OECD 2006) recognize that the impact of corrupt practices stretches far beyond the specific misbehaviour of the actors involved. Its repercussions sweep across entire populations and results in unfinished roads, crumbling schools and crippled health systems.

This chapter draws on the work of Transparency International and first considers the ethical and moral challenges that engineering-construction industry personnel face, before explaining the role that TI has in fighting corruption in construction projects. The economic and social costs of corruption are examined, and reference to TI's Corruption Perception Index (CPI) provides a backcloth to the international dimension of corruption. Global anti-corruption initiatives are discussed, and the involvement of Export Credit Agencies in supporting projects exposed to corruption is covered. A case study project (Lesotho Highlands Water Project) is used to expose the complex web of relationships that is often necessary to disguise corrupt behaviour. The chapter concludes by returning to the topic discussed in the following section, by reinforcing the argument that corrupt actions are, foremost, an issue of personal morality. Corporations and governments exist only through the decisions and actions of individuals and groups. Owing to the inevitable need to limit the length of the chapter, a deeper analysis of relevant topics (e.g. the World Bank's and United Nations' role in combating corruption) has not been undertaken.

Ethical dilemmas for the construction industry

Construction industry personnel who knowingly engage in corrupt practices break the codes of ethics established by professional bodies such as the Institution of Civil Engineers (ICE). Kang *et al.* (2004) note that the ethical characteristics of these codes include terms such as trustworthiness, honesty, integrity, dignity and fitness. Other characteristics such as 'to be incorruptible, act impartially and be transparent and accountable' are mentioned by Bowen *et al.* (2007) in their study of ethical behaviour in the South African construction industry. These authors argue that unethical behaviours can emanate from competing ethical choices. That is, individual consciences and pledges to the likes of the ICE form a backcloth to an industry that is characterised by economic and management pressures. These pressures can be the catalyst for unethical behaviour.

In 2007, the *New Civil Engineer* journal cited a senior UK construction industry figure who appears to play 'devil's advocate' with regard to the apparent ethical dilemmas noted above. He discusses payments of commission to foreign officials to win work:

> Many people would not accept that it [is] corruption by the standards and values of that particular society in which you are doing business. As an engineer you have to decide what to do; will you have no work for the UK and let the French win it?
>
> (Armitt 2007: 10)

This perspective does have some legitimacy and Bottelier (1998) found that the actual occurrence of corruption and financial fraud is determined by many factors, including cultural and moral values. However, in a subsequent issue of the *New Civil Engineer* journal, Ferguson (2007: 16), Deputy Director General of the ICE, warns against this line of reasoning by quoting the ICE's rules of professional conduct. Rule 1 emphasises that those members found guilty of involvement in bribery and corruption, direct or indirect, would most likely be expelled from the ICE.

However, it appears that many individuals in the infrastructure, construction and engineering sectors fail to recognise corrupt situations or, if they do recognise them, fail to appreciate the risks of becoming involved in corruption (Stansbury and Stansbury 2007: 4). Moreover, engagement in what might be perceived as 'low level' corruption fits the analogy of progressing from soft to hard drugs. Bottelier (1998) warned that 'small corruption, when tolerated or condoned by society, creates uncertainty and often leads to big corruption'. In the United Kingdom, under the Anti-Terrorism Crime and Security Act 2001, UK-registered companies and UK nationals can be prosecuted in the United Kingdom for any act of bribery committed either in the United Kingdom or overseas.

In common with other professional engineering bodies, the American Society of Civil Engineers (ASCE 2007) has challenged engineers to become signatories towards an initiative aimed at 'zero-tolerance' towards bribery. Interestingly, in June 2007, the register was updated with several new names associated with the Society of Afghan Engineers. Given Afghanistan has an unenviable ranking on Transparency International's 2007 Corruption Perception Index (discussed later in this chapter) this will be a test for all stakeholders engaged in construction and development work. However, in signing up to the ASCE code, the Afghan engineers are demonstrating their commitment in the fight to remove corruption and no doubt it is also a prerequisite for securing future investment from the likes of the World Bank.

Transparency International

Transparency International, founded in 1993, plays a lead role in improving the lives of millions around the world by building momentum for the anti-corruption movement. TI has a global network including more than 90 locally established national chapters. They bring together relevant players from government, civil society, business and the media to promote transparency in elections, in public administration, in procurement and in business. TI notes that, although the devastating impact of corruption is often obvious, an example from the engineering-construction sector shows that sometimes corruption's impact is not immediately apparent – faulty buildings, built to lower safety standards because a bribe passed under the table in the construction process, which collapse in an earthquake or hurricane (TI 2007a). TI has five global priorities in the fight against corruption (politics, public contracting, private sector, international anti-corruption conventions, and poverty and development) and this chapter uncovers the engineering-construction sector's involvement in corrupt activities, detection and prevention over these five themes. Its initiative to prevent corruption on construction projects is being led by TI (UK) and has three primary objectives:

- to raise awareness of the damage and risks caused by corruption in construction projects;
- to develop anti-corruption actions and tools;
- to promote the implementation of anti-corruption actions and tools in construction projects.

TI (UK) has developed business tools, reports, actions and information which can help prevent corruption on construction projects, and these are disseminated through a dedicated webpage (http://www.transparency.org. uk/programmes/CICE.htm) that provides the following guidance:

- Project Anti-Corruption System (PACS) (Construction Projects)
- Anti-Corruption Training Manual
- Anti-Corruption Reports and Tools
- Anti-Corruption Action Statement.

Types and methods of corruption

Although distinct types of corruption are associated with the construction industry, Stansbury and Stansbury (2007: 5) have concluded that there is no international definition of this word. They note that an array of corrupt activities (bribery, extortion, fraud, deception, collusion and money-laundering) are considered to constitute criminal coerced corruption', and in reflecting Armitt's comments discussed earlier in this chapter, note that constructors may feel compelled to offer a bribe during tendering, often as a means to level the playing field with competitors who they believe will also have done this. Further reference to bribery is noted by a senior advisor to the World Bank:

> When an official accepts, solicits, or extorts a bribe or when a private agent offers a bribe to circumvent public policies and processes for competitive advantage and profit. Public office can also be abused for personal gain even if no bribery occurs, through patronage (cronyism) and nepotism, underpricing of State assets, collusion to divert public resources, or outright theft.
>
> (Bottelier 1998)

Large international construction-engineering projects typically have lengthy gestation and construction phases and Transparency International (2005: 1) note the propensity for corrupt activity at every stage in these undertakings.

Previous exposure of corruption has shown that it can take place during planning and design, prequalification and tender, project execution and operation and maintenance phases. In particular, TI acknowledges that some projects would not have passed the planning stage without the motivation of corruption. Moreover, the awarding of construction contracts is often tainted with corruption (particularly due to international pressure from Governments) as are the operation and maintenance of projects after construction is finished.

Cost of corruption

It is evident that the cost and the resultant impact of corruption within the global construction industry can be viewed from a buyer's or a seller's perspective as well as that of societal stakeholders. The American Society of Civil Engineers (2005) adopted TI's conservative estimate that ten per cent of the global construction turnover – US$3.9 trillion – is lost through bribery, fraud and corruption; i.e. about US$390 billion is diverted annually from projects that provide water, pollution control, electricity, roads, housing and other basic human needs.

Given the apparent magnitude of corruption within the international construction sector, it is perhaps no surprise that Transparency International (2005a) has published a special focus report on *Corruption in Construction and Post-conflict Reconstruction* with one chapter devoted to the 'cost of corruption'. Authored papers within this chapter present diverse definitions that include both social and economic factors. Sinha (2005) reports on the mysterious murder of an Indian civil engineer whose 'whistle-blowing' included a letter to the Indian Prime Minister's Office about corruption on a road project where he worked. His requests for confidentiality were ignored. Collier and Hoeffler (2005) examine the true total cost of corruption in infrastructure projects and find that it brings about lower current living standards, with the poorest hit hardest, and slower growth. Furthermore, expensive and low-quality infrastructure may inflict costs on society that are far in excess of the money directly wasted in the process of provision. Bosshard (2005) argues that corruption in the development planning process results in projects with serious environmental and social impact costs. Lewis (2005) examines the collapse of buildings in earthquake zones and cites examples from Italy and Turkey to illustrate how the marriage of corrupt contractors and corrupt building inspectors and other public officials had resulted in ignored building codes, lax enforcement and the absence of on-site inspection, which are deadly when they occur in earthquake-prone areas.

From a client's perspective, the inclusion of bribes at project inception and execution stages can add considerable sums to the overall capital cost of a project and therefore act as 'opportunity costs' that are denied use in future developments. Transparency International (2005) considers that bribes amounting to five per cent of total project cost are quite low, and that bribes can be in excess of 30 per cent. They argue that in such cases corruption may result in losses and reputational risks for the funder in that additional costs caused by corruption can make the project uneconomic, which may result in non-payment to the funder. Moreover, TI believes that many major cost overruns that are blamed on management inadequacies or changes in design may, in fact, be due to corruption. TI notes (2005: 2) the impact of corruption on other parties:

- Corruption can disadvantage the developing world by resulting in projects which are unnecessary, unsuitable, defective or dangerous; which require overly complex components; which are overpriced or expensive to operate and maintain; or which are delayed.
- Corruption can disadvantage construction and engineering companies by resulting in wasted tender expenses, tendering uncertainty, increased project costs, economic damage, blackmail, criminal prosecutions, fines, blacklisting and reputational risk.
- Corruption can disadvantage individual directors and managers by resulting in criminal prosecution, fines, imprisonment, loss of professional status, disqualification from office, and loss of employment.

Perception of corruption

Bribe Payers Index (BPI)

The BPI survey focuses on the supply side and was first undertaken in 1999 and then in 2002 and 2006. The BPI examines the propensity of companies from 30 leading exporting countries to bribe abroad. Companies from the wealthiest countries generally rank in the top half of the Index, but still routinely pay bribes, particularly in developing economies (Transparency International 2006b). In the 2002 BPI survey, respondents were asked to rate different business sectors – how likely is it that senior public officials in this country (respondent's country of residence) would demand or accept bribes, for example, for public tenders, regulations and licensing, in the following business sectors? The public works/construction sector was perceived to have the highest level of corruption and most likely to pay the highest bribes (Transparency International 2002).

Corruption Perception Index (CPI)

Transparency International has its own index of measuring the perception of corruption. The Corruption Perception Index (CPI), introduced in 1995, focuses on the demand-side and ranks countries in terms of the degree to which corruption is perceived to exist among public officials and politicians. It is a composite index, a poll of polls, drawing on corruption-related data from expert and business surveys carried out by a variety of independent and reputable institutions. The CPI reflects views from around the world, including those of experts who are living in the countries evaluated. The CPI focuses on corruption in the public sector and defines corruption as the abuse of public office for private gain. The surveys used in compiling the CPI ask questions that relate to the misuse of public power for private benefit, for example, bribery of public officials, kickbacks in public procurement and embezzlement of public funds, or questions that probe the strength of anti-corruption policies, thereby encompassing both administrative and political corruption (TI 2007b).

The 2007 CPI ranked 180 countries and concluded that a strong correlation between corruption and poverty is prevalent. Forty per cent of those countries scoring low (i.e. Iraq, Afghanistan and Somalia), indicating that corruption is perceived as rampant, belonged to low-income countries as classified by the World Bank. However, the study implicates 'bribe payers' as well as 'takers', and this is reflected in the case study examined later in this chapter.

Global insights

The anti-corruption programmes within the construction industries in two countries, China and Tanzania, are now examined. Burke (2007) reveals that China has a long history of relations with Tanzania and that privatisation and deregulation of Tanzania's construction industry have been accompanied by a steady increase in the number of Chinese companies entering the country. Given that both countries have experienced growth within their construction sectors, the pressure to share best practice in CSR principles and, in particular, anti-corruption techniques should be paramount. This is particularly so, if both countries are to continue seeking funding assistance from the World Bank, government agencies and NGOs. However, TI's Bribe Payers Index survey (2006a) found that, in the case of China and other emerging export powers, efforts to strengthen domestic anti-corruption activities have failed to extend abroad.

China

In contrast to Tanzania, China is a country experiencing rapid industrial growth over the past decade. The volume of government expenditure on public procurement alone in China jumped from 3.1 billion Yuan (US$0.4 billion) in 1998 to 150 billion Yuan (US$18.7 billion) in 2003. The country's rapid transition has resulted in huge investments in construction, which in turn has bred widespread corruption (Transparency International 2004). A study of value management (VM) in Chinese construction, a concept, unlike corruption, that is designed to 'add value' to projects, highlights the problem. Shen's and Liu's (2004: 17) study of VM practice acknowledged that the Chinese construction market is in a period of transition but corrupt practices cloud the transparency of the construction market and hinder the dissemination of VM practice throughout the industry. Zou's (2006) study of corruption in the Chinese construction industry found that current anti-corruption practices are reactive rather than proactive, and recommended, amongst other things, the promotion of an ethically clean construction culture to mitigate the scourge. Bristow (2007) points to the punishment for being found guilty of corruption in China, citing the case of a local party secretary being executed following the collapse of a pedestrian bridge in Sichuan Province in 1999. The collapse led to the death of 40 people, and it was later discovered that the politician had accepted a bribe from a childhood friend in exchange for a bridge-building contract.

It appears that the Chinese government and the construction industry are actively seeking to resolve this problem and no doubt improve the country's and its construction industry's position on TI's Corruption Perception/Bribe Payers Indexes. Indeed, Stansbury's (2005) visit to Beijing found that the Chinese government is making the reduction of corruption in the construction industry a priority. Furthermore, Transparency International's (2007c) national contact in China has established a 'Promoting Transparent Procurement and strengthening Corporate Responsibility' project. Funded by the UK embassy in Beijing, the initiative included a workshop featuring a case study that focused on 'Integrity in Chinese Construction Sector'.

Tanzania

Africa is constantly accused of having the highest corruption across the globe in everything from day-to-day politics to rigged procurement procedures. However, in the eastern part of the continent, Tanzania held the spotlight for one of the most innovative anti-corruption measures in the history of the continent. In 1995, the country's President Benjamin Mkapa created a commission to fight corruption and employed former Prime Minister Joseph Warioba as its chairman. The commission was given the task of investigating corruption and preparing a country report on the state of the corruption (the *Warioba Report* 1996). The need for the report is perhaps emphasised in the title of a paper by Langseth and Michael (1998): 'Are bribe payments in Tanzania "grease" or "grit"?' The report investigated all types of corruption from what was categorised as 'petty type', such as police bribes, to the very large-scale 'grand type' that involved public tenders and corrupt procurement practices. Doig and Riley (1998) refer to the report and note that grand-scale corruption had been identified in the procurement of goods and in the award of large public contracts, in particular, in road building and public construction. An assessment of 24 construction contracts had found substantial cost overruns, and corruption was considered to be contributory to this outcome.

The Warioba report has proved to be extremely useful in the country's efforts against corruption and conveys the exemplary efforts made by the country against the crime. The results were highlighted in the report as national priorities against corrupt practices; some of these highlights are listed below:

- *Rule of law*: To create conditions, which can restore confidence in the judiciary services and law enforcement agencies.
- *Financial discipline*: To reduce siphoning of public funds and increase revenue collection to enable financial social services.
- *Procurement*: To enforce strict adherence to and transparent administration of tendering procedures.
- *Public awareness*: To create awareness of how corruption harms the economy and ultimately transforms the fabric of society.
- *Public service*: To recognise that public officers are not 'masters' but 'servants' accountable for their actions (who therefore deserve a fair remuneration package).
- *Media*: To report corrupt elements without fear or favour and to publicise the harm they do to the innocent, the poor and the weak in Tanzania.

Commentators who have sought to measure the success of the recommendations in the report provide a mixed view. Heilman and Ndumbaro (2002) argue that, though the Mkapa administration has taken partially successful steps to control corruption, these efforts have not fundamentally undermined the supporting environment for corruption in the country. In the context of the construction industry, Debrah and Ofori (2005) reviewed the training needs of construction professionals in Tanzania. They found that the existing programmes are fragmented and lack overall strategy and continuity, coordination and sustainable funding. They recommend the establishment of an industry-specific training levy for professionals. However, they warn that in an emerging/developing-country environment there is the need to ensure that both the training programme and the administration of finance are not derailed by bureaucratic bottlenecks and other constraints, such as corruption. Moreover, an overview of the construction industry points to a growth in the number of registered contractors (100 in 1997 to 1400 in 2005) and the difficulty in enforcing the mandatory registration and operating requirements. Mkono and Ache (2006) found that these regulations have been hampered by poor monitoring and weak enforcement by the regulatory bodies, which has permitted many contractors to ignore several laws, although the government's capacity and willingness to police and punish offenders is expected to increase.

Anti-corruption schemes and a business integrity management system

The International Federation of Consulting Engineers (FIDIC) has undertaken a sustained programme of anti-corruption initiatives including the formation of a Business Integrity Management System (BIMS) developed by Engeli and Pieth (2000). They note the propensity for consulting engineers to become embroiled in corruption:

> If corrupted élites have created a network of corruption, they have been supported by exporting companies and investors from the North too weak to refuse to participate in 'the game'. And – it must be said – sometimes consulting engineers have tolerated or even participated in these dealings.
>
> (Engeli and Pieth 2000)

FIDIC have sought to influence its members by reminding them that corruption creates unnecessary waste in the procurement of projects. Furthermore, it undermines the values of society, breeds cynicism and demeans the individuals involved. They specifically outlaw corrupt behaviour through their code of ethics:

> The consulting engineers shall neither offer nor accept remuneration of any kind which in perception or in effect either (a) seeks to influence the process of selection or compensation of consulting engineers and/or their clients or (b) seeks to affect the consulting engineer's impartial judgment.
>
> (FIDIC 2001)

The BIMS is intended to promote integrity during the procurement of projects and has been specifically designed to be suitable for alignment with an organisations' quality management system such as ISO 9001: 2000. Engeli and Pieth (2000) note that FIDIC had already established ground rules for integrity with its code of ethics and specific policy statement on corruption in 1996. However, they recommended the establishment of 'integrity manuals' that are based on a shrewd and concrete analysis of procurement procedures that went wrong or, at least, on a detailed risk analysis of the hypothetical with participation of professionals working in the (geographically or sectorially) sensitive areas.

The FIDIC BIMS is intended to provide an organisation with a set of tools and a process-based approach towards the management of integrity within an organisation. In their 2007 survey, FIDIC found that, to date, there have been 78 firms reporting the implementation of a BIMS that follows FIDIC guidelines (FIDIC 2007a). They note that of the 78 companies, 31 were from South Africa, 7 from Japan and 5 from Denmark. The others were scattered across the globe; noticeably absent were consulting engineers from the UK (FIDIC 2007b).

Amongst the companies that have adopted a BIMS are Canadian consultant Acres International Limited (taken over by Hatch Energy in June 2004). Hearne's (2005) interview with John Ritchie, a civil engineer and Director of Corporate Affairs at Acres, probed Ritchie on the company's court-fine regarding its involvement in corruption (bribery) on the Lesotho Highlands Water Project (see case study in this chapter). She notes that he has worked closely with FIDIC in the development of BIMS and has travelled extensively as an advocate for the system. In the article, Ritchie acknowledges that some NGOs have criticised the BIMS for lacking credibility owing to its voluntary approach; however, he argues:

> If they're only going through the motions, they aren't protecting their firm. If the fact that it's the right thing to do isn't enough for them, they need to realise that society's morals are changing and that being out of step can severely damage their business. I emphasize the need to train your people and make it very clear that corrupt practices won't be tolerated. I emphasize the need to keep very close tabs on any outside representatives you hire. This is a real change in the industry.
>
> (Ritchie 2005)

Hatch Energy (2007), in reflecting on Acres' involvement, claims that the Lesotho corruption case demonstrates the value of operating an effective BIMS, and the need to have strong defences against bribery allegations.

The role of export credit agencies

Research (Goldzimer 2003) into the role of Export Credit Agencies (ECAs) reveals widespread criticism of their involvement in projects where corrupt activities have been exposed with accusations that they lack transparency and accountability in their operations. Goldzimer also notes that many NGOs have now begun to grapple with ECAs after discovering that they had become the principal financiers of the projects which local communities in developing countries were battling against because of environmental or social impacts, corruption, or other ills. He refers to instances where NGOs successfully campaigned to stop or delay certain ECA-backed projects, such as the Maheshwar dam in India and the Ilisu dam in Turkey. The latter project was to see Friends of the Earth's (FOE's) involvement, and FOE has also criticised the manner in which the UK's Export Credit Guarantee Department (ECGD) is 'heavily' influenced by businessmen connected with companies that benefit from its guarantees (FOE 1999). This situation can lack transparency and encourage engineering-construction companies to develop self-fulfilling prophecies by searching out commercial opportunities in developing countries, which are not socially and economically beneficial to the host country, and pressuring their ECA to back the proposal. It is not uncommon for a parallel activity to involve bribery of a government official in the country targeted and thus a win-win scenario all round for the corporation(s) concerned.

The propensity for this is noted by the United Kingdom's National Audit Office (NAO 1992: 7) which found that the Samanalawewa Dam in Sri Lanka was approved in advance of the expected need for increased power supplies; and the commercial case for aid and trade support was weak. This was also the case with the Philippines' mini hydro project. In these two cases the administration had carried out only limited appraisals. The NAO concluded that the projects had not been fully justified, and the Philippines' mini hydro project will not, in the event, secure the full expected benefits.

It would appear that Goldzimer's description of the role of ECAs is based on such assumptions:

An export credit agency is an agency of – or backed by – a government. Usually overseen by the finance, trade, or economics ministry, an ECA uses taxpayers' money to make it cheaper and less risky for domestic corporations to export or invest overseas. Almost all industrialized nations have at least one ECA. Like department stores that provide credit so people without cash will buy the stores' products, rich countries (through their ECAs) provide loans and credit to developing countries, so that they will buy the rich country's exports. The results include debt for poor countries and increased sales and foreign investment opportunities for multinational corporations based in wealthy countries.

(Goldzimer 2003)

Drew (2002) notes that the United Kingdom's ECGD, in common with other OECD Export Credit Agencies, is under a legal and political obligation to introduce measures to combat bribery and corruption under the Convention. She considers that requirements for ECAs to obtain no-bribery declarations from applicants and to provide for financial sanctions against companies that have been found to have engaged in bribery are especially important. However, she concluded that if the ECGD is to meet its objectives and be effective in fully implementing the OECD Convention, then it must take action that is based on a wider interpretation of the requirements of the Convention and in line with the OECD's own monitoring brief.

Hawley (2003) argues that one of the major reasons why the ECGD has ignored corruption is that some of the best opportunities for British exports are in countries with the most serious corruption problems. She reviews several engineering-construction ECGD-assisted projects, including the Lesotho Highlands Water Project (see case study that follows this section) and provides extensive evidence to show where the ECGD has failed to address issues of corruption. She concluded that repeated failures point to an institutional culture within the ECGD that verges on gross irresponsibility in its handling of public funds. She calls on the ECGD to readdress the issue of corruption, if the United Kingdom's reputation on tackling corruption is not to be further tarnished. Indeed, Coates and Reale (2002) find that under ECGD guarantees, many of the risks are transferred from the private sector to the public sector. Projects that clearly fail, or that involve substantial amounts of bribery and corruption, or that supply arms to repressive regimes may end up being paid for by the British taxpayer.

Hawley (2003: 55) documents the case against the UK's ECGD as:

- A persistent failure to take notice of corruption allegations and a deep reluctance to investigate them.
- Inadequate investigatory procedures.
- An unwillingness to pass on corruption allegations to the appropriate external investigatory authorities.
- Inadequate due diligence regarding the potential for corruption in the projects it backed, coupled with complete disregard for international concerns about corruption.
- Inadequate vetting of UK companies and inadequate due diligence regarding consortia, partners and agents used by UK companies in countries in which they supported projects.
- Lack of openness and accountability regarding whether it had backed certain projects.

Case study: Lesotho Highlands Water Project (LHWP)

The LHWP is a multi-purpose project constructed in a series of phases (Phase IA completed in 1998; Phase IB in 2002; Phase II & III under development) and involves the construction of a series of dams, tunnels, pumping stations and hydroelectric works within the highlands region in the Kingdom of Lesotho, a country land locked by the Republic of South Africa (RSA). The primary objectives of the scheme are to supply parts of the RSA with water, generate hydroelectric power for Lesotho and provide regional social and economic development, water supply and irrigation in Lesotho. McCully (2001: 251) argues that these ventures are unusually susceptible to corruption because the amount of money at stake is greater than with most other construction projects. Furthermore, hydropower is a vital source of energy but, as in all large infrastructure projects, significant corruption can occur from the policy and planning stage through construction to the actual electricity production. Corruption invariably reduces the benefits from a project, at the same time increasing the human, economic and ecological damages (Water Integrity Network 2008).

A core feature of the Phase IA involved the design and construction of the main impounding reservoir and dam at Katse (Nthako and Griffiths 1997). The entire project is expected to cost US$8 billion by the time of its completion in 2020 (South African Reporter 2004). It is evident that the project has already made a contribution to Lesotho's GDP as well as other societal benefits, albeit these have come under scrutiny. Indeed, Earle and Turton (2005a) find that whilst some reports refer to the development and the promise of a better life for the citizens of the two developing nations, others point to a web of corruption and deceit on an international scale. This appears to reflect Whitelegg's (2000: 31) argument that the process of developing the built environment is a powerful determinant of who gains and who loses in such circumstances.

The House of Lords (2005) notes that in 1991 the Lesotho Highlands Development Authority (LHDA) engaged a consortium of seven companies from the United Kingdom, South Africa, Italy, Germany and France to construct Phase IA, the Katse Dam, in Lesotho. All the implicated companies are from countries that have signed the OECD Convention, which makes it a crime to bribe public officials abroad (Transparency International 1999a). Collectively, the contracting companies were referred to as the Highlands Water Venture. The contract was made on 15 February, 1991, and was concluded on the standard FIDIC Conditions of Contract (4th edition) with terms and additions. The contract was governed by the law of Lesotho.

In the High Court of Lesotho, key players in the LHWP were accused of taking part in bribe payments made by bidding companies to Mr Ephraim Sole, Chief Executive (CE) of the LHDA. Through an extensive international network of bank accounts and 'agents', bribes were paid to Sole, who was able materially to influence the tendering procedure and outcome through his unchecked power as CE of the LHDA (IPOC 2007). The degree to which Sole had influence over the tendering process and his ability to make preferential awards, without open competition, to the contractors who offered bribes is implicit, and Earle and Turton (2005a: 5) argue that he wielded considerable power and influence. How far this control extended along the supply chain is less clear. However, in a review of the tendering procedures adopted by the LHDA, Nthako and Griffiths (1997: 8) refer to the employment of a six-member team of renowned international experts who managed the tender evaluation process and issued recommendations as to award of construction contracts.

The role of offshore banks in jurisdictions where banking secrecy is considered normal has been criticised by Transparency International (2007b: 2) which notes that global financial centres play a pivotal role in allowing corrupt officials to move, hide and invest their illicitly gained wealth. Transcripts of Sole's Swiss and South African bank accounts were released, albeit Judge Cullinan (2002) commented that these were uncovered only because Sole's local bank accounts had been released by a court order during an earlier civil trial. Earle and Turton (2005a) note that a forensic examination of the bank accounts, undertaken by PriceWaterhouseCoopers (PWC), revealed a movement of funds from one of the accused companies, to an agent and then on to Sole's account. It was no wonder, then, that most of the contractors and consultants working on the project attempted to block the release of Sole's Swiss and South African accounts. Figure 7.1 shows the three agents (Cohen, Du Plooy, and Bam) and the number of transactions between the accused

(Continued)

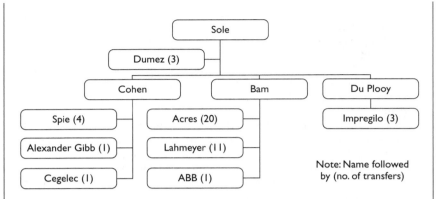

Figure 7.1 Flow of funds from international companies to LHWA chief executive.
Source: Earle and Turton (2005b).

companies and Sole. Earle and Turton (2005a: 12) note that Canadian contractor Acres used a representation agreement (RA) with the agent, and that the prosecution in the court case argued that this was a façade intended to provide a cover of legitimacy should the matter ever be investigated. The RA usually sets out a list of services that the agent is to perform for the contractor. Accordingly, they usually have a 'no duck–no dinner' clause – if the agent is not successful in assisting the contractor to secure the contract being tendered for, then the agent receives no payment. Hildyard (2000) directs partial blame for this problem at the United Kingdom's ECGD which supported the project through a £66 million contribution and argues that it failed to vet the corruption records of the companies bidding for contracts, particularly given that, he asserts, these companies were no strangers to allegations of corruption. This form of corruption is described below:

Bribery to obtain main contract award

A contractor who is tendering for a project is approached by an agent who claims that he will be able to assist the contractor to be awarded the project. They agree that, if the contractor is awarded the project, the contractor will pay the agent a commission of 5% of the contract price. The agent is appointed under a formal agency agreement which states that the agent will carry out specified services. However, the fee being paid to the agent is grossly in excess of the market value of the legitimate services which the agent is committed to provide. The agent intends to pay part of the commission to a representative of the project owner to ensure that the contractor is awarded the contract. Although the contractor does not actually know that the agent will use the commission for that purpose, the contractor thinks it likely that this will be the case due to the significant disparity between the value of the legitimate services to be carried out by the agent and the amount of the fee. The contractor is awarded the contract. The contractor pays the agent the commission. The agent pays the representative of the project owner a bribe out of the agent's commission. The cost of the commission (and therefore of the bribe) is included in the contract price. The project owner therefore pays more than it would have done had there not been a bribe. The project owner is unaware that one of its representatives has been bribed.

(Stansbury and Stansbury 2007: 18)

An engineer's responsibility to society

In sentencing Ephraim Sole, Judge Cullinan (2002) noted that Sole held a Bachelor of Civil Engineering degree from a university in Canada. Interestingly, Darroch (2005) notes that it was a Canadian engineering company, involved in two contracts within the LHWP, that was the first company to be found guilty of bribing Sole and that the company had benefited from the bribe to the detriment of its competitors. The company appealed but failed, and was later debarred by the World Bank for a period of three years.

In his defence, Sole submitted that his good actions outweighed the bad, and, furthermore, the scale of bribery was in some manner consistent with the size of the project. Judge Cullinan (2002) noted that the accused's Swiss and South African bank accounts were credited over a nine-year period with payments from as many as 12 consultants/contractors. The transactions indicated activities involving 48 corrupt or fraudulent transactions, including the acceptance of 46 bribes. Sole was sentenced to 18 years' imprisonment with a subsequent appeal unsuccessful. Judge Cullinan also commented on Sole's responsibility to society and that his position as CE of such a prestigious project would have provided a source of immense pride and commitment to most men. He contrasts Sole's actions with those of the contractors' workforce who were working in the alternating heat and biting cold of the mountains of Lesotho and who were exposed to risks during potentially dangerous tunnelling operations. The health and safety aspects were of particular concern during the construction of the Katse Dam, and Develay *et al.* (1998: 29) recollect that the local labour force was largely unfamiliar with heavy civil engineering construction methods. Moreover, although safety training courses were given to foremen, they found that old habits were difficult to change and that the conflict of safety and progress remained an ongoing problem.

Despite the LHWP receiving various plaudits (e.g. ICE 1997) for its technological, societal and project management achievements, it has attracted further criticism on several issues related to CSR. For example:

- An environmental action plan (EPA) was developed for the LHWP. This concerned compensation, natural environment and heritage, public health and rural development and was used for monitoring the impact and management of the social and environmental aspects. However, the

(Continued)

EPA was finalised only in May 1991, after the start of construction on major project elements, and its implementation continued to lag behind the construction (Nthako and Griffiths 1997). Indeed, the completion of documentation to ensure completeness for QA purposes rather than as a management tool to mitigate foreseeable problems has received attention from McCully (2001: 54). He argues that the employment of Environmental Impact Assessments (EIA) has become a bureaucratic formality and part of a self-fulfilling prophecy whereby governments and funders rubber stamp projects that they have already decided to approve. According to Hildyard (2000), the LHWP was approved on the basis of a deeply flawed and inadequate environmental impact assessment.

- The workforce employed during construction are considered to have been a contributing factor in introducing and spreading HIV/AIDS within the country (Lerer and Scudder 1999: 116; McCully 2001: 86).

Conclusions

To most observers it is evident that corruption has been a fact of life since the earliest times of mankind (Engeli and Pieth 2000), like a plague, affecting entire societies. Unfortunately, corruption is part of human nature, emanating from greed and disregard for the consequences and damages to fellow human beings, society at large and the natural environment. However, valiant efforts are being undertaken to curb corruption through personal, corporate, social, political, religious, regional, national and even international interventions.

These efforts have met with varying degrees of success. Currently, there are various bodies including Transparency International, multilateral organisations such as World Bank (WB) and International Monetary Fund (IMF), international professional societies as well as consumer organisations which have highlighted the curse of corruption and are trying to combat it at different levels. Specifically in the construction industry, 'procurement acts' and procedures have emerged in almost all countries and anti-corruption legislations being formulated and implemented. Governments have also, through 'disclosure' requirements, made it difficult for corrupt money to be invested and banked. However, it is also sad to note that corruption is getting very sophisticated and the amount of corrupt money is increasing exponentially.

Some developed countries and some developing countries have managed to reduce corruption through political, economical, social or religious awareness. Specifically, developed countries in Scandinavia and Australasia have managed to curb corruption through social and economic policies.

Some of the Middle Eastern countries also have managed to reduce corruption through strict adherence to religious tenets.

However, these efforts are very limited in scope, and until and unless human beings are forced to understand that corruption is bad for everyone including themselves, the success will always be limited. Corruption is a 'moral' issue besides being a financial, economical and criminal issue. The 'morals' can only be taught voluntarily and also be accepted by the recipients voluntarily. This can be encouraged through the dissemination of CSR values and associated topics such as professional codes of ethics within college and university courses, whether they be engineering, management, law or finance courses.

Organisations have to unite to produce a sustainable programme to continue to educate on, expose and fight corruption. The whole process of combating corruption in the construction industry can be summarised in the acronym 'ACCESS':

A = ANNIHILATE
C = CORRUPTION in
C = CONSTRUCTION through
E = EDUCATION and
S = SOCIAL
S = SINCERITY

Given that this chapter has drawn heavily, in places, on Transparency International and its initiative Preventing Corruption on Construction Projects, it is fitting that the concluding thoughts are given to Peter Eigen, Chairman of Transparency International (TI) and Neill Stansbury, Project Director for Construction & Engineering at TI-UK:

> Corruption in large-scale public projects is a daunting obstacle to sustainable development. Corruption in procurement plagues both developed and developing countries. When the size of a bribe takes precedence over value for money, the results are shoddy construction and poor infrastructure management. Corruption wastes money, bankrupts countries, and costs lives.
>
> (Eigen 2005b)

> Corruption in construction projects can be avoided if all parties put into place the necessary preventive measures. This requires coordinated international action by governments, banks, export credit agencies, project owners, contractors and other relevant parties.
>
> (Stansbury 2005)

References

American Society of Civil Engineers (2005) *Combating Corruption*. Online. Available at: http://www.asce.org/pressroom/news/policy_details.cfm?hdlid=508 (accessed 28 May 2007).

American Society of Civil Engineers (2007) *Combating Corruption in Engineering and Construction*. Online. Available at: http://content.asce.org/global/principles/Charter.html (accessed 2 December 2007).

Armitt, J. (2007) Armitt defends overseas 'commission' payments, *New Civil Engineer*, 21 June: 1.

Bosshard, P. (2005) The environment at risk from monuments of corruption, *Transparency International Global Corruption Report 2005, Special Focus, Corruption in Construction and Post Conflict Resolution, Part 1: Corruption in Construction, the Costs of Corruption*: 19–23.

Bottelier, P. (1998) Corruption and development, Remarks for *International Symposium on the Prevention and Control of Financial Fraud*. Online. Available at: http://web.worldbank.org/WBSITE/EXTERNAL/NEWS/0,,contentMDK:20060366~menuPK:34474~pagePK:34370~piPK:42770~theSitePK:4607,00.html (accessed 15 December 2007).

Bowen, P., Akintoye, A., Pearl, R. and Edwards, P.J. (2007) Ethical behaviour in the South African construction industry, *Construction Management and Economics*, 25: 631–648.

Bristow, M. (2007) *China's Construction Projects 'Rushed' The Collapse of a Bridge in China has Put the Spotlight on the Nation's Many Construction Projects*, BBC News 24. Online. Available at: http://news.bbc.co.uk/1/hi/world/asia-pacific/6945972.stm (accessed 12 December 2007).

Burke, C. (2007) China's entry into construction industries in Africa, *China Report*, 43 (3): 323–336.

Coates, B. and Reale, D. (2002) Export credits: what is the public policy aim? *World Development Movement NGO Seminar on Export Credit Reform*, House of Commons, London. Online. Available at: http://www.thecornerhouse.org.uk/item.shtml?x=51996#index-01-00-00-00 (accessed 10 October 2007).

Collier, P. and Hoeffler, A. (2005) The economic costs of corruption in infrastructure, *Transparency International Global Corruption Report 2005, Special Focus, Corruption in Construction and Post Conflict Resolution, Part 1: Corruption in Construction*: 12–19.

Corner House (2000) Underwriting corruption Britain's role in promoting corruption, cronyism and graft, *Corner House*. Online. Available at: http://www.thecornerhouse.org.uk/item.shtml?x=51990 (accessed 20 January 2008).

Cullinan, B.P. Judge (2002a) *In the Matter Between Rex vs. Masupha Ephraim Sole*. In the High Court of Lesotho, Before the Hon. Mr Acting Justice B.P. Cullinan on the 4th June Sentence.

Cullinan, B.P. Judge (2002b) *Sentence Handed Down in Rex v. Masupha Ephraim Sole*. In the High Court Lesotho, Odiousdebts, June 4. Online. Available at: www.odiousdebts.org/odiousdebts/publications/SoleSentence.pdf (accessed 20 August 2007).

Darroch, F. (2005) Case study: Lesotho puts international business in the dock, In *Transparency International (2005a) Global Corruption Report 2005, Special Focus, Corruption in Construction and Post Conflict Resolution, Chapter 2, Corruption in Practice*: 31–36.

Debrah, Y.A. and Ofori, G. (2005) Emerging managerial competencies of professionals in the Tanzanian construction industry, *International Journal of Human Resource Management*, 16 (8): 1399–1414.

Develay, D., Hagen, R.J. and Bestagno, R. (1998) Lesotho highlands water project – design and construction of Katse dam, *Proceedings of the Institution of Civil Engineers; Civil Engineering, Special Issue, No. 1*, Lesotho Highlands Water Project, 120: 14–29.

Doig, A. and Riley, S. (1998) Corruption and anti-corruption strategies: issues and cases from developing countries, *Corruption and Integrity Improvement Initiatives in Developing Countries*, UNDP, Chapter 3. Online. Available at: http://mirror.undp.org/magnet/Docs/efa/corruption/Chapter03.pdf .

Drew, K. (2002) *The UK Export Credit Guarantee Department, Corruption and the Case for Reform*, Public Services International Research Unit, Unicorn, A Global Unions Anti Corruption Network. Online. Available at: http://www. againstcorruption.org/reports/2002-06-C-UKECGD.doc (accessed 19 August 2007).

Earle, A. and Turton, A. (2005a) *No Duck No Dinner: How Sole Sourcing triggered Lesotho's Struggle against Corruption*. Online. Available at: www.acwr.co.za/pdf_files/07.pdf (accessed 20 December 2007).

Earle, A. and Turton, A. (2005b) Corruption on the Lesotho highlands water project – a case study, *Proceedings of World Water Week in Stockholm* , Drainage basin Management – Hard and Soft Solutions in Regional Development, Stockholm International Water Institute, Stockholm.

Eigen, P. (2005a) Introduction, in *Global Corruption Report 2005*, Special focus, Corruption in Construction and post conflict resolution, Transparency International, http://www.transparency.org/publications/gcr/download_gcr/download_gcr_2005#download (accessed 14 October 2007).

Eigen, P. (2005b) Quoted in, *Latest News*, A world built on bribes? *Corruption in Construction Bankrupts Countries and Cost Lives, says TI Report*, www.transparency.org/news_room/latest_news/press_releases/2005/16_03_2005_gcr_relaunch (accessed 12 September 2007).

Engeli, G. and Pieth, M. (2000) Developing an integrity programme for FIDIC: a private sector initiative to prevent corruption in IFI-funded public procurement, *Background Paper*, Commissioned by the World Bank and prepared for the Annual Meeting of FIDIC 2000 in Honolulu.

Ewins, P., Harvey, P., Savage, K. and Jacobs, A. (2006) *Mapping the Risks of Corruption in Humanitarian Action*, Overseas Development Institute and Management Accounting for NGOs (MANGO), A report for Transparency International and the U4 Anti-Corruption Resource Centre.

Ferguson, H. (2007) The ICE says no to bribery, *New Civil Engineer, Letters*, 28 June: 16.

FIDIC (2001) Guidelines for business integrity management in the consulting industry (BIMS) guidelines test ed (2001), FIDIC, Switzerland, http://212.74.172.248/bookshop/prod_page.asp?ProductCode=FI-QI-G-AA-10 (accessed 10 October 2007).

FIDIC (2007a) *FIDIC BIMS (Business Integrity Management System) Survey 2007*. Online, International Federation of Consulting Engineers, Available at: www1.fidic.org/about/bims07/ (accessed 10 December 2007).

FIDIC (2007b) 2007 Singapore Conference, *Winning with Integrity Management*. Online, International Federation of Consulting Engineers, Available at: http://www1.fidic.org/conference/2007/talks/wed/bims/ochoa_bims.pdf (accessed 10 December 2007).

Friends of the Earth (1999) *Export Credit Agency, Least Green of the Lot!* Online. Available at: www.foe.co.uk/resource/press_releases/19990802120100.html (accessed 5 June 2007).

Goldzimer, A. (2003) Worse than the World Bank? Export credit agencies – the secret engine of globalization, Institute for Food and Development Policy, *Backgrounder*, Winter, 9 (1). Online. Available at: www.foodfirst.org/pubs/backgrdrs/2003/w03v9n1.html (accessed 2 June 2007).

Hatch Energy (2007) *Case Study BIMS Implementation and Results: Acres International*, Hatch Energy.

Hawley, S. (2003) Turning a blind eye corruption and the UK export credits guarantee department, June 2003, *The Corner House*. Online. Available at: www. thecornerhouse.org.uk/pdf/document/correcgd.pdf (accessed 24 November 2007).

Hearne, B. (2005) *Building Capacity to Cut Graft*, Ethical Corporation, 30 March. Online. Available at: http://www.ethicalcorp.com/content.asp?ContentID=3601 (accessed 29 December 2007).

Heilman, B. and Ndumbaro, L. (2002) Corruption, politics, and societal values in Tanzania: an evaluation of the Mkapa administration's anti-corruption efforts, *African Journal of Politics*, 7 (1): 1–20.

Hildyard, N. (2000) The Lesotho highland water development project, what went wrong? *Global Policy Forum*. Online. Available at: www.globalpolicy.org/nations/corrupt/lesotho.htm (accessed 6 June 2007).

House of Lords (2005) *Opinions of The Lords of Appeal For Judgment in The Cause Lesotho Highlands Development Authority (Respondents) V. Impregilo Spa And others (Appellants)* [2005] UKHL 43.

ICE (1997) Lesotho highlands water project, *Proceedings of the Institution of Civil Engineers; Civil Engineering*, Special Issue, No. 1, 120.

Information Portal on Corruption in Africa (2007) *Case Studies, Lesotho Highlands Water Project*. Online. Available at: www.ipocafrica.org/content/view/31/66/ (accessed 10 December 2007).

Kang, B., Price, A.D.F., Thorpe, A. and Edum-Fotwe, F.T. (2004) Developing a systems approach for managing ethics in construction project environments, *Proceedings of Twentieth Annual ARCOM Conference*, 1–3 September, Heriot Watt University, 2: 1367–1375.

King, M. (2002) Trinidad & Tobago: from airport corruption to the collapse of government, *Transparency International Global Corruption Report*, Regional Reports – Central America, Mexico and the Caribbean, 79.

Langseth, P. and Michael, B. (1998) Are bribe payments in Tanzania 'grease' or 'grit'? *Crime, Law & Social Change*, 29: 197–208.

Laurenson, J. (2007) Corruption, quote taken from web based video, Organization for Economic Co-operation and Development. Online. Available at: http://www.viewontv.com/oecd/161007_corruption/index.php (accessed 10 October 2007).

Lerer, L.B. and Scudder, T. (1999) Health impacts of large dams, *Environmental Impact Assessment*, 19, 113–123.

Lewis, M. (2004) Public procurement and corruption in Trinidad and Tobago, *Journal of Construction Procurement*, 10 (1): 4–15.

Lewis, J. (2005) Earthquake destruction: corruption on the fault line, *Transparency International Global Corruption Report 2005, Special Focus, Corruption in Construction and Post Conflict Resolution, Part 1: Corruption in Construction, Costs of Corruption*: 23–30.

McCully, P. (2001) *Silenced Rivers: The Ecology and Politics of Large Dams*, London: Zed Books Ltd.

Major Projects Association (2006) Corruption-damage, *Risk Perception*, Summary of Seminar 125 held at The Royal College of Pathologists, London, 26 April.

Mkono, N.E. and Ache, P. (2006) Tanzania: construction investment, *International Financial Law Review*. Online. Available at: www.iflr.com/?Page=10&PUBID =33 &ISS=22257&SID=643197&TYPE=20 (accessed 23 September 2007).

National Audit Office (1992) Overseas aid: water and the environment, *Report by the Comptroller and Auditor General*, 6 May, London: HMSO.

Nthako, S. and Griffiths, A.L. (1997) Lesotho highlands water project – project management, *Proceedings of the Institution of Civil Engineers; Civil Engineering*, Special Issue, No. 1, Lesotho Highlands Water Project, 120: 3–13.

Organisation for Economic Co-operation and Development (2006), *The OECD Fights Corruption*, Paris, http://www.oecd.org/dataoecd/31/51/37393705.pdf (accessed 20 August 2007).

Ritchie, J. (2005) Quoted in, Building capacity to cut graft, *Ethical Corporation*, Bernadetta Hearne, 30 March. Online. Available at: http://www.ethicalcorp.com/content.asp?ContentID=3601 (accessed 29 December 2007).

Shen, Q. and Liu, G. (2004) Applications of value management in the construction industry in China, *Engineering, Construction and Architectural Management*, 11: 9–19.

Sinha, R. (2005) Blowing the whistle on corruption: one man's fatal struggle, *Transparency International Global Corruption Report 2005, Special Focus, Corruption in Construction and Post Conflict Resolution*, 9–12.

South African Reporter (2004) *Africa's Biggest Water Project*. Online. Available at: www.southafrica.info/doing_business/economy/infrastructure/sa-lesothowaterproject.htm (accessed 02 October 2007).

Stansbury, N. (2005) A world built on bribes? Corruption in construction bankrupts countries and cost lives, says TI report, Quoted in, *Latest News*, 2005. Online. Available at: www.transparency.org/news_room/latest_news/press_releases/2005/16_03_2005_gcr_relaunch (accessed 12 September 2007).

Stansbury, C. and Stansbury, N. (2007) *Anti-Corruption Training Manual: Infrastructure, Construction and Engineering Sectors, International Version*, Transparency International UK. Online. Available at: www.transparency.org/tools/contracting/construction_projects/section_b (accessed 10 January 2008).

Transparency International (1999a) *International Construction Companies Bribe Top Official in Large Dam Project for South Africa*. Online. Available at: www.transparency.org/news_room/latest_news/press_releases/1999/1999_08_06_safrica (accessed 23 September 2007).

Transparency International (1999b) Hong Kong, The airport CORE program and the absence of corruption, *Report by a Mission of Transparency International Comprising Peter Rooke and Michael H. Wiehen*. Online. Available at: http://unpan1.un.org/intradoc/groups/public/documents/APCITY/UNPAN013116.pdf.

Transparency International (2000) Press Release: *How to Build a Mega Airport Without Corruption*. Online. Available at: http://www.transparency.org/news_room/latest_news/press_releases/2000/2000_01_28_hongkong_airport (accessed 10 May 2008).

Transparency International (2002) *Bribe Payers Index, 2002*, Explanatory Notes and tables. Online. Available at: www.transparency.md/Docs/TI_BPI2002_en.pdf (accessed 12 June 2007).

Transparency International (2004) *China Moves Against Bribery in Construction Sector*. Online. Available at: www.transparency.org/news_room/latest_news/press_releases/2004/2004_05_13_china_moves_constr (accessed 23 September 2007).

Transparency International (2005) *Preventing Corruption on Construction Projects*, Risk Assessment and Proposed Actions for Banks, Export Credit Agencies, Guarantors and Insurers. Online. Available at: http://www.transparency.org.uk/ programmes/UKACS/ECGD/TI.PREVENTING.CORRUPTION.RISK.FUNDERS. pdf (accessed 22 February 2007).

Transparency International (2006a) *Report on the Transparency International Global Corruption Barometer 2006*, Policy and Research Department, International Secretariat, Germany www.transparency.org/content/download/12169/115654/version/1/file/Global_Corruption_Barometer_2006_Report.pdf (accessed 5 September 2007).

Transparency International (2006b) *Bribe Payers Index 2006*, Analysis Report, Policy and Research Department, International Secretariat. Online. Available at: www.transparency.org/news_room/in_focus/bpi_2006#pr (accessed 5 September 2007).

Transparency International (2007a) *About Transparency International; What is Transparency International?* Online. Available at: http://www.transparency.org/about_us (accessed 12 December 2007).

Transparency International (2007b) *Corruption Perception Index, Press Kit: Persistent Corruption in Low-income Countries Requires Global Action*. Online. Available at: www.transparency.org/policy_research/surveys_indices/cpi/2007 (accessed 12 December 2007).

Transparency International (2007c) *Systematic Corruption in the Construction Sector Brings Chinese and International Experts Together*. Online. Available at: www.transparency.org/news_room/latest_news/press_releases/2007/2007_08_23_construction_china (accessed 12 February 2008).

Transparency International (2007d) *Policy Paper: Poverty, Aid and Corruption*.

Warioba Report (1996) *The Report of the Presidential Commission of Inquiry Against Corruption*, Dar Es Salaam, Government of Tanzania.

Water Integrity Network (2008) *Global Corruption Report 2008 – Corruption in the Water Sector*. Online. Available at: www.waterintegritynetwork.net/home/learn/win_news/global_corruption_report_2008_corruption_in_the_water_sector (accessed 2 January 2008).

Whitelegg, J. (2000) Building ethics into the built environment, in Fox, W. (ed.), *Ethics and the Built Environment*, UK: Routledge, Chapter 3: 31–43.

Zou, P.X.W. (2006) Strategies for minimizing corruption in the construction industry in China, *Journal of Construction in Developing Countries*, 11 (2): 15–30.

Chapter 8

Cartels in the construction supply chain

Steve Male

Introduction and background to anti-competitive behaviour

The central focus of this chapter is collusion in construction industry supply chains, namely, the development of cartels that act against the normal operations of competitive behaviour in the industry, in short, corrupt practices. The chapter adopts a theoretical and investigative commentator approach to its content, raising and debating issues surrounding the presence of construction cartels and anti-competitive behaviour.

Transparency International (2005a), a global network organisation committed to fighting corruption and corrupt practices, distinguishes between *bribery*, where one party can bribe another party using either cash or non-cash inducements, and *fraud*, where one party is attempting to deceive another party. Bribery is normally initiated as a corporate act with the recipient being an individual. In the case of fraud the perpetrator of the deception will be attempting to extract payments or some other form of advantage from the other party. Equally, this may involve preventing the other party receiving a payment or an advantage. An example of the former, noted by Transparency International (TI) in the construction sector, is paying a bribe to win a project. Furthermore, TI highlights that bribery will inevitably involve a degree of fraud; for example, the bribe whilst covert has the intent, in the preceding example, of being seen externally and visibly as a legitimate arm's-length transaction. Additional examples are where contractors collude covertly during a project tendering process to increase the contract price or where clients/owners claim delays or defects with a view to withholding payments. Fraud may or may not involve bribery. Price-rigging through the use of cartels is a form of anti-competitive behaviour and hence fraud. TI (2007b) estimates that due to the size and scope of the construction sector globally, corruption may run at $3200bn per annum. This figure indicates that corruption, either in the form of bribery or fraud, is endemic in the international construction industry.

In the United Kingdom, anti-competitive behaviour is prohibited under Chapters I and II of the Competition Act 1998 and may be prohibited under Articles 81 and 82 of the European Union Treaty. These laws prohibit anti-competitive agreements between businesses and the abuse of a dominant position by a business. Businesses that infringe competition law may face substantial financial penalties of up to 10% of their worldwide turnover (OFT 2007a). Cartels are seen as a particularly damaging form of anti-competitive activity, where their purpose is to increase prices by removing or reducing competition. As a result they directly affect the purchasers of the goods or services, whether they are public or private businesses or individuals. Cartels are also seen to have a damaging effect on the wider economy, as they remove the incentive for businesses to operate efficiently and to innovate. Detecting and taking enforcement action against the businesses involved in cartels form one of the main enforcement priorities for the United Kingdom's Office of Fair Trading (OFT), which locates construction within part of the infrastructure and knowledge economy (2007b). In the context of anti-competitive behaviour, the OFT defines a *'business'* or *'undertaking'* as any entity engaged in economic activity irrespective of its legal status. This includes companies, partnerships, Scottish partnerships and individuals operating as sole traders.

Set against this background and depending on the scale of economic development in a country, construction normally accounts for some 8–15% of a nation's GDP. The Department for Business, Enterprise and Regulatory Reform (BERR 2007), formerly the Department of Trade and Industry (DTI), notes that the UK construction industry currently contributes 8.2% of the nation's gross value added and employs approximately 2.1 million people in over 250,000 firms. UK construction output is the second largest in the European Union, and its output is ranked in the global top ten. Depending on the stage of the economic cycle, construction contributes between 45% and 55% of fixed capital formation. Construction is, therefore, a major UK industry, is an enabler for other sectors of the UK economy and provides essential infrastructure for economic progress and social development. Equally, its relative importance to other industries is often underplayed. The construction industry includes the construction, refurbishment and maintenance of all facilities, namely, civil, building and power and process-engineering structures. Construction is also an important barometer of economic health, both for investment in fixed capital assets and for assessment of the stage of economic development.

The foregoing sets the parameters around which the operation of cartels in construction will be explored. The next section explores the nature of anti-competitive behaviour.

What is anti-competitive behaviour and a cartel?

Chapter I of the United Kingdom's Competition Act 1998 and Article 81 of the European Union Treaty provide a blanket prohibition on agreements between undertakings that affect trade between EU member states and that have as their objective or effect the distortion of competition within the market. Chapter II of the United Kingdom's Competition Act 1998 and Article 82 of the EU Treaty govern the abuse by an undertaking of a possible dominant position in the market and are therefore concerned with monopolies and oligopolies, namely, the control of the market by a small number of suppliers. It applies where one or more undertakings have a dominant position on a relevant product or geographical market and abuse that position which affects trade between member states.

At its most simple level, a cartel is an agreement between businesses not to compete with each other, and can often be verbal. Typically, cartel members may agree on (OFT 2007c):

- *price fixing* – manipulating the price they will charge or the discounts/credit terms they will offer their customers for goods or services;
- *bid-rigging* – deciding who should win a contract in a competitive tender process;
- *output quotas/restrictions* – limiting the levels of products or services supplied to a market in order to increase the price;
- *market sharing* – choosing which customers or geographic areas they will supply, or preventing competitors (e.g. foreign competitors) from entering the market.

This is seen as 'hard core' behaviour of cartel operations and may also include tie-ins and information exchanges. In some cartels several of these elements may be present, and there is often an international dimension to cartel activity. The boundaries of anti-competitive behaviour and cartel operations essentially revolve around limiting the supply or production of goods or services that would otherwise be available under a more freely operating competitive market situation, and in this sense cartel operations are seen as a clear abuse of a competitive marketplace.

Cartels can occur in almost any industry and can involve goods or services at the manufacturing, distribution or retail level. Some sectors may be more susceptible to cartels than others because of their structure or the way in which they operate; cartels may be more likely to exist in an industry where

- there are few competitors;
- the products have similar characteristics (which leaves little scope for competition on quality or service);
- communication channels between competitors are already established (e.g. trade associations);
- the industry is suffering from excess capacity or there is a general recession.

Equally, the fact that these conditions are not present does not rule out the possibility that a cartel is operating. Conversely, the fact that an industry shows some or all of these characteristics does not automatically mean that some form of cartel operates, but it may be a possibility (OFT 2007d). The next section will explore the nature and the structure of the construction industry.

Industry structure and markets in construction

The terms *industry* and *market* are in common use in construction almost interchangeably. Analytically, however, they have distinct meanings. An *industry* is a supply-side concept. It is an arbitrary boundary within which firms compete with each other to produce related or similar products. There are five major forces determining industry structure, and these jointly establish the profit potential in an industry (Porter 1980). An awareness of their joint influences is important for understanding the competitive environment facing firms and the individual firm in an industry. *Buyers* and *suppliers* are two of the determinants of an industry's structure and have similar effects in that if they are powerful, profit margins can be pushed down in an industry. Buyers (or clients) as a group are particularly important in construction since they and their advisers, in the form of the construction professionals, can dictate the rules of competition for contractors, especially through the choice of procurement path. In the case of suppliers, construction is a highly interconnected industry through material inputs from other industries. Therefore, where industrial concentration of other industries may be high, the opportunity for suppliers to influence input costs could be considerable. *Threat of entry* is concerned with the likelihood of new competitors entering the industry. This is dependent on the presence or absence of entry barriers. It is often claimed that entry and exit barriers to the construction industry are low. However, others have suggested that there are subtle forms of entry barriers present in construction that have no direct counterpart in manufacturing. This stems directly from the heterogeneity and characteristics of the industry and the nature of the project-based market structures within it. *Substitute products or services*, according to Porter (1980), are those that undertake the same function. Finally, the *extent of competitive rivalry* is determined by the degree of mutual dependency or interaction between competitors and the likelihood of this setting off retaliatory strategic moves between them.

The structure of an industry directly impacts on the nature of competition between firms in that industry and the competitive strategies available to them (Porter 1980). The construction industry in the United Kingdom has traditionally been viewed as being subdivided into the civil engineering and building industries. However, by their nature, many construction contracts will involve both aspects of building and civil works. Equally, Groak (1994) argues convincingly that construction should not be considered an industry at all and contends that construction studies should no longer focus on an 'industry' paradigm, but should embrace potentially a technological paradigm whose impact is greater at project and production levels.

A *market*, a demand-side concept, has an economic and social dimension. It organises an exchange relationship between buyers and sellers to determine price (Male 1991). Kay (1993) also distinguishes between markets and industries, introducing the term 'strategic groups' into the equation using the work of Porter (1980, 1985):

- A market, defined by demand conditions, is based on consumer needs and is characterised by the 'law of one price'. A market is bounded by the ability of the consumer to substitute one product for another.
- An industry, determined by supply conditions, is based on production technology and is defined by the markets chosen by firms. Industries are determined by the manner in which production is organised.
- A strategic group, defined by the strategic choices made by firms in terms of whom they decide to compete against, is based on a firm's distinctive capabilities and market positioning. Group membership is determined subjectively by management within different competing firms.

According to Kay, the important issue for any firm is its choice of markets in terms of product and geography; the membership of a strategic group and industry follows from that choice. Hence, industries are relatively stable and are based around production capabilities and the outputs therefrom; markets are transient, are determined by consumer needs and are bounded by price and substitute products (Langford and Male 2001). There are two distinct types of market structures identified in construction with different economic forces operating in each type (Male 1991):

1. Construction to contract – a project-based business environment – which involves a company in constructing a facility to a customised design where the roles and responsibilities of the constituent parties are contractually defined. The method of price determination is the reverse of that in manufacturing in that the contractor determines price prior to production. Under this form of market structure the final product is pre-demanded by the client.
2. Speculative construction, which involves anticipating, responding to or creating demand and where the product is developed and sold to a wider market of consumers. A typical example in construction is speculative housebuilding.

Property developers could be viewed as a hybrid of types 1 and 2 since their initial activity is formed under type 2, but later results in type 1 activity. Different skills are required for each type of market structure. In the first instance, the emphasis is more on managerial and technical skills. In the second, entrepreneurial activity involves market forecasting of a different type, with market research, the assembly of financial packages and land banks (Male 1991).

The concept of the industry and the forces shaping it, as identified by Porter (1980), are not without problems in construction when considering the operation of cartels; these can operate at industry structure level around general price fixing or at market level through the project-based environment of construction-to-contract and the tendering/bidding process. The next section adopts a greater project focus.

The demand and supply chain system in construction

The choice of procurement strategy brings together the demand and supply sides of the buyer–producer relationship (Male and Mitrovic 2005). Demand and supply chain systems in construction comprise (Croner's 1999) the following:

- Professional service firms, typically comprising the designers and other professional consultants.
- Construction and assembly firms, comprising contributors to the on-site manufacturing process at a specific site location. These firms are involved in fitting, installation, assembly, repair, on-site and off-site labouring and fabrication.
- Materials and products firms, delivering the materials, products and hired plant required for the on-site manufacturing process. These form part of the main contractor's supply chain.

Transparency International (2005) highlights the fact that construction projects involve a large number of participants that are linked together through complex contractual structures which probably contributes to the widespread nature of corruption in construction. Complex contractual structures occur as a result of procurement and supply chain options.

In terms of procurement strategies that link supply chains together, typically, the primary procurement routes that now exist in UK construction are as follows:

- Traditional, where design is split from construction.
- The management forms: construction management and management contracting, acting as overlays on the former for more complex projects. Project management could be included within this grouping, although it is essentially an issue of a client's agent commissioned to manage the interface between the client and the construction industry and not a distinct procurement route as long as any of these mentioned here remain available.
- Design and build, and its variants such as turnkey, package deal, and early contractor involvement.
- Prime contracting and its variants (e.g. NHS Procure 21, and alliances with technology clustering).
- Private finance initiative (PFI)/public–private participation (PPP).

Prime contracting and PPP/PFI procurement routes, with a greater emphasis also on supply chain integration and the operational phase, require between four and five consortia and their supply chains to bid through three successive tendering stages, with potentially up to 9–12 consortia involved in some form of pre-qualification process. Regardless of whether it is PFI or prime, both go through similar bid processes. Bidding in PFI has been seen as a costly exercise, with sums of up to £0.5m quoted as the cost of bidding for the early stages of bidding, and between £2m and £3m often quoted as the cost of bidding for projects in the £50m–£200m size range, towards the final stages. For much larger and very complex projects, the cost of tendering can be in the tens of millions. The cost of bidding under such types of procurement mechanisms is prohibitive for many firms in the industry; it can also be a significant drain on financial resources for even the largest of firms should they have periods of unsuccessful tendering. Equally, consortia are formed from amongst the ranks of the same large contractors, consultants and suppliers. The exorbitant cost of bidding on such large projects, either within the United Kingdom or at international level, can conceivably create a climate for corrupt practices to creep in. In this vein, Transparency International (2005) notes many case-study examples derived from real experiences of corrupt practices throughout the project life cycle and into the operational phase of projects – particularly on international projects where the traditional, design and build/turnkey or FDBOT routes may predominate.

To summarise, competition in construction is affected by market structure, within the context of a broader industry structure. At an industry level, the evidence indicates that resulting from the five competitive forces noted by Porter that drive competitive industrial structures, each has been the subject of corrupt practices; cartels, in particular, operate within the supplier dimension where there is intense competitive rivalry between firms. When supplier monopolies and oligopolies operate at the industry structure level, general cartel price-fixing structures can come into play. Under project-based construction to contract, the rules of competition are set by the client and the client's advisers in the form of procurement, tendering and contractual strategies and arrangements. Flanagan and Norman (1989) use the term 'contestable markets' to describe these oligopolistic situations operating in bid/tendering situations where there are a few competitors who have been selected from all those available in the marketplace. Pre-qualification and selective tendering procedures set up oligopolistic market structures in the bidding situation and can lead to price and bid-rigging.

The next section explores supply chain management in construction.

Supply chains in construction: perspectives on terminology

Due to competitive pressures, a number of recent and significant trends have resulted in a need for organisations to become more effective and efficient with a consequent influence on the way they are managed, including the purchasing function and its relationship to supply chain management. Organisations, especially the larger ones, have needed to become more adaptable, responsive and flexible to changes in the business environment. There is an increased focus on understanding core competencies and what should or should not remain within the organisation or be outsourced. This is also linked to a greater appreciation of the opportunities that can accrue from more cooperative ways of working between suppliers and customers resulting in cultural changes within firms. The consequence of this is that firms are more willing to consider working closely with suppliers and customers to create a more integrated production process that goes beyond legally defined organisational boundaries. The impact has been that firms now have to work out their role and positioning within a wider configuration of organisations, with a consequent impact on organisational structuring, team and personal relationships.

This different way of thinking about supply and purchasing has resulted in a series of new terms being adopted and used to describe the domain. The supply chain concept has emerged having an underlying implication of a sequencing of interdependent activities that are internal or external to the firm, as a legal entity (Christopher 1998; Saunders 1997). This can encompass single-location activities for fairly simple firms to multi-site, geographically dispersed locations, often located internationally. The idea of the 'chain' has been extended to include analogies with rivers, and 'upstream' and 'downstream' terminology has become infused into the language, often to include the concept of a 'supply pipeline'. Further extensions to the concept have included viewing a supply chain in network terms, seeing it as a series of connected, interdependent organisations cooperatively working together to transform, control, manage and improve materials and information flows from suppliers to customers and end users. One implication of the supplier network is that the boundaries of different supply chains might overlap, and particular suppliers might become nodes within a more complex web pattern of suppliers. There is also a clear link between the resource procurement and transformation processes of the supply chain and the value chain or value system (Johnson and Scholes 1999).

The manufacturing industries have provided the main thrust for theoretical and practical development in supply chain management, where the objectives are seen as synchronising customer-service requirements with materials' flows from suppliers. Supply chain management relies on trust and cooperation to work effectively and is concerned with both internal and external integration of the supply chain through internal process linkages within the firm and the relationships between suppliers and customers. For the supply chain to work as an integrated system, the management of both materials and information is required (Gattorna and Walters 1996; Saunders 1997). It also involves managing the upstream and downstream business relationships between customers and suppliers to deliver superior customer value at lower cost to the supply chain as a whole (Christopher 1998).

'Partnering' philosophies and policies are typical, but an increasingly important view is that different types of relationship may be equally appropriate in managing the supply chain. Supply chain management, as a relatively new concept, is inherently interrelated with sources of competitive advantage that stem from pitting supply chain against supply chain (Christopher 1998). As Saunders (1997) points out, there has been a shift in perspective from viewing the procurement of supplies as an operational activity into one that is now seen as strategic and linked to the long-term survival of a firm. The foregoing further sets the context within which corrupt practices could operate. Figure 8.1 indicates the supply chain relationships operating in construction.

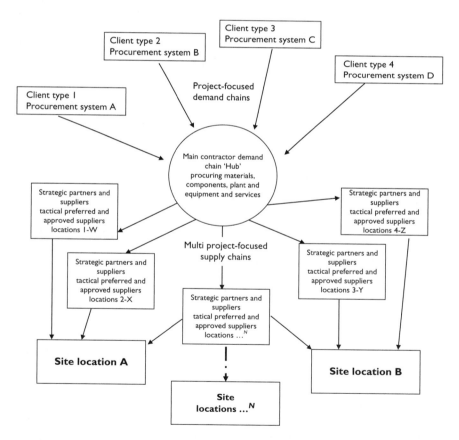

Figure 8.1 Construction demand and supply chain system.
Source: Adapted from Langford and Male (2001).

Corruption, collusion or cartel structures in construction

Corruption in construction

A recent survey conducted by the Chartered Institute of Building (CIOB) concluded that amongst the sample surveyed across a variety of sectors in the industry ($N = 1404$) there was considerable variation in the way corruption was perceived. Many respondents had no direct experience of corrupt practices, although 41% of the sample indicated they had been offered a bribe on at least one or more occasions; 34% indicated they had come across cartel operations. The sample was close to evenly split on the extent to which they thought corruption was common in UK construction. There was a series of interesting views on what is considered corrupt practice:

- Cover pricing was seen by 63% of the sample as very (18%) or moderately corrupt (45%).
- Bribery to obtain planning permission was seen by 73% of the sample to be very (56%) or moderately corrupt (17%).
- Employing illegal workers was seen by 84% of the sample as very (52%) or moderately corrupt (32%).
- Concealment of bribes was seen by 79% of the sample as very (52%) or moderately corrupt (17%).
- Collusion between bidders for market-sharing reasons was seen by 76% of the sample as very (45%) or moderately corrupt (31%).
- Bribery to obtain a contract was seen by 75% of the sample as very (57%) or moderately corrupt (18%).
- Leaking information to a preferred bidder was seen by 79% of the sample as very (39%) or moderately corrupt (40%).
- Production of fraudulent timesheets and invoices was seen by 84% (very 53% and moderately 31%) and 76% (very 54% and moderately 22%) of the sample, respectively, as corrupt.
- False or exaggerated claims against a contractor to withhold or reduce payments to a contract was seen by 79% of the sample as very (44%) or moderately (35%) corrupt, whilst 81% saw inclusion of a false extra cost to a claim as being very (42%) or moderately (39%) corrupt.
- Bribes to win operation and maintenance contracts by the building contractor was seen by 72% of the sample as very (53%) or moderately (19%) corrupt.

The CIOB concluded that one of the key issues to emerge from the findings is the lack of clear definition surrounding corruption and corrupt practices, noting that respondents clearly indicated that corruption did exist, but to what extent was not clear. Equally, whilst there was a clear consensus on what corrupt practices were, that consensus broke down over the extent to which certain practices were seen as more or less corrupt, as indicated above. Other perceptions that came across were indications that some practices were just seen as the way the industry does business, for example, the use of non-cash versus cash incentives to gain commercial advantages. The CIOB notes that it can become difficult to draw the line in the area of corporate entertainment where, for example, personal relationships and increased levels of team-working and networking are seen by some as team-building activities, whilst others see these as corrupt practices. The CIOB's final conclusion is that the responses gave a clear view that the current situation was unacceptable and that improvements were needed.

As noted earlier, Transparency International (2005) notes that many examples of corrupt practices may occur at different stages of the project life cycle, some of which are examples of bribery and some of fraud, a number covering the operation of cartels. A number of the areas noted by TI were also covered in the CIOB survey. The next section explores cartels in construction.

Collusion and anti-competitive behaviour

Collusive and other concerted practices between competitors, who should be competing and not coordinating their behaviour, are likely to infringe the prohibition on anti-competitive agreements set out in the UK Competition Act. In a number of EU and UK competition cases, the tender processes by which a significant amount of work was won in the sector have been found, in certain circumstances, to facilitate such collusion between competitors (Dundas and Wilson 2005). Collusive tendering or bid-rigging occurs when two or more firms agree not to submit competitive prices for the supply of goods or services or agree in advance the price that each is to tender. The effect of such arrangements is to increase the price which consumers, namely individuals, organisations or government agencies have to pay for goods and services. Firms may also try to achieve the same effect by other means, such as dividing their sales areas between them and agreeing not to sell in each other's designated area (postcode selling), or agreeing not to poach each other's customers, thereby enabling each firm to set prices, knowing that the other will not undercut it.

Different forms that bid-rigging can take include:

- *Bid-suppression*: where one or several competitors refrain from tendering or withdraw already submitted tenders so that a competitor's tender will be accepted.
- *Complementary or cover bidding*: competitors agree to submit tenders that are too high or that contain special conditions that will be unacceptable to the buyer. This method gives the appearance of genuine competition.

- *Bid rotation*: competitors agree to take turns as to who will be best placed to win the tender i.e. the competitor whose turn it is to win puts in the lowest offer.
- *Awarding subcontracts*: in exchange for not bidding or for submitting an obviously unacceptable bid, competitors agree to share amongst them the illegal gains, e.g. the award of a lucrative subcontract for agreeing to participate in the collusive tendering scheme.

In the absence of a formal subcontracting relationship or a formal joint venture established to combine expertise, management, or finances to strengthen a bid, there is no reason why undertakings invited to participate in a tender process should need to communicate with one another in relation to a tender. However, although bid-rigging operations are often very sophisticated to avoid detection, there are certain signs that can be seen, particularly where contracts are awarded regularly; these include (OFT 2005) the following:

- Certain suppliers unexpectedly decline an invitation to bid.
- An obvious pattern of rotation of successful bidders is evident.
- An unusually high margin between the winning and unsuccessful bids.
- All bid prices drop when a potential new bidder has come onto the scene, that is, one who is not a member of the cartel.
- The same supplier is found to be the successful bidder on several successive occasions for a particular area or for a particular type of contract.
- One or more suppliers has made a series of bids although they consistently fail to win a contract.

The European dimension to cartels in construction

The European Commission is to set up a new Directorate devoted exclusively to cartel enforcement (Foundation World 2005). This was announced by the European Competition Commissioner at a Brussels conference entitled 'Taking Competition Seriously – Anti-Trust Reform in Europe' organised jointly by the Commission and the International Bar Association. The creation of the Directorate reflects the determination to make the fight against cartels a top priority. The prohibition of mergers does not stem from opposition to large companies. In merger control, it is not the size of company that matters, but whether sound economic analysis proves that the merged entity will dominate its competitors and ignore customers. It is seen that the only sustainable option for long-term economic success is based on national and European firms being able and keen to compete at home. These companies will be the ones that have proven themselves by winning in open and fair domestic and European markets and have the innovative edge and resourcefulness required to compete and succeed at global level.

Baer *et al.* (2005) add to this when they indicate that the Commission has been intensifying its pursuit of illegal cartels in recent years, although 2003 did see a reduction in the number of cartel decisions taken and total fines imposed. Investigative activities are likely to remain high now that the Commission has freed resources previously designated to consider applications for the exemption of individual agreements under Article 81(3) of the EC Treaty. Under the modernisation of EC competition law, national competition authorities of member states will assume greater responsibilities and a higher level of cooperation. This is likely to lead to significantly higher levels of cartel investigation and enforcement at both EU and national levels. Institutional changes in the European Commission and in many member states' authorities have also provided the necessary means for such increased activity. The Commission also has available to it new investigative powers under the 'modernisation regulation', EC Regulation 1/2003, which came into effect on 1 May 2004. These, together with the increased powers devolved on national competition authorities, enable the Commission to increase its focus on trans-border cartels.

The modernisation legislation allows for a greater exchange of information between the new and closer network that has been established between the Commission and national competition authorities in the European Union. However, under the rules governing the new network, an application for leniency to a competition authority in the network is not to be considered as an application for leniency to any other member of the network. It is expected that multiple and simultaneous applications will need to be made to members of the enforcing network. The information voluntarily submitted by a leniency applicant will be transmitted to another member of the network only with the consent of the applicant, and once the applicant has given its consent it cannot be withdrawn.

When a national competition authority in the European Union deals with a cartel, it must inform the European Commission and may also inform other EU national competition authorities. The Commission has an equivalent obligation to inform the member states' network. When the case has been initiated as a result of a leniency application, the information submitted to the network may not be used by other members of the network as the basis for starting an investigation on their own behalf.

Cartels in the European construction supply chain

The activities of construction firms in Europe have been the subject of intense scrutiny from the European Commission over the last 20 or so years (CMC Cameron McKenna 2004). The production of beams, pipes, bars, tubes, cement and roofing felt have all been the subject of cartels. The Commission has detected and punished 17 construction cartels, representing well over a quarter of all cartels, and has imposed a staggering €1.5bn in fines on the participants including the biggest individual fine of nearly €250m on Lafarge in the plaster board cartel.

At the end of October 2001, the Competition Authority launched an investigation into the asphalt sector in Sweden, suspecting the existence of a possible cartel (Petterson and Thunstrom 2004). Following inspections carried out at the premises of a number of construction companies, the Competition Authority announced that the construction companies had allegedly rigged bids submitted in public procurements and shared the market geographically between them. The Competition Authority suspected that contracts of collusive nature existed between 11 construction companies, at least from the time of the introduction of the Competition Act in 1993, and until the Competition Authority launched its investigation. National competition authorities in Norway and Finland are currently undertaking similar investigations in the asphalt sector in their respective countries. The Competition Authority brought action at the Stockholm District Court in March 2003, proposing a total fine for the 11 construction companies of approximately SKr1600m. The amount of the fine is the highest ever imposed by the Competition Authority.

In November 2002 Brussels antitrust regulators fined 4 building materials' companies, including Lafarge and BPB, a total of €478m for a six-year conspiracy to fix the price of a widely used construction product (Guerrera *et al.* 2002). France's Lafarge was the hardest hit, receiving a fine of €250m – the third-largest penalty ever levied by the Brussels authorities on a single company – with its UK rival BPB getting a penalty of €139m. The German group Knauf has been ordered to pay €86m, while Gyproc of Belgium was fined €4m. A four-year inquiry concluded that the companies conspired to fix the price of plasterboard in France, the United Kingdom, Germany and the Benelux region by agreeing to stop a price war. The total penalties for the plasterboard cartel rank second only to 2006's €855m fines on eight chemicals' companies for fixing the price of vitamins. The decision to impose heavy fines – first reported in The Financial Times – highlights the European Commission's determination to crack down on price-fixing conspiracies.

At the time, Lafarge and BPB said they would appeal against the decision in Europe's second-highest court, denying that the arrangements were illegal and claiming that the fines were disproportionate. Lafarge said the fine was equivalent to 2.5 times its annual sales of plasterboard in the United Kingdom and Germany – the only two markets it had been asked about by the Commission. It added that plasterboard prices in the two countries were lower in 1998 than when Lafarge entered the market.

Mario Monti, competition commissioner, said the heavy fines showed the Commission had a 'zero-tolerance' policy against price-fixing agreements. He said the cartel focused on a market worth €1.2bn at the time of formation of the cartel and affected the vast majority of Europe's

370 million consumers, as plasterboard was widely used in do-it-yourself and construction.

It was also reported in 2004 that the European Commission over a three-year period conducted investigations into several markets in the construction sector (Blake Lapthorn Linnell 2004):

- Fines totalling €85.04 m (about £59.5m) were imposed on eight businesses when the Commission discovered a cartel in Italian concrete reinforcing bars.
- An investigation into the activities of the industrial copper tubes sector found a cartel operated by 75% of the suppliers of these tubes used mainly in the air-conditioning and refrigeration industry. The cartel members met at meetings of the trade association formed for promoting quality standards within the industry.

Six cement producers were fined in April 2004 a sum totalling €660m as Germany's cartel office imposed its largest-ever sanction following a price-fixing probe (Wassener 2005). Germany's Heidelberg-Cement received the largest fine, though the €251.5m charge was less than the €400m predicted by some observers. The company has not admitted any liability in the case and, while it has made provisions against the fine, it plans to appeal. Dyckerhoff and Schwenk, both based in Germany, received fines of €95m and €142m respectively. France's Lafarge was fined €86m and Alsen, a Hamburg-based subsidiary of Switzerland's Holcim, received a €74m fine. Readymix, a unit of British building-materials group RMC, which had been quick to cooperate with the authorities, received the smallest fine, of €12m. Thirty companies operating in Germany were raided in 2004/2005 by officials seeking evidence of price-fixing and agreements over regional delivery quotas. Thirteen investigations against small- and medium-sized companies are still in progress. Ulf Boge, cartel office president, said the fact that cement buyers and users had been 'massively damaged' by collusion between the companies justified the high fines, though the final punishment reflected the level of cooperation from the companies. Separately, Holcim, the world's second-largest cement company, said Italian antitrust investigators had launched a probe into the concrete industry and had seized documents from some of its units.

In September 2004, the Commission imposed fines totalling €222.3m (about £155.6m) when eight companies were found to have been involved in a cartel for water, heating and gas tubes used in plumbing. Through the cartel the manufacturers had been involved in sharing production volumes and the market, the setting of standard terms of supply and price-fixing. They had exchanged information on sales, orders, market-shares and prices. Such was the organisation that it involved regular meetings and even the use of code names.

The UK dimension

The Competition Act 1998 and the Enterprise Act 2002

The Enterprise Act 2002 has introduced a criminal offence for individuals who dishonestly engage in cartel agreements (Baer *et al.* 2005). The OFT competition enforcement branch is responsible for cartels and works across all market sectors, conducting on-site investigations and, when appropriate, liaises closely with the relevant authorities in the European Commission and across the world (OFT 2005). Actions against cartels to break them apart is conducted because they allow businesses to achieve greater profits for less effort to the detriment of consumers and the economy as a whole, and, for the purchasers of their goods or services, this means higher prices, poorer quality, and less or no choice.

A cartel offence under the Enterprise Act 2002, which came into force in June 2003, provides for liability where an individual dishonestly agrees with one or more others to make, implement or cause to be made or implemented certain types of agreement between undertakings. This offence might typically arise where directors and senior managers of two or more undertakings enter into a cartel on behalf of their respective firms.

In 2004, the UK national competition authority – the Office of Fair Trading – published new guidelines on how it will use its criminal investigation powers, including covert surveillance and listening devices. The OFT has also agreed upon a memorandum of understanding with the United Kingdom's Serious Fraud Office (SFO) on how they will cooperate in investigating and prosecuting individuals for participating in cartels.

As at the Spring of 2005, no individual had been prosecuted under the Enterprise Act, nor has an individual been disqualified from acting as a company director under the powers granted by the Act. Nevertheless, the OFT has been active in investigating cartels and in enforcing the civil law against them. The OFT has also imposed fines on roofing companies for operating a collusive tendering arrangement in flat-roofing services. One company was granted total immunity from fines and another a 50% reduction, under the leniency scheme run by the OFT.

There is no possibility in UK law for private prosecutions for a cartel offence. The Enterprise Act does, however, provide for civil claims to be made by consumers, and by approved consumer organisations on behalf of groups of named consumers. The Act also provides the Competition Appeal Tribunal (the specialist tribunal that hears appeals from decisions of the OFT) with the power to hear applications for damages following a decision that a company has infringed UK or EC competition law. In 2004, the first such cases were brought. These concerned actions by purchasers of vitamins and were brought against members of the vitamins cartel. These actions have not yet been heard.

UK courts may also award damages to an applicant who suffers harm as a result of the infringement of EC competition law. In 2004 damages were awarded by the UK courts for the first time in *Crehan v Inntrepreneur (Court of Appeal (Civil Division), [2004] EWCA 637 2004)*. The Court accepted a claim by an applicant in relation to harm suffered as a result of the operation of an agreement entered into by the applicant and the defendant. The Court was willing to contemplate such an application because the applicant was in a 'markedly weaker [position] than the other party such as seriously to compromise or even eliminate his freedom to negotiate the terms of his contract and his capacity to avoid the loss or reduce its extent'. Damages were awarded even though the applicant would not have proved his case had it been brought purely under UK competition law. The Court accepted that there was an obligation to award damages under the EU 'principle of effectiveness', which prevents national laws from removing a right to damages as a result of a breach of EC competition law. The judgement has been appealed to the House of Lords, but there is little doubt that the current judgement will encourage competition-based claims in UK courts, especially where EC law can be pleaded.

The Office of Fair Trading

Competition law is enforced in the United Kingdom principally by the OFT. However, in certain industries such as gas and electricity, the sector regulators have been given 'concurrent powers' to apply and enforce competition law (OFT 2005).

In addition, the OFT and SFO are under the Enterprise Act 2002 responsible for carrying out investigations of directors suspected of involvement in the so-called hard-core cartel offences which carry punishments including disqualification as a director, unlimited fines and imprisonment of up to five years if found guilty.

The OFT and cartels in the UK construction supply chain

The construction industry is a target sector for the OFT. A previous OFT Chairman, Sir John Vickers, indicated in 2004 that '[e]vidence on cartel activity in the construction sector...is mounting. It is understood that some of the new cartel cases involve public sector contracts and a range of construction services. Some are believed to have stemmed from the initial investigation, but they are not limited to the West Midlands'.

Again, in December 2004 *The Financial Times* indicated that the construction and housing industry was to come under heavy scrutiny from the competition watchdog, amid concerns that there may be cartels operating across the sector (Sherwood 2004). The OFT's hit list of five areas it was most worried about also included the healthcare sector and credit markets, along with government-imposed trading restrictions and mass-marketing scams. The OFT can also refer cases to the Competition Commission, the regulator that conducts more detailed analysis of mergers and markets that could pose competition problems.

OFT cartel investigators consistently believe they must step up enforcement activity in the construction sector and property-related services such as estate agencies. Again, the former OFT Chairman, Sir John Vickers, warned that the OFT would be looking for whistle-blowers to lift the lid further on construction cartels. He said: 'Our leniency arrangements are there for people to come forward with information about cartel activity and the construction sector might be well advised to listen to that'. Vickers pointed to the experience of the Dutch authorities, which had uncovered a string of cartel cases after whistle-blowers came forward. He said the Dutch experience would not 'read through' directly to the United Kingdom, but it was an example of how the combination of tough anti-cartel powers and offers of leniency could hit price-fixers in the sector.

Yet again, in February 2004, it was indicated that the OFT has the construction industry firmly in its sights, having identified it as one of its priority areas in the OFT draft annual plan for 2005–2006 (Martineau Johnson 2004); approximately 30% of all the cartels the OFT was investigating involved the UK construction industry.

In November 2004, the SFO confirmed that it has launched a criminal investigation into the award of building contracts by an NHS hospital in Nottingham. The case was initially brought to the attention of the OFT after the hospital became aware of a pattern emerging vis-à-vis the award of construction contracts. It is believed that this is the first time the OFT has referred a case for criminal investigation under the Enterprise Act. It is understood that the OFT is conducting a parallel investigation under the Competition Act (Dundas and Wilson 2005).

The Financial Times in 2005 noted that the OFT was poised to break a string of cartels in the construction industry (Sherwood 2005). A number of whistle-blowers were understood to have alerted the OFT to anti-competitive behaviour by a series of companies, in an apparent vindication of the regulator's policy of offering lenient punishments for cartel members who come forward with information. Some cartels were expected to be hit with potentially heavy fines; another set of investigations, which could provide a second wave of cases, is in process. The investigations also raised the possibility of the first criminal cartel prosecutions or director disqualifications, whilst a criminal cartel offence was introduced by the Enterprise Act 2002. However, OFT officials at that stage were keeping open the option of criminal prosecutions, in addition to the prospect of levying civil fines. News of the cartels came as an appeal tribunal upheld the OFT's finding that two roofing contractors, with seven others, had agreed to fix prices by collusive tendering for repair and maintenance contracts in the West Midlands. The Competition Appeal Tribunal agreed with the OFT's £35,922 fine against Apex Asphalt and Paving Co although it reduced an £18,000 fine against Richard W. Price to £9000. The OFT had imposed fines on the nine companies involved ranging from zero, for the company that blew the whistle on the cartel, to £80,550.

In addition, in January 2005 at a conference in Edinburgh, the OFT took the unprecedented step of announcing that it intends to continue targeting the construction and housing sector in a bid to rid it of anti-competitive behaviour. Indeed, it was already investigating a cartel in the Scottish construction industry, and it is clear that the OFT believes that anti-competitive behaviour is rife within the industry (Dundas and Wilson 2005). An extract from the evidence gathered by the OFT in a construction cartel under investigation and disclosed at the OFT Annual Plan Stakeholder Event, Edinburgh, highlights the problem, '. . . I think it was only three or four contractors and there was a discussion over the phone . . . well, there's no point in us all . . . cutting our throats to pick up this work and ending up with one contractor with all the work at low prices, let's see if we could share this about and not be silly about it . . .'. The OFT has cited three reasons for its decision to target construction and housing:

- a significant proportion of the current workload of the cartel branch is in the construction and housing sector;
- the sector is of prime importance in many large public procurement exercises;
- the OFT continues to receive complaints from consumers.

Unlike the professions and utilities, the construction and housing sector has largely escaped external regulation and is effectively self-regulated by market forces.

Mechanisms to address cartels

Whistle-blowing

Since the coming into force of the Competition Act 1998, the OFT has shown a willingness to exercise its powers of search and seizure against companies thought to be involved in anti-competitive practices which, as noted earlier, if proven, will attract fines of up to 10% of the undertaking's (company's) turnover for each year of the infringement, to a maximum of three years. The Public Interest Disclosure Act 1999 provides for a system of leniency for whistle-blowers, and this is likely to see more whistle-blower cases undertaken by the OFT.

Fines and penalties

In September 2004, the UK engineering company IMI was fined €44.98m (£30.5m) for taking part in a 12-year-long cartel activity with European competitors to inflate the cost of plumbing equipment (Castle 2004). The IMI group, made up of IMI plc, IMI Kynoch Ltd, and Yorkshire Copper Tube Ltd, was one of eight companies running an elaborate cartel to inflate the prices for water, heating and gas tubes and squeeze out cheaper rivals. Announcing the fine, the European Commission said the companies had been operating a well-structured, classic cartel, with code names, meetings in anonymous airport lounges and the clear objective to avoid competition. The Commission quoted notes from one of the companies involved, taken at the cartel's first meeting of European firms in 1989 in Zurich, which said, 'The objective is to keep the prices in the high price level... if possible to increase even more'.

The illegal arrangement ran between June 1988 and March 2001; IMI joined in September 1989. Although IMI sold its copper tube and fitting business in 2002, it is still held responsible for the findings of the Commission. Copper piping used in housebuilding and by industry accounts for a major use of the metal, second only to cabling. The market, including plastic-coated copper plumbing tubes, was worth around €1.15bn in 2000. The cartel was finally exposed by one of the eight participants, Mueller Industries, based in the United Kingdom, France and the United States, which blew the whistle in exchange for escaping a fine. The other seven firms, including Sweden's Boliden and Finland's Outokumpu, were fined a total of €222.3m. IMI's penalty was the second-largest after Societa Metallurgica Italiana's €67.1m fine. Mario Monti, the European Commissioner for Competition, said: 'Because of the companies' illegal behaviour, European consumers paid more for plumbing replacement work or when buying a house. Today's decision illustrates the relentless fight against cartels by this Commission'. Outokumpu and Boliden indicated that they would appeal against the fine.

In November 2004, the OFT imposed fines totalling in excess of £2.4m (reduced to £1.7m on leniency) on five suppliers of IG desiccant, a chemical used in double glazing manufacture (Dundas and Wilson 2005). The OFT found that the main UK supplier, UOP Limited, had entered into an anti-competitive agreement with four of its UK distributors – UKae Limited, Thermoseal Supplies Limited, Double Quick Supplyline Limited (DQS) and Double Glazing Supplies Group plc (DGS) – to maintain and/or fix the resale price of IG desiccant. In March 2004, nine roofing contractors were fined a total of £330,000 for fixing the price of repair, maintenance and improvement (RMI) services for flat roofing in the West Midlands through collusive tendering. Amongst the projects affected by the collusive tendering were a number of schools, a community library, a shopping centre and a car park. The contractors' behaviour meant that buyers were unable to obtain competitive prices locally when tendering for RMI flat-roofing services. In appeals brought by two of the contractors, the level of fines was upheld.

Directors who have committed a breach of UK or EU competition law may become the subject of a Competition Disqualification Order (CDO) imposed by the court on application by the OFT. The result of a CDO is that it is a criminal offence to be a director of a company involved in a breach of UK or EU competition law or take on certain other roles relating to its management. The maximum period of disqualification is 15 years.

Leniency

To encourage cartel participants to come forward, i.e. whistle-blowing on their fellow cartelists as noted above, the OFT will grant total immunity from financial penalty to the first party to come forward before the OFT has commenced an investigation into the cartel activity, subject to certain conditions being met (e.g. refraining from any further involvement in the cartel, other than that directed by the OFT). A party that is not the first to come forward with information or that does not fulfil these conditions may still benefit from a reduction in financial penalty. It appears that a number of recent (and existing) investigations have been prompted by members of cartels who wish to escape fines by blowing the whistle in this way. The double-glazing cartel case, noted above, shows how leniency for information providers can help expose cartels. The fines before and after leniency were:

- UOP – £1,540,000 (reduced to £1,232,000 on leniency)
- Thermoseal – £279,000 (reduced to £139,000 on leniency)
- DQS – £109,000 (not granted leniency)
- DGS – £227,000 (not granted leniency)
- UKae – £278,000 (reduced to zero on leniency).

In a West Midlands roofing cartel, Briggs Cladding & Roofing Limited was granted 100% leniency, and Howard Evans (Roofing) Limited received a 50% reduction on the basis of leniency.

More recently, the OFT noted in a press release (OFT 2007e) that it was looking to fast-track its cartel and bid rigging investigations into the UK construction industry by reducing its penalties to those firms that have not applied for leniency but may be willing to cooperate. The release notes that over a two-year period, the OFT had raided the premises of 57 companies in England, of which 37 had applied for leniency. The OFT evidence collated during this activity suggested that bid-rigging had been going on in thousands of tenders with a combined value of an estimated £3bn. Equally, in the roofing sector of construction, the OFT had imposed aggregated financial penalties in excess of £4m.

Conclusions

The construction industry globally is an important sector for any national economy in terms of employment and its contribution to Gross Domestic Product. Equally, the creation and maintenance of the built environment is important for sustained economic as well as social well-being and development in the developed and developing world. The often complex nature of construction projects, with myriad organisations, people, stakeholders and contractual structures, provides an important opportunity for building and infrastructure projects to be delivered on time, to budget and specification with no untoward events or incidents happening. On the other hand, the characteristics of construction projects, both nationally and internationally, and the often highly competitive nature of the industry, also provide opportunities for corrupt practices to occur, involving either bribery or fraud or some combination of both.

This chapter has provided a brief overview of corrupt practices and how and when they can occur. The evidence suggests that corrupt practices appear common globally and within the United Kingdom, and the OFT has targeted construction as one of five sectors which it wishes to investigate. Ongoing investigations over a sustained period have uncovered numerous examples of collusive and anti-competitive practices. The CIOB survey confirms the worrisome trend. However, that same survey also raises issues surrounding the definition of corruption, and alludes to the fact that some practices are seen as the way that business is conducted in the industry. The Transparency International (2002) report, using its bribery index across 17 industrial sectors, rates public works/construction as having the highest perceived incidence of bribery, followed in second place by arms and defence. Both sectors are again seen as the ones where the biggest bribes are likely to be paid, with construction again coming out on top.

The chapter has focused on the incidence of cartels in the construction supply chain. It is clear that cartels can operate first at the industry structure level, focusing on setting general minimum price levels. In the materials supply sector these are more likely to operate within strategic groups that compete intensely with each other. Second, cartels operate at a project level through bid or price-rigging and are likely, based on the evidence presented here, to be located within contracting, but given the increasing competition within the provision of consultancy services, they cannot be excluded from the consulting sectors. Transparency International classes this as fraud. In addition, where collusive practices occur it may or may not involve bribery to keep the deception covert. In the United Kingdom, the OFT has adopted the strategy of civil legal proceedings and setting fines against cartels, and legislation also exists for it to move to criminal investigations.

The work produced by Transparency International and other evidence presented here indicate that the operation of cartels, either at industry or at project levels, is a corporate act and raises issues over appropriate corporate governance mechanisms, ethical conduct and also an increased requirement for overall corporate responsibility. One other clear conclusion is that the evidence presented here does not add to the image of the industry, even though these activities represent only a small percentage of the regular initiation, design and delivery of numerous construction projects nationally and internationally. Unfortunately, the industry already has an inappropriate image owing to its health and safety record and being perceived at large as one that is both dangerous and dirty. The presence of corrupt practices adds further to that already tarnished image.

References

Baer, W., Frazer, T. and Gyselen, L. (2005) Cartel prosecution around the world, *The Global Counsel Competition Handbook*, The Practical Law Company.

BERR (2007) Department for Business, Enterprise and Regulatory Reform. Online. Available at: http://www.dti.gov.uk/sectors/construction/index.html (accessed 24 October 2007).

Blake Lapthorn Linnell (2004) Talking to your competitors – good business or high RISK? *Construction Issues*, Issue 2, October. Online. Available at: http://www.bllaw.co.uk/pdf/CON_1004_construction%20issues.pdf (accessed 4 November 2007).

Castle, S. (2004) British firm fined £30m for its part in plumbing cartel, *The Independent*, London, 4 September.

Chartered Institute of Building (CIOB) (2006) *Corruption in the UK Construction Industry*, Chartered Institute of Building.

Christopher, M. (1998) *Logistics and Supply Chain Management: Strategies for Reducing Cost and Improving Service*, 2nd edn, London: Financial Times & Pitman Publishing.

CMC Cameron McKenna (2004) *Competition Law Survival for Construction Firms*, CMC Cameron McKenna. Online. Available at: http://www.law-now.com/law-now/sys/getpdf.htm?pdf=compsurvconst.pdf (accessed 4 November 2007).

Croner's (1999) *Management of Construction Projects*, July: 541–543.

Dundas and Wilson (2005) *EU and Competition*, http://www.dundas-wilson.com/news/bulletins.php (accessed 24 June 2008).

Flanagan, R. and Norman, G. (1989) Pricing policy, in Hillebrandt, P.M. and Cannon, J. (eds), *The Management of Construction Firms*, London: Macmillan.

Foundation World (2005) *Commission Increases Resolve on Cartel Enforcement*, News, 13/3/05. Online. Available at: http://www.foundationworld.org.uk/jsp/effc.jsp?lnk=022&id=122 (accessed 4 November 2007).

Gattorna, J.L. and Walters, D.W. (1996) *Managing the Supply Chain: A Strategic Perspective*, Basingstoke: Macmillan Business.

Groak, S. (1994) Is construction an industry? *Construction Management & Economics*, 12 (4): 287–293.

Guerrera, F., Mallet, V. and Smyth, L. (2002) Plasterboard cartel fined euros 478m for price fixing, *Financial Times*, London, 28 November.

Johnson, G. and Scholes, K. (1999) *Exploring Corporate Strategy*, Hemel, Hempsted: Prentice Hall.

Kay, J. (1993) *Foundations of Corporate Success: How Business Strategies Add Value*, Oxford: Oxford University Press.

Langford, D. and Male, S.P. (2001) *Strategic Management in Construction*, Oxford: Blackwell Publishing.

Male, S.P. (1991) Strategic management in construction: conceptual foundations, in Male, S.P. and Stocks, R.K. (eds), *Competitive Advantage in Construction*, Oxford: Butterworth-Heineman.

Male, S.P. and Mitrovic, D. (2005) The project value chain: models for procuring supply chains in construction, *Joint RICS Cobra 2005, CIB and AUBEA Conference*, Queensland University of Technology, Brisbane, Queensland, Australia, 4–8 July.

Martineau Johnson (2004) *Competition and Procurement Brief*, February, www.martineau-johnson.co.uk (accessed 24 April 2005).

OFT (2005) *The Office of Fair Trading*. Online. Available at: www.oft.gov.uk (accessed 4 November 2007).

OFT (2007a) *Cartels*. Online. Available at: http://www.oft.gov.uk/oft_at_work/ enforcement_regulation/Cartels/ (accessed 24 October 2007).

OFT (2007b) *Infrastructure and Knowledge Economy*. Online. Available at: http://www.oft.gov.uk/oft_at_work/markets/infrastructure/ (accessed 24 October 2007).

OFT (2007c) http://www.oft.gov.uk/oft_at_work/enforcement_regulation/Cartels/ (accessed 24 October 2007).

OFT (2007d) *What is a Cartel?* Online. Available at: http://www.oft.gov.uk/advice_ and_resources/resource_base/cartels/what-cartel (accessed 4 November 2007).

OFT (2007e) OFT closes door on cartel leniency in construction bid rigging cases in England. Press Release 50/07 (22 March 2007).

Petterson, T. and Thunstrom, M. (2004) New Regulations for Swedish Competition Law, Mannerheim Swartling Advokatbyra.

Porter, M.E. (1980) *Competitive Strategy: Techniques for Analysing Industries and Competitors*, New York: The Free Press.

Porter, M.E. (1985) *Competitive Advantage: Creating and Sustaining Superior Performance*, New York: The Free Press.

Saunders, M. (1997) *Strategic Purchasing and Supply Chain Management* (2nd edn), London: Financial Times & Pitman Publishing: 44.

Sherwood, B. (2004) Competition watchdog homes in on areas of concern OFT, London: Financial Times, 2 December.

Sherwood, B. (2005) OFT set to break building industry cartels, London: Financial Times, 25 February.

Transparency International (2005) *Preventing Corruption on Construction Projects: Examples of Corruption*, Report, March.

Transparency International (2007) *Construction, Engineering, and Post-Disaster Construction*.

Wassener, B. (2005) Cement cartel is fined euros 660m, *Financial Times*, London, 15 April 2003.

Part IV

Sustainable development

Chapter 9

The evolution of sustainable development

Martin Sexton, Peter Barrett and Shu-Ling Lu

Introduction

A central principle of corporate social responsibility is that firms should treat their stakeholders in an ethical fashion and that this behaviour should embrace environmental, as well as economic and social considerations. The purpose of this chapter is to provide a theoretical exploration of the concept of sustainable development in its broadest sense and, in so doing, encourage researchers and practitioners to locate and progress with their corporate social responsibility work within a robust 'sustainable development' framework.

There is an increasing appreciation that Earth's ecological systems cannot indefinitely sustain present trajectories of human activity. The nature and scale of human activity is exceeding the carrying capacity of the Earth's resource base, and the resultant waste and pollution streams are exceeding the assimilative capacity. The contribution of the built environment and construction activity to this unsustainable human activity is substantial, and Lenssen and Roodman argue that:

> responsibility for much of the environmental damage occurring today – destruction of forests and rivers, air and water pollution, climate destabilization – belongs squarely at the doorsteps of modern buildings.
> (Lenssen and Roodman 1995: 95)

The prevailing 'vision', which is arguably preventing a sustainable future, is the failure to appreciate and embrace the reality that human well-being is a derivative function, secondary to the well-being of the Earth, and that ecological processes provide the biophysical context for human existence. Human activity and the natural world are thus viewed as being on a collision course.

The 'urgent and radical reform' to meet this challenge was influentially envisioned and contextually defined by the World Commission on

Environment and Development (WCED) as 'development which meet the needs of the present without compromising the ability of future generations to meet their own needs' (WCED 1987: 8).

This concept is particularly pertinent for the construction industry, as the construction industry has always played a major role in producing the built environment that society has required, and has played an important part in the development of the human race. This role has never been as important as it is now, when there is a growing consensus that appropriate corporate social responsibility strategies and actions are needed to ensure sustainable built environments and construction activity.

Model of societal–ecological system interaction

Description of model

Figure 9.1 presents a systems model of social system and ecological system interaction. The rationale and operation of the model is described below.

The finite *biosphere* supra-system represents the Earth and encompasses all the elements of both the social and ecological systems. The *ecological* system contains sources and sinks. *Sources* are energy and natural resources, which make up *natural capital* and which are *utilised* (or *invested* in for future utilisation) by the *economic* system (a subsystem of the *social* system). The economic system serves, and is nurtured by, the ongoing development of *human capital* production and consumption. A distinction

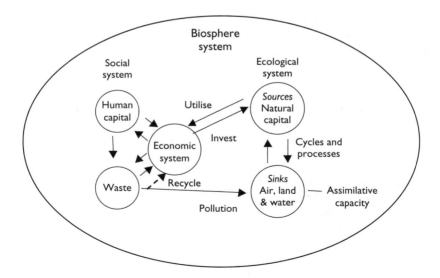

Figure 9.1 Model of interaction between ecological and social systems.

is made between exhaustible (or non-renewable) and renewable natural capital. Exhaustible natural capital (such as minerals and fossil fuels) consists of an initial stock which, from a human time perspective, is only very slowly renewed. Renewable natural capital (such as fish, forests, groundwater), in principle, is reproduced within the human time perspective although, increasingly it is becoming exhausted. The *sinks* are physical components of the ecological system (air, land and water) for the assimilation of materials and energy, which are transferred from the economic system back to the ecological system as *pollution* (from both production and consumption *waste* which has not been *recycled*). The source and sink functions are related in the sense that a higher extraction of resources, such as oil or coal, will mean more pollution and waste and increased pressure on the assimilative capacity of the ecosystem. The sources and sinks of the ecological system are linked by the natural services provided by the natural capital system (such as the maintenance of essential climatic and ecological cycles and processes), the quality of which is essential for supporting economic production and welfare. The system model is dynamic, with the composition and interaction changing through time, either because of natural system disturbance or because of internal ecological mechanisms.

The ecological system has a limited resource-creating capacity for the substances that the social system extracts and a limited assimilation capacity for the pollution and waste that society returns to nature. When the societal influence exceeds these capacities of nature, damage occurs. Sustainability, in the system terms set out in this model, is thus achieved when resource extraction from the ecological system occurs within the carrying capacity of the resource base and when waste transfer to the physical components of the ecological system does not exceed the assimilative capacity of the particular ecosystems.

This model thus clearly identifies the key issues as the organisation of production and consumption of the social system, the quantity and quality of ecological-system functions, and the dynamic interaction between the social system and the ecological system – in summary, the model captures the thesis that humans are dependent upon ecological systems, for 'without the services provided by natural ecosystems, civilisation would collapse and human life would not be possible' (Ehrlich 1986: 239).

At present, it is argued that the organisation of, and interaction between, the social and ecological systems is not sustainable and, unless rearranged, will lead to a permanent breakdown in human time-span terms, of suprasystem resilience (the ability of the system to stay in dynamic balance) and integrity (the ability of the system to support services of value to humans). The next section will present evidence to support this claim by first, examining the present state of socio-ecological system interaction at a global level, and, second, identifying the built environment and construction industry contribution to this interaction.

Present state of socio-ecological system interaction

Global situation

Mounting evidence shows that the ecosystems of Earth cannot sustain current levels of economic activity, let alone increased levels. By the year 2025, the world population is projected to total about 8.3 billion people, or about 45 per cent more than the estimated current population of 5.7 billion. By 2050, the global population could be about ten billion (United Nations 2004). The population increase, coupled with a five-fold increase in global economic activity since 1950, is elevating the consumption levels of natural capital and the production of pollution and waste in excess of the replenishing rate of the ecological system's sources and the assimilation capacity of its sinks, respectively. For example, at the present rates of consumption, world reserves of oil, natural gas, coal and all minerals are predicted to be substantially depleted by the end of this century.

Built environment and construction industry contribution

The contribution of the built environment and construction to these trends is substantial. Between 1971 and 1992, primary energy use in buildings worldwide grew on average two per cent annually. The built environment accounts for about a third of total world energy consumption, including 26 per cent of fossil fuels, 45 per cent of hydropower and 50 per cent of nuclear power. It is estimated that between 13 and 30 per cent of all solid waste deposited in landfills worldwide comprises construction and demolition waste (Bossink and Brouwers 1996). The construction industry, including building-material production, is probably the greatest consumer of natural resources, using between 17 and 50 per cent of the extracted resources, such as water, wood, minerals and fossil fuels. According to the Worldwatch Institute, building construction consumes 40 per cent of the raw stone, gravel and sand used globally annually, and 25 per cent of the virgin wood. Buildings also account for 16 per cent of the water used annually worldwide (Roodman and Lenssen 1995).

Summary

These global trends, to which the built environment and the construction industry are substantial contributor is, have fuelled the inevitable conclusion that 'the major cause of the continued deterioration is the unsustainable pattern of consumption and production, particularly in industrialized countries' (Agenda 21, Chapter 4). A diverse range of commentators increasingly argues that there is a need for a fundamental reconceptualisation of the interaction between social and ecological systems. The *Brundtland Report* framed the challenge by saying:

the time has come to break out of past patterns. Attempts to maintain social and ecological stability through old approaches to development and environmental protection will increase instability.

(WCED 1987: 21)

It is argued that two principal elements are needed to bring about and maintain such a reconceptualisation: an envisioning, motivating portfolio of goal orientations which can direct and shape the transition; and a conceptual framework to locate and integrate stakeholders' diverse policies and actions to generate the ability for appropriate, complementary progress. The portfolio of goal orientations has been loosely captured in the term 'sustainable development'. What this term means is discussed more fully in the next section.

Sustainable development

What is it?

The concept of sustainable development was contextually defined by WCED as quoted in the 'Introduction' section of this chapter (WCED 1987: 8), and, in its broadest sense, this influential definition has been widely accepted by many firms, institutions and governments across the globe. The goals embedded within sustainable development serve several important functions (described below) that vary according to the perspective of sustainable development advocated:

- *Focus.* A given view of sustainable development will generate a distinctive set of goals that serve as guidelines for action, directing and channelling efforts and activities of relevant stakeholder participants. In this regard, a clear view of sustainable development provides focus for activity by prescribing what 'should be' done. This crucial role is set out, for example, in the need for

> establishing a vision of sustainable development and clear goals that provide a practical definition of that vision in terms that are meaningful for the decision-making unit in question.
>
> (Hardi and Zdan 1997: 1)

It has been stressed that there is a need for an appropriate hierarchy of goals: aims at the general level (e.g. preserving and improving environmental quality); qualitative goals at the intermediate level (e.g. preserving the ozone layer); and specific quantitative targets at a more specific level (e.g. reduction of car pollution levels in a given city) (OECD 1997). Further, the lack of clear focus of this kind, for example, underpins the observation that more attention is needed on how sustainable development can be translated into concrete goals and criteria at the level of sectors, regions and projects (van Pelt *et al.* 1990).

- *Constraints.* To the extent that a given set of sustainable development goals prescribes what 'should be' done, they also serve to prescribe what 'should not be done'. A given view of sustainable development that commits itself to certain goals reduces the amount of discretion it has to pursue other outcomes. The concept of 'accessibility space', for example, argues that the range of sustainable development trajectories available is restricted by a raft of physical, human and time constraints that vary depending on the goals being pursued (Bossel 1999).
- *Source of motivation and legitimacy.* Goals also provide a source of motivation and legitimacy for relevant stakeholders by justifying their activities. The work of the President's Council on Sustainable Development (1996: 4) in the United States, for example, 'gave [people] credibility to continue innovative projects for which they did not yet have widespread support'. Similarly, it has been noted that a variety of organisations and institutions, with very different interests and objectives, utilise the notion of sustainable development to justify or rationalise particular strategies and actions as being in the global interest (Harvey 1996).
- *Measures of performance.* To the extent that sustainable development goals are clearly stated and understood, they offer a seedbed of appropriate measures or indicators for evaluating performance. This need, for example, is expressed in the argument that,

> if we genuinely embrace sustainable development, we must have some idea if the *path* we are on is heading towards it or away from it. There is no way we can know that unless we know what it is we are trying to achieve – i.e. what sustainable development means – and unless we have indicators that tell us whether we are on or off a sustainable development path.
>
> (Pearce 1988: 22)

In summary, a clear understanding of different sustainable development perspectives will make more transparent the differing objectives, criteria and constraints guiding action, along with the underpinning sources of motivation and legitimacy driving and protecting the various sustainable development trajectories being pursued. There is thus a clear tension between the normative need for establishing a clear understanding of sustainable development from which consistent and coherent goals and actions can be stimulated and the reality of multiple, often discordant, views of sustainable development obstructing what these goals and actions should be. To try and better understand this tension, and thus tease out some guidance on how progress can be made to its resolution, there is a need to understand why different stakeholders have such divergent, often incompatible, aspirations of sustainable development that can severely hinder progress at policy and operational levels. Two key strands will be followed. First, some of the principal components of the ideas that are generally shared by the majority of sustainable development perspectives will be identified. Second, the complex role of people's worldviews in shaping the focus and composition of these sustainable development components will be discussed.

Principal elements of sustainable development

A number of recurring elements which flavour, to varying degrees, the majority of the definitions of sustainable development can be articulated. For the purposes of contextualising these elements, sustainable development is viewed as:

> *Endurable, appropriate progress*, built on *socio-ecological system principles*, that are temporally and spatially *equitable* in its focus and *participatory* in its formulation and implementation.

Each of the components is discussed below.

- *Endurable, appropriate progress*. Most definitions of sustainable development appreciate that development must be within the carrying and assimilation capacities of the Earth (see 'socio-ecological system principles' below) and that it must be distributed fairly across spatial and temporal dimensions (see equitable below). The term 'development' is generally viewed as progress in the quality of life through social and cultural progress, rather than the more traditional goal of increasing economic activity. Progress does not rule out growth but it certainly dictates the type of growth which is desirable. This view of development is consistent with the post-materialistic thesis that argues that societies are changing their cultural values towards 'quality of life' issues, away from material consumption and away from economic distribution conflicts (Inglehart 1997).
- *Socio-ecological system principles*. The majority of sustainable development perspectives appreciate that the production and consumption demands of the social system must not exceed the carrying capacity of the resource base and that resultant waste and pollution flows must not exceed the assimilative capacity of the ecological system.
- *Equitable*. Fair distribution of benefits from development across intergenerational, intragenerational and spatial dimensions is a central consideration in most conceptions of sustainable development. Commentators contend that the resource use of each present generation is

depriving the right or possibility of future generations using the same resource (Pearce and Atkinson 1995). Intergenerational equity draws upon this tension to argue that the opportunity for quality of life must not diminish for future generations, requiring that future generations should have access to the same resource base as existing generations (Solow 1992; Weiss 1989).

Intragenerational equity is generally conceived as the elimination of poverty (Barbier 1987; Burayidi 1997; Dasgupta 1993); almost one quarter of the global population lives in absolute poverty. The rationale behind this principle is that poverty has an adverse impact upon the environment and, thereby, jeopardises welfare and resources along with intergenerational equity, since natural capital will be diminished for future generations. Implicit within the notion of equity is spatial equity: sustainable development cannot be achieved in one nation or region at the expense of another nation or region (Bhaskar and Glyn 1995; Pearce *et al.* 1989).

- *Participatory.* This facet of sustainable development is closely linked to intragenerational equity. The essence of the argument is that if there is to be positive discrimination in favour of poorer groups and minorities, then such groups have to be closely involved in defining their own needs and engaging relevant decision-making authorities and processes. This consistent strand of sustainable development resonates strongly with the minority-issue literature which encourages participatory approaches to social problems (Rahman 1993).

The common principal elements of sustainable development have been outlined. In any given conceptualisation of sustainable development, however, the emphasis on and the combination of these elements will differ, which will, in turn, produce different goals and policies. To understand why different stakeholders have different conceptualisations of sustainable development, it is critical to understand how they perceive the environment, their relationship with it, and their interactions with each other. These perceptions are very much shaped and filtered through stakeholders' 'worldviews'. The concept and role of worldviews will now be discussed.

Sustainable development and worldviews

The nature and role of worldviews

Worldviews are understood to be 'the constellations of beliefs, values and concepts that give shape and meaning to the world a [stakeholder] experiences and acts within' (Norton 1991: 75), providing a system of co-ordinates or a frame of reference in which everything presented to us by our diverse experiences can be placed. It is a symbolic system of representation that allows us to integrate everything we know about the world and ourselves into a global picture, one that illuminates reality as it is presented to us.

(Aerts *et al.* 1994: 39)

Worldviews play a major role in complex decision-making, particularly in complex, ambiguous and subjective issues (Jolly *et al.* 1988). Stakeholders' worldviews are thus critical in helping them determine which elements of the sociological and ecological systems are important to heed when formulating objectives, policies and actions (Gary and Belbington 1993). Research has supported the view, for example, that stakeholders' values, beliefs and attitudes shape their environmental norms (Stern and Dietz 1994).

Interaction and understanding (though not necessarily mutual acceptance) of worldviews are thus required to develop a discourse of shared terms and language that are needed in order for analysis, debate, negotiation and problem-solving to occur (Dryzek 1997). The need for dialogue of this nature is firmly located within the relevant literature, with it being argued that the basic causes of conflict between stakeholders are the differences in their knowledge and values (Dorcey 1986), and that these shape the way information is gathered, perceived and acted upon by these various groups (Simmons 1993). Thus, it is argued, for example, that in order to incorporate all the appropriate components of sustainable development, the identification of criteria and indicators of sustainable development must not only be approached by scientific means, but also include perceptions and values set by society as a whole (Cairns *et al.* 1993; Young 1997) and by individual stakeholder groups (Schwartz and Thompson 1990; Thompson *et al.* 1990; Vreis 1989). (This understanding, in part, has focused attention on the need to create and manage a participatory dimension to sustainable development, to ensure that all relevant stakeholders are closely involved in defining their own needs and engaging relevant decision-making authorities and processes.)

To investigate the concept and role of worldviews, numerous commentators have categorised similar worldviews into groups and located these groups along continua or in frameworks. Such continua can be usefully bundled together to form two 'worldview' umbrella clusters: the currently dominant 'neoclassical' worldview, and the 'ecological' worldview espoused to varying degrees in the sustainable development movement. (The argument that the 'neoclassical' worldview is currently dominant is supported in the discussion below.) This process enables a more integrated discussion to take place, with otherwise fragmented ideas being interwoven to develop a more holistic, systemic understanding of stakeholder worldviews. Further, the discussion will follow three interdependent lines of enquiry:

- The different positions engaged by the neoclassical and ecological world-views on the *relationship between human beings and the environment* will be examined.
- This relationship will provide the ethical context that motivates and legitimises the opposing standpoints articulated by the neoclassical and ecological worldviews on the *relationship between the firm and the environment*.
- The differing views taken by the neoclassical and ecological worldviews on the *interaction between social capital and ecological capital*. The nature and scale of this interaction is a key determinant of whether system interaction is sustainable or not.

Relationship between human beings and the environment

Neoclassical worldviews adopt the anthropocentric ethic, namely, there is a fundamental dualism between human beings and the natural environment (Pauchant and Fortier 1990). This ethic grants moral standing exclusively to human beings and considers non-human natural entities and nature as a whole to be only a means for human ends.

In contrast, ecological worldviews reject the anthropocentric premise that human beings occupy a privileged place in the biosphere. Rather, they adopt an ecocentric ethic that morally enfranchises, to varying degrees, living and non-living things. Commentators argue that the anthropocentric-based neoclassical worldview must be recognised and eradicated before fundamental changes can take place towards an ecocentric nurtured ecological worldview (Oelschlaeger 1991).

The anthropocentric ethic is, however, the dominant ethic at present (Midgley 1994). Indeed, the Rio Declaration at the Earth Summit asserted the claim that 'human beings are at the centre of our concerns' (United Nations Conference on Environment and Development 1992). This appreciation of the neoclassical worldview dominance which provides significant insights into what guides and motivates the relationship between the firm (taken to be the vehicle for stakeholder influence) and the ecological system is discussed in the following section.

Relationship between the firm and the environment

The neoclassical worldview legitimises, through its anthropocentric ethic, the means whereby rational, self-interested agents can optimise and exploit the social system and ecological system for their own end. It has been commented on, for example, that this worldview shapes the observation that 'traditional organizations serve only their own ends. They are, and indeed are supposed to be, selfish' (Trist 1981: 43); firms are more likely to pursue an economically advantageous course of action when confronted with a choice between environmental preservation and economic development (Axlerod 1994). In particular, the dominant drive would seem to be towards profits and profit maximisation. This is justified by neoclassical economists:

> few trends could so thoroughly undermine the very foundations of our free society as the acceptance by corporate officials of social responsibility other than to make as much money for their stockholders as possible.
>
> (Friedman 1963: 133)

Further, neoclassical economic welfare arguments largely ignore intergenerational equity issues, tending towards utilitarian assessments that celebrate aggregate growth.

The anthropocentric ethic generates 'simple thought' (Morin 1992), which produces organisational policies and actions that have difficulty understanding and perceiving that they are nested within a broader biosphere (Bateson 1972). Such firms do not give adequate consideration to how their activities will have an impact on, alter, or interfere with the complex behaviour of the biosphere's constituent social and ecological systems (Dunlap and Catton 1993). Indeed, commentators have (perhaps cynically) concluded that even

> marginalist reformers...[do not]...consider the dominant ideology of present forms of capitalism and they lack the imagination and creativity to develop the real strategies which will bring about the fundamental change which is needed....They merely scratch the surface of the problem and quickly paper over the cracks with industry-centred and profit-centred solutions.
>
> (Welford 1995: 2–3)

It is increasingly apparent that neoclassical economics does not reflect social, economic and environmental realities in a world of limited resources (Friend 1992). In its most basic form, neoclassical economics treats nature as an infinite supply of physical resources (i.e. raw materials, energy, soil and air) to be used for human benefit, and as an infinite sink for the by-products of the consumption of these resources, in the form of various types of pollution and ecological degradation. This throughput aspect of the flow of resources from ecological system sources into the economic system and the flow of wastes back into the ecological system does not enter into economic thinking, as it is believed to be infinite in extent (Daly 1989). Thus, there is no explicit biophysical 'environment' to be managed, since it is irrelevant to the economy. Externalities highlight what can be termed 'market failure'; that is, the market does not capture the full environmental implications of social system–ecological system interactions (Rees 1990). The neoclassical worldview thus generates a market that consumes and substitutes ecological capital for social capital, and this adverse interaction has become a major contributor to current environmental problems (Welford and Gouldson 1993).

In contrast, ecological worldviews argue that firms and industries as a whole need to take a much broader view of the business environment to embrace (a) the ecology of the planet Earth; (b) the world economic, social and political order; and (c) the immediate market, technological and socio-political context of organisations (Davis 1991; Stead and Stead 1992). The above factors are systemic – interconnected and interdependent – and need a new kind of systematic, or ecological, thinking to be understood and solved (Callenbach 1993).

This discussion has drawn upon the neoclassical ethic to explain its role in legitimising and motivating firms to exploit the ecological system in an unbalanced fashion. The key issue is the degree to which firms substitute social capital for ecological capital in their exploitative endeavours. This issue is discussed in the following section.

Relationship between social capital and ecological capital

The clear implication from the previous discussion on the interaction between the firm and the environment is that the fundamental assumption in neoclassical worldview states that substitutions can be made between social and ecological capital. The diversity of sustainable development worldviews on this issue can be fruitfully located along a 'weak' sustainability–'strong' sustainability continuum (Pearce *et al.* 1989; Pezzy 1992). Both are based on the concept that humanity should live on the 'interest' of its ecological capital, preserving the capital for future generations (Daly and Cobb 1990). The ecological capital comprises source and sink resources.

Neoclassical worldviews tend towards 'weak' sustainability, contending that resources (both in the ecological system and in the social system) are substitutes for others (solar energy for oil, for example) and allow substitutions as long as the combined social and ecological capital is not diminished. Neoclassical worldviews assume a high level of resource substitution, particularly through technological development and the price mechanism that increases resource cost as it becomes relatively scarcer (Dasgupta and Heal 1979; Solow 1974).

In contrast, ecological worldviews embrace 'strong' sustainability. Under strong sustainability, both ecological and social capital should be independently maintained in physical/biological terms (Brekke 1997). The motivation for this view is either the recognition that ecological resources are essential inputs into the social system that cannot be substituted by social capital, or the ecocentric ethical acknowledgement of environmental integrity and rights in nature. In either case, it is understood that environmental components are unique and that environmental processes may be irreversible (over relevant time horizons) (Pearce and Atkinson 1995).

Summary and worldview framework

This discussion of neoclassical and ecological worldviews has shown two contrasting ways of perceiving and understanding the interaction between social system and ecological system. The argument has been developed that the concept and operationalisation of sustainable development is located within different stakeholders' worldviews, within which ethical positions guide, shape and legitimise firm behaviour, and the scale and form of interaction between social system and ecological system. All stakeholders operate to a greater or lesser extent in keeping with the neoclassical worldview, although the ecological worldview is emerging as a viable and necessary alternative. Further, it is clear that the current diversity of worldviews is unlikely to change, except in focused areas, and that this should ideally be appreciated and accommodated, rather than viewed as a source of debilitating confusion.

The neoclassical and ecological worldview matrix, shown in Figure 9.2, is proposed as a simple, but effective, typology which allows the worldviews embodied in definitions of sustainable development to be categorised.

The framework categorises different definitions of sustainable development along a sociological continuum from 'neoclassical worldview sustainability' (DPS – dominant product/service sustainability) to 'ecological worldview sustainability' (EBS – ecosystem benefit sustainability). The different categories are discussed below:

1. *Dominant product sustainability* (DPS) results in a narrow range of ecosystem products defined as economically valuable by existing markets. The rationale is economic efficiency rather than aesthetic value. Economic gain or provision of a vital product justifies sustaining the dominant product.

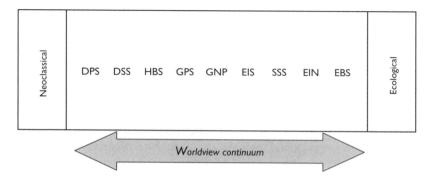

Figure 9.2 Worldview framework.

2. *Dependent social systems sustainability* (DSS) is orientated towards specific human social systems, such as communities, occupations or families, that depend on an ecosystem and its products. The rationale represents a value judgement that asserts an anthropocentric priority of designated social systems.

3. *Human benefit sustainability* (HBS) maintains the flow of diverse human benefits that result from intensive resource management. In contrast to dominant product sustainability, this type emphasises a greater range of resource products and contributions to the larger society rather than to targeted resource-dependent social systems. Resources are valued on both economic and non-economic criteria. The rationale represents the idea that ecological systems should be managed to yield the maximum good for the greatest number of people.

4. *Global product sustainability* (GPS) emphasises the flow of unique or increasingly valuable natural resource commodities produced by local ecosystems for the international market. The dominant rationale is that nations and their ecosystems are encouraged to produce specialised goods for the global market place. This rationale attempts to balance the diverse needs of international consumers with the ability of local ecosystems to produce unique or increasingly valuable natural resource products for the global village.

5. *Global niche preservation* (GNP) sustains some specific local ecosystems judged as integral to the larger goal of sustaining the entire Earth. This global perspective has led to wilderness preservation, marine sanctuary protection, and efforts to identify and safeguard endangered species. The dominant rationale is that both ecosystems and human populations occupy interdependent global niches, and that humans have no right to destroy ecosystems.

6. *Ecosystem identity sustainability* (EIS) is orientated towards a general land use or ecosystem type, such as forest, desert, estuary or wetland. The dominant rationale is a long-term commitment to sustaining resources within a broad land use. Implicit is the principle that it is better to sustain existing ecosystem identity than to convert to a radically different pattern or use.

7. *Self-sufficient sustainability* (SSS) supports long-term natural resource ecosystem integrity, as characterised by relatively balanced, self-sustaining ecosystems. Such ecosystems, needing little human intervention, may nonetheless yield products for human use. However, because of the less intensive management, sustainable output levels are likely to be significantly lower than under human benefit sustainability. The rationale is an ecocentric ethic that asserts that humans have no right to intervene in ecological system evolution. A secondary rationale focuses on the lack of scientific knowledge about how ecosystems function; allowing ecosystems to operate without human intervention assists in clarifying how complex ecosystems sustain themselves.

8. *Ecosystem insurance sustainability* (EIN) is concerned with ecosystem diversity. Specific ecosystems, plant species or animal species are divided into two categories: the first continues to supply traditional products for use, whereas the second is protected in a more natural condition as a genetic storehouse. The dominant rationale is of ecosystem disaster, occurring either cataclysmically or through the gradual reduction of ecosystem diversity because of human intervention.

9. *Ecosystem benefit sustainability* (EBS) focuses most strongly on ecological systems rather than social systems. Natural ecosystems as free from human intervention as possible are targeted, even if their condition falls below the threshold of self-sufficient sustainability. The principal assumption is that nature exists for its own benefit rather than for humans, and that nature has its own intrinsic value.

This framework can assist those in developing corporate social responsibility strategies to enable different stakeholders to better understand each other's particular needs and aspirations, thereby creating the necessary common foundation and language to facilitate the development of 'win-win' solutions that engage and motivate all relevant stakeholders.

Holistic, systemic framework

Introduction

The discussion to date has explored the significant influence of stakeholder worldviews on the *goals* of sustainable development. Goals set out a broad vision to which different stakeholders aspire, but this in itself is insufficient to make any substantial or coherent progress. Goals provide an essential starting point, but need to be translated into, and operationalised by, appropriate indicators so that progress towards these goals can be measured and guided. It is argued that before this can be done, there is a critical need for an appropriate holistic systems-orientated framework to locate and integrate stakeholders' diverse policies and actions to generate the *ability* for appropriate, complementary progress.

The next section will first discuss the need for a conceptually rigorous, but practice-orientated framework that facilitates the identification and integration of key sustainable development indicators.

The need for an appropriate framework

The interrelations between the social system and the ecological system are extremely complex and systemic in nature. There is a need, therefore, to use a framework that provides direction, consistency and coherence in the progression of sustainable development goals and indicators. The contribution that such a framework will make is expressed in the claim that:

- in the longer term, [it will help] to develop a more sustainable construction industry, embracing all aspects of manufacture, design, construction, use and disposal of the built environment.

- in the shorter term, [it will help] to clarify the actions required to improve the sustainability of construction.

(DETR 1998: 1)

Dynamic PSR model

Systemic nesting of scales

The first important task is to contextualise the framework within an appropriate portfolio of scales. A key question for sustainable development, for example, is over what space is sustainable development to be achieved, and over what time period? Spatial boundaries can be determined: global, national, regional, and so on, but it must be appreciated that these boundaries are socially or politically contrived, and are, in actual fact, systemically interlinked. It has been argued, for example, that the specific regional, environmental and economic structure determines the sensitivity of a region to external environmental and economic forces (Siebert 1995). Similarly, the timescale over which sustainable development occurs differs depending on whichever system is under consideration.

The key issue being made here is that appropriate deliberation should be made on what point of a given scale is sustainable development being considered, and what the implications of interactions between multiple scales are. The primary consequence of this nested context is that any management decisions will affect several scales (higher and lower levels) (Boureron and Jensen 1994). Therefore, it has been argued that ecosystem patterns and processes need to be studied at varied spatial and temporal scales or within 'ecological time frames' (Reichman and Pulliam 1996).

Drawing upon these spatial and temporal scale debates, Figure 9.3 presents a framework (Barrett *et al.* 1998) that can infuse the Dynamic PSR model with the required systemic focus and linkage across a range of pertinent scales. Moving from Level A to Level D involves increasing spatial areas and time frames, as well as increasing complexity and effort, and need for collaboration and integration with third parties outside the industry. This framework identifies the different scales (and the linkages within and between them) that need to be actively investigated and managed for the progress of sustainable development.

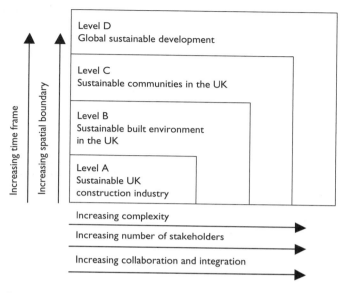

Figure 9.3 Systemic nesting of scales.

Basic framework

The dynamic PSR model explicitly links pressures, states and responses in a cycle. This is shown in Figure 9.4. Intuitively, it makes sense that pressures create states, that in turn demand responses, which in turn have an effect on the original pressures.

Further, the definitions of the PSR boxes are modified to capture the learning and improvement dimensions to the model:

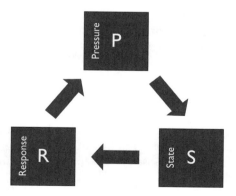

Figure 9.4 Basic rearrangement of PSR framework.

- *Pressure* – drivers for change, from a range of possible sources, such as: regulation, markets, social, technical. Pressures are viewed as ranging from strong to weak.
- *State* – the level of understanding and willingness of relevant actors within the industry to act, viewed as ranging from 'unaware' to 'aware, but not active' to 'aware and active'.
- *Response* – actions taken in practice, viewed as ranging from zero (passive) to positive and negative actions on either side.

Units of analysis

It is crucial, if the model is to make sense, that it is used in such a way that, at any one time, the same stakeholder's perspective is used for P, S and R and that the issue or objective in question is also kept constant. The focus on stakeholder and/or issue can be difficult due to the intrinsic variety of stakeholder perspectives on and ecological complexity of sustainable development; but any slippage on this makes it inconsistent with the proposed cause–effect cycle of the model.

The need for consistency on stakeholder/issue may be considered restrictive, but it is strongly proposed that the *same framework* can be and should be used flexibly at different levels of abstraction. For example, a study could be done of the construction industry as a whole (stakeholder) in relation to environmental issues generally. For example, a study on waste minimisation (issue) from a contractor's perspective (stakeholder) could be supported by the framework. The key point being made here is that by keeping a consistent framework, particular stakeholders can make sense of their situation (e.g. in relation to their supply chain partners) *and* the possibility of combined analyses is opened up. For example, the impact of a particular regulation could be followed through a number of exercises to understand different responses by different parts of the industry. This approach has the advantage of flexibility and consistency. It can be empowering for particular groups of stakeholders and enable strategic syntheses to be developed, extending to international comparisons. The possibility of infinite applications can be addressed at a strategic level by choosing key issues and stakeholders to focus upon.

Gap analysis

The operationalisation of the *dynamic PSR model* is fruitfully achieved through viewing the model as a gap analysis framework. Interrogation of the model reveals two categories of gaps. Those related to P, S and R and those related to the relationship between P, S and R. These gaps are shown in Figure 9.5 and defined in broad terms in Table 9.1 (Barrett *et al.* 1998).

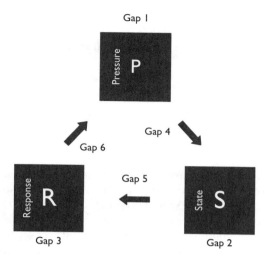

Figure 9.5 Framework for change.

Table 9.1 Gaps in knowledge and understanding and their implications.

Gap	Lack of knowledge about . . .	Generic questions raised
1	*Pressures,* in terms of drivers for change	What are the relevant drivers for the given issue and how strong are they from the point of view of the players?
2	*States,* in terms of players' level of understanding, willingness to act	What is the profile of the players' level of understanding and willingness to act on the given issue?
3	*Responses,* in terms of actions taken by players	What is the profile of the players' responses to the given issue ranging from passive to positive or negative?
4	The relationship between *pressures* and *states*	Is there a mismatch between the strength of drivers and the level of understanding and readiness of the players to respond on the given issue?
5	The relationship between *states* and *responses*	Is there a mismatch between the level of understanding and readiness to act of the players' and their actual actions, both positive and negative?
6	The relationship between *responses* and *pressures*	Is there a mismatch between players' actions and the original intentions of the drivers for change?

The learning and improvement cycle dimension of the dynamic PSR model provides a *mechanism* for *systemic understanding* to guide decision-making and action, and the gap analysis dimension provides a *process* to drive and support the necessary *effective change* for sustainable construction.

Conclusion

In this chapter, the substance of the sustainable development challenge, which is increasingly embedded in corporate social responsibility endeavours, has been explored, identifying, in particular, the importance of appreciating and accommodating diverse stakeholder worldviews, and the need to develop and operate a system-orientated framework to guide decision-making and action. The corporate social responsibility agenda is located predominantly within a neoclassical context that celebrates short-term profit generation and hedonistic client satisfaction. This context will inevitably constrain the motivation and capability of the majority of stakeholders – policy-makers, firms and clients alike – to bring about sustained, meaningful corporate social responsibility. Far greater international and national policy-driven enforcement is arguably the key way forward – the early shoots of which are beginning to emerge as a result of the global warming agenda gaining legitimacy. But whether these early shoots can withstand the entrenched ambient conditions of neoclassicalism is far from certain. This chapter culminated in the presentation of the dynamic PSR model as a potentially fruitful framework to develop appropriate corporate social responsibility strategies.

References

Aerts, D., Apostel, L., De Moor, B., Hellemans, S., Maex, E., van Belle, H. and van der Veken, J. (1994) *Worldviews: From Fragmentation to Integration*, Brussels: VUB Press.

Axlerod, L.J. (1994) Balancing personal needs with environmental preservation: identifying values that guide decisions in ecological dilemmas, *Journal of Social Issues*, 50 (3): 85–104.

Barbier, E.B. (1987) The Concept of Sustainable Development, *Environmental Conservation*, 14 (2): 101–110.

Barrett, P.S., Bootland, J., Cooper, I., Gilham, A. and Jenkins, O. (1998) *Report for the Construction Research and Innovation Strategy Panel – Sustainable Construction Theme Group: Research and Innovation for Sustainable Construction*, London: CRISP.

Bateson, G. (1972) *Steps to an Ecology of Mind*, New York: Ballantine.

Bhaskar, V. and Glyn, A. (eds) (1995) *The North and South: Ecological Constraints and the Global Economy*, Tokyo: United Nations University Press.

Bossell, H. (1999) *Indicators for Sustainable Development: Theory, Method, Applications: A Report to the Balaton Group*, Winnipeg, Canada: International Institute for Sustainable Development.

Bossink, B.A.G. and Brouwers, H.J.H. (1996) Construction waste: quantification and source evaluation, *Journal of Construction and Engineering Management*, 122: 1.

Boureron, P.S. and Jensen, M.E. (1994) An overview of ecological principles for ecosystem management, in Jensen, M.E. and Bourgeron, P.S. (eds), *Ecosystem Management: Principles and Applications*, Pacific Northwest Research Station.

Brekke, K.A. (1997) *Economic Growth and the Environment: On the Measurement of Income and Welfare*, Cheltenham: Edward Elgar.

Burayidi, M. (1997) Environmental Sustainability of third world economic development: constraints and possibilities, *Environmental Sustainability*, 9 (2): 31–42.

Cairns, J., McCormick, P. and Neiderlehner, N. (1993) A proposed framework for developing indicators of ecosystem health, *Hydrobiologica*, 263: 1–44.

Callenbach, E. (1993) *EcoManagement – The Elmwood Guide to Ecological Auditing and Sustainable Business*, San Francisco: Berret-Koehler.

Daly, H.E. (1989) Steady-state versus growth economics: issues for the next century, *Proceedings of the Hoover Institute Conference on Population, Resources and Environment*, Stanford University, 1–3 February.

Daly, H.E. and Cobb, J.B. (1990) *For the Common Good: Redirecting the Economy Towards Community, the Environment and a Sustainable Future*, London: Greenprint.

Dasgupta, P.S. (1993) *An Inquiry into Well-being and Destitution*, Oxford: Oxford University Press.

Dasgupta, P.S. and Heal, G.M. (1979) *Economic Theory and Exhaustible Resources*, Cambridge: Cambridge University Press.

Davis, J. (1991) *Greening Business*, Oxford, England: Basil Blackwell.

Department of Environment, Transport and the Regions (1998) *Opportunities for Change: Sustainable Construction*, London: DETR.

de Vreis, H.J.M. (1989) *Sustainable Resource Use: An Inquiry into Modelling and Planning*, Groningen, The Netherlands: University Press.

Dorcey, A.H.J. (1986) *Bargaining in the Governance of Pacific Coastal Resources: Research and Reform*, Westwater Research Centre, Vancouver, University of British Columbia.

Dryzek, J.S. (1997) *The Politics of the Earth: Environmental Discourses*, Oxford: Oxford University Press.

Dunlap, R.E. and Catton, W.R. (1993) The development, current status, and probable future of environmental sociology: toward an ecological sociology, *Annuals of the International Institute of Sociology*, 3.

Ehrlich, P.R. (1986) *The Machinery of Nature*, New York: Simon & Schuster.

Friedman, M. (1963) *Capitalism and Freedom*, Phoenix Books, Chicago: University of Chicago Press.

Friend, A.M. (1992) Economics, ecology and sustainable development: are they compatible? *Environmental Values*, 1: 157–170.

Gary, R. and Belbington, K.J. (1993) The global environment and economic choice, in Adams, D.K. (ed.), *Environmental Issues: The Response of Industry and Public Authorities*, Halifax, England: Ryburn, 21–35.

Hardi, P. and Zdan, T. (1997) *Assessing Sustainable Development: Principles in Practice*, Winnipeg, Manitoba: International Institute for Sustainable Development.

Harvey, D. (1996) *Justice, Nature and the Geography of Difference*, Cambridge, MA: Blackwell.

Inglehart, R. (1997) *The Silent Revolution: Changing Values and Political Styles Among Western Publics*, Princeton, NJ: Princeton University Press.

Jolly, J., Reynolds, T. and Slocum, J. (1988) Application of the means-end theoretical for understanding the cognitive bases of performance appraisal, *Organizational Behavior and Human Decision Processes*, 41: 153–179.

Lenssen, N. and Roodman, D.M. (1995) Making better buildings, in Brown, L.R. (ed.), *State of the World 1995*, London: Earthscan, 95–112.

Midgley, M. (1994) The end of anthropocentrism? in Attfield, R. and Belsey, A. (eds), *Philosophy and the Natural Environment – Royal Institute of Philosophy Supplement*, Cambridge: Press Syndicate, 36: 103–112.

Morin, E. (1992) The concept of system and the paradigm of complexity, in Maruyama, M. (ed.), *Context and Complexity: Cultivating Contextual Understanding*, New York: Springer-Verlag, 125–136.

Norton, B.G. (1991) *Toward Unity Among Environmentalists*, New York: Oxford University Press.

OECD (1997) *Sustainable Consumption and Production: Clarifying the Concepts*, Paris: OECD.

Oelschlaeger, M. (1991) *The Idea of Wilderness: From Prehistory to the Age of Ecology*, New York: Yale University Press.

Pauchant, T. and Fortier, J. (1990) Antropocentric ethics in organizations, strategic management and the environment, in Shrivastava, P. and Lamb, R. (eds), *Advances in Strategic Management*, Greenwich: Jai Press, 6: 99–114.

Pearce, D. (1988) Optimal prices for sustainable development, in Collard, D., Pearce, D. and Ulph, D. (eds), *Economics, Growth and Sustainable Environments*, New York: St. Martin's Press, 58.

Pearce, D.W. and Atkinson, G. (1995) Measuring sustainable development, in Bromley, D.W. (ed.), *The Handbook of Environmental Economics*, Oxford: Blackwell. 166–181.

Pearce, D., Markandya, A. and Barbier, E.B. (1989) *Blueprint for a Green Economy*, London: Earthscan Publications.

Pezzy, J. (1992) *Sustainable Development Concepts: An Economic Analysis*, World Bank Environment Paper No. 2, Washington, DC: World Bank.

President's Council on Sustainable Development (1996) *Sustainable America: A New Consensus for Prosperity, Opportunity, and a Healthy Environment for the Future*, Washington, DC: PCSD.

Rahman, A. (1993) *People's Self-development: Perspectives on Participation Action Research – A Journey Through Experience*, London: Zed Books.

Rees, J. (1990) *Natural Resources: Allocation, Economics and Policy*, 2nd edn, London: Routledge.

Reichman, O.J. and Pulliam, H.R. (1996) The scientific basis for ecosystem management, *Ecology Applications*, 6 (3): 694–696.

Roodman, D.M. and Lenssen, N. (1995) *A Building Revolution: How Ecology and Health Concerns are Transforming Construction*, Worldwatch Paper 124, Washington, DC: Worldwatch Institute.

Schwartz, M. and Thompson, M. (1990) *Divided We Stand: Redefining Politics, Technology and Social Choice*, New York: Harvester Wheatsheaf.

Siebert, H. (1995) *Economics of the Environment: Theory and Policy*, Berlin: Springer-Verlag.

Simmons, I.G. (1993) *Interpreting Nature: Cultural Constructions of the Environment*, New York: Routledge.

Solow, R.M. (1974) Intergenerational equity and exhaustible resources, *Review of Economic Studies*, 41: 29–45.

Solow, R.M. (1992) Sustainability: our debt to the future, *USA Today*, September, 40.

Stead, E.W. and Stead, J.G. (1992) *Management for a Small Planet: Strategic Decision Making and the Environment*, Newbury Park: Sage Publications.

Stern, P.C. and Dietz, T. (1994) The value basis of environmental concern, *Journal of Social Issues*, 50 (3): 65–84.

Thompson, M., Ellis, R. and Wildavsky, A. (1990) *Culture Theory*, Boulder, USA: Westview Press.

Trist, E.L. (1981) The sociotechnical perspective: the evolution of sociotechnical systems as a conceptual framework and an action research program, in Van de Ven, A.H. and Joyce, W.F. (eds), *Perspective on Organization Design and Behavior*, New York: Wiley, 19–75.

United Nations (2004) *World Population Prospects: The 2004 Revision*, Paris: United Nations.

United Nations Conference on Environment and Development (1992) *Agenda 21*, Rio de Janeiro: UNCED, 3–14 June.

van Pelt, M.J.F., Kuyvenhaven, A. and Nijkamp, P. (1990) Project appraisal and sustainability: methodological challenges, *Project Appraisal*, 5 (3): 139–158.

Weiss, E.B. (1989) *In Fairness to Future Generations: International Law, Common Patrimony, and Intergenerational Equity*, Tokyo: United Nations University.

Welford, R.J. (1995) *Environmental Strategy and Sustainable Development: The Corporate Challenge for the Twenty-first Century*, London: Routledge, 2–3.

Welford, R.J. and Gouldson, A.P. (1993) *Environmental Management and Business Strategy*, London: Pitman.

World Commission on Environment and Development (WCED) (1987) *Our Common Future*, New York: Oxford University Press.

Young, J.W.S. (1997) A framework for the ultimate environmental index: putting atmospheric change into context with sustainability, *Environmental Monitoring and Assessment*, 46: 135–149.

Chapter 10

The alternative eco-building movement and its impact on mainstream construction

Tom Woolley

Introduction

In this chapter it is argued that an alternative ecological building movement exists in the United Kingdom and represents a small but significant contribution to concepts of social responsibility within building construction. While corporate statements from large companies and public sector bodies may make claims of ethical behaviour, the more radical alternative movement provides a deeper green model that challenges the status quo. The term 'ecological' is used here in an effort to distinguish genuinely environmentally responsible approaches to construction from terms like 'sustainable' and 'green'. These labels are being increasingly applied to almost every branch of conventional construction practice and are therefore in danger of becoming meaningless as they are used as marketing tools rather than representing a genuine change in practice. Even 'eco' or 'ecological' is being devalued.

As 'sustainability' has become a policy and marketing requirement for the construction industry, some architects, contractors and developers who are trying to change their practice have embraced it whole heartedly. On the other hand, much of what is claimed as green or sustainable is little more than 'greenwash', an advertising device to present conventional practice as environmentally acceptable. For those who are offering genuinely ecological and alternative solutions, wider acceptance of sustainability is a mixed blessing. 'Greenwash' makes it difficult for clients and the public to distinguish bogus claims from genuine ones. This raises ethical questions for both mainstream construction industry and the growing radical fringe. Assessing sustainability and establishing standards and guidelines assume a new importance because defining sustainable construction is necessary to distinguish 'genuine' from 'greenwash' as well as complying with a raft of government policy initiatives and new legislation.

Architects, builders and their clients who decide to follow a genuine ecological building approach do so largely for reasons of social and environmental responsibility despite additional costs and other obstacles, and how they do this will be examined in more detail later in this chapter. Clients, particularly in the public sector, are increasingly looking for statements and evidence from their consultants and contractors that the latter have not only adopted a CSR statement but that this includes a commitment to environmental responsibility. However, there is still a great deal of confusion as to what this should involve.

Perhaps the most significant innovations in ecological building have been in the housing sector and also in buildings for community, environmental and voluntary groups. Increasingly, however, examples can be found of large-scale commercial buildings, such as large warehouses, offices and health and school buildings, where highly innovative environmental measures have been adopted.

The principles of alternative building

The United Kingdom has a small but significant alternative movement within the construction industry. It is alternative in the sense that it espouses ethical intentions and a different approach to conventional construction methods. This movement encompasses a wide range of activities from suppliers, who manufacture in the United Kingdom or import ecological building materials from Germany, Austria and Switzerland, builders' merchants, construction companies and self-builders who construct 'natural' buildings in the countryside. Most ecological building materials and methods have to comply with normal regulations and standards, but many of the self-build ecological pioneers have built without normal planning and building approvals (Rosen 2007).

> The job of shaping the built environment comes with a responsibility beyond the wants of the paying client, and beyond our personal wants as well. May the wisdom that we bring to our practice include an understanding of the effects of our building designs and materials choices on all beings now alive and their descendants.
>
> (Elizabeth and Adams 2000)

Elizabeth and Adams present the sustainability principle as it applies to alternative construction in a relatively poetic language. Embodied in this statement is a principle of social responsibility that goes beyond normal professional standards. To put this idea into practice, it is necessary for exponents to consider the 'downstream' impacts of any decisions they make such as pollution, waste streams and health issues for building occupants. 'Upstream' impacts must also be taken into account in which the source of materials, transportation, manufacturing energy use and pollution are investigated. Normal architects and builders rarely consider these issues because they are preoccupied with getting the job done, within budget and on time. Where the timber has come from, for example, is of little interest, and little attention is given to the source of other materials or pollution issues during design and specification. An example of this lack of concern is the suggestion by environmental groups like Greenpeace that the United Kingdom is one of the biggest importers of illegally logged hardwood. Despite the government supporting the concept of timber certification, Greenpeace has exposed the use of illegal hardwood in government buildings in Westminster (Greenpeace 2007). This suggests that even where social responsibility statements or green procurement policies are in force, getting the materials on site with little regard for source is the general order of the day.

The alternative ecological building movement starts from a concern about the environmental impact of buildings and the construction industry and offers a range of alternatives that are seen as having beneficial effects. These alternatives cover a very wide range of organisations from those that are barely distinguishable from conventional construction to those which can be argued to have introduced significant new practices. Many of these practices go beyond what is required by legislation and normal standards. For instance, ecological building exponents have been building to a standard of energy efficiency, which goes well beyond the minimum standards of the building regulations. They frequently refer to using green materials such as planted roofs, sustainable timber, natural insulation materials and recycled materials.

Many green architects and building materials' suppliers talk of creating 'healthy' buildings even though there are no UK standards that define healthy buildings. Indoor Air Quality (IAQ) standards have been introduced into the England and Wales building regulations recently, but these do not go far enough to ensure that buildings are healthier (Yu and Crump 2007). Ecological builders and architects may try to achieve good IAQ through a sense of social responsibility rather than because of stringent requirements. They may specify paints, glues and finishes free of volatile organic compounds and other toxic emissions.

The origins of the ecological building movement

While many of the ideas and inspiration for this movement began in the 1970s and 1980s with books like *Radical Technology* (Alexander 1979; Boyle and Harper 1976; Papanek 1984), most of the organised activity developed in the 1990s. The Ecological Design Association (EDA) was one of the first organisations; founded by David Pearson and others, it began to produce a magazine, *Ecological Design*, in the late 1980s, and Pearson published a number of beautifully illustrated and influential books (Pearson 1989). The EDA was largely made up of interior and product designers, with some architects. Shortly afterwards, a group of builders and energy experts set up the AECB (AECB 2007) which now calls itself the Sustainable Building Association. The EDA has disappeared as an organisation, but the AECB now has over 1000 members which includes architects, local authorities, materials' suppliers and so on, representing most facets of the construction sector. The Scottish Ecological Design Association (SEDA 2007) is also still strong and active.

The Association of Community Technical Aid Centres (ACTAC), a network of professionals providing advice to community and voluntary groups, flowered for over a decade in the 1980s and, in the 1990s, organised a series of 'Green-Building Fairs' and published the *Green-Building Digest*, now incorporated into the *Green-Building Handbook* (Woolley and Kimmins 2000; Woolley *et al.* 1997). ACTAC has also disappeared, but several community technical aid centres still exist in Belfast, Manchester and elsewhere.

The influence of the ecological building movement

It would be easy to dismiss this alternative movement as ephemeral and marginal, as organisations tend to come and go with little obvious influence on mainstream building activities or public policy. For instance, the UK Government's Sustainable Building Task Group, chaired by Sir John Harman and Victor Benjamin, published a report entitled 'Better Buildings, Better Lives' in May 2004. The ministerial response to this stated the importance of consultation with stakeholders and listed a number of organisations that should be involved in this. None of the alternative organisations like the AECB were mentioned at this time:

> Bodies not established by government, but which are concerned with sustainable buildings include:
>
> - Constructing Excellence (www.constructingexcellence.org)
> - Strategic Forum for Construction (www.strategicforum.org.uk)
> - Sustainability Forum
> - Construction Industry Council (CIC), and, in particular, its Sustainable Development Committee (www.cic.org.uk)
> - Construction Clients Group (CCG)
> - CRISP (Construction Research and Innovation Strategy Panel) (www.crisp-uk.org.uk)
> - Sustainability Alliance of Professions (www.sustainabilityalliance.org.uk)
> - Water Regulations Advisory Scheme (www.wras.co.uk)

Non-government bodies that government ministers recognised as stakeholders in developing a sustainable construction policy were listed, but the list did not include alternative organisations such as the AECB. Subsequent consultation about a proposed 'Code for Sustainable Building' largely ignored the alternative movement and instead focused on getting mainstream construction to buy into sustainability measures. This led eventually to the 'Code for Sustainable Homes', which is discussed later.

The active members of organisations like the AECB are largely small businesses, though bigger companies and public sector bodies have joined. Research, funded by the Engineering and Physical Sciences Research Council (EPSRC), into the alternative building movement indicated a tendency that most participants were individuals or small businesses with a wide variety of views and interests (Woolley and Caleyron 2002).

Many are struggling to make a living and establish their businesses as viable concerns. Any ethical principles they may espouse can be an impediment to commercial success, as they are competing with cheaper products. As a collective force, they are relatively weak and lacking in influence. Their interface with the main environmental NGOs is tenuous, for instance. While there has been some collaboration between the AECB and Friends of the Earth on energy policy, bodies like Greenpeace and WWF tend to adopt their own campaigning policies, and these only intersect with the green-building movement from time to time. However, sustainable construction has not been a significant campaigning issue for many environmental NGOs, though they embrace aspects of it occasionally. WWF have been the most interested with their 'Million Sustainable Homes' campaign, for instance, and active support for certified timber.

The influence of the ecological building movement

Given the frequently repeated statement that buildings and the construction industry are responsible for over half of carbon emissions, at least a third of landfill waste and a quarter of all resource consumption, it is strange that more coordinated pressure has not been exerted on this sector to improve environmental performance. The main reasons for this are the complexity of the issue and the difficulty of identifying simple solutions. First, the construction sector is very fragmented. Many clients for buildings are building for the first time and on a steep learning curve, and sustainability seems like a problem too far, when there is so much else to deal with. Although some public sector bodies have begun to adopt sustainable construction policies, these are limited by the influence of vested interests in the construction industry. There is very little consensus on the exact nature of sustainable construction, and this creates confusion in the minds of those who want to address the issues.

Second, much of the pioneering work in alternative construction has come from a strongly anti-establishment movement of people, who grew out of a mixture of 'hippie', squatter and back-to-the-land movements of the 1970s and the 1980s (Rosen 2007). This disparate movement of people can be seen as mounting a challenge to normal global capitalist models of society through the personal activities of its members such as living in tepees in Welsh valleys, travelling around the country, attending the summer solstice celebrations at Stonehenge or helping out on organic farms. There is little coordinated activity to influence government policy. This approach, often characterised as 'lentils and sandals', is perceived by some as an obstacle to the advancement of sustainable building into the 'mainstream'. Public perception of buildings of straw and earth creates the impression that sustainable building can be embraced only by a handful of long-haired self-builders in the countryside and is not relevant to meeting housing or other mainstream building needs in the cities. Those trying to sell ecological materials into mainstream construction can be tarred with the same brush even if they are businessmen in suits!

Third, UK government policy and research have barely addressed the issue of alternative ecological building materials and methods. Much of the research and literature on this subject is from the Bau-Biologie movement in Germany (Bau-Biologie 2007) or work in the United States and Australia. This means that official bodies do not provide support or advice on the use of alternatives, and this discourages those who might specify or use them.

The Bau-Biologie movement has undoubtedly advanced understanding of the science of ecological building and is often referred to by the manufacturers of ecological building products in Germany. However, many of its exponents also subscribe to 'new-age' mystical concepts of building design and pseudo-sciences such as Feng shui and geomancy. This can undermine the scientific credibility of green-building ideas for those who prefer a more rational approach (Woolley 1998).

Self-build in the countryside

Some self-builders using straw-bale walls or cob (mud and straw walls) have gone ahead with eco-building, without seeking normal planning and building control approval, often because they would not get such approval. Although the number of such people living a fully ecological lifestyle, off the land and local resources, is quite small, their influence on a wide range of the population who still have jobs and drive cars is nevertheless more significant than might be expected. During the compilation of the book *Natural Building* (Woolley 2006a), which documents many alternative ecological building techniques, the majority of illustrated examples was chosen from the countryside, as this is where the greatest scope for experimentation seems to have existed.

The aspirations of *The Good Life* on TV in the 1970s have gone deep into the psyche of British society, and thus many aspire to a greener lifestyle and living off the land. More recently, a number of organisations like Lammas (2007), Chapter 7 (2007) and Land for People (2007) are trying to establish eco-villages in the countryside that will demonstrate ways of living in the post-peak oil era, where land is used more efficiently. Currently in the United Kingdom only about one per cent of the population get their employment from the land, but it has been suggested that this will have to change significantly as fossil fuel resources run out (Heinberg 2003, 2004). Richard Heinberg presents some of the most cogent arguments in his book *Power Down* that alternatives must be developed for a post-carbon economy and that where people live is a key factor. While these new UK rural initiatives are attempting to go through the planning system and to be fully approved by regulatory bodies, this has not been an easy process, as planning policy has not been adjusted to take account of the fundamental changes envisaged by Heinberg and others.

Rural planning policies have been largely concerned with protecting the countryside from development and this has forced up property prices as wealthy people buy second homes or retire to picturesque country villages. This has led to the decline of local schools, transport and facilities in rural areas, and local agricultural workers cannot afford to live in the countryside any longer. Although many who want to live an alternative lifestyle may have been able to buy land in the countryside, planning restrictions mean that it is almost impossible for them to build houses to live in.

As a result, there has been a growth of illegal buildings and settlements, often hidden in woodlands, built by people who feel that planning restrictions aimed at voracious housing developers should not apply to them. They argue that as they are living and working in harmony with the land, different policies should be applied. So compelling is this argument that many planning authorities are now agreeing to housebuilding in the countryside where the applicants can prove that their occupation such as farming and woodcraft requires them to be on the land 24 hours a day. Low-impact development policies have been adopted by some rural-planning authorities, though they set a very high standard for people to be 75 per cent self-sufficient within two years.

In the vast majority of cases, these new rural dwellers subscribe fully to ethical principles of sustainable construction, and many innovative forms of building have emerged as a result. Using local green and round wood timber, straw, mud and earth, wattles and shingles, genuinely zero impact buildings can be built. Often these are completely off the grid, relying on renewable energy for power and hot water and on-site ecological forms of sewage treatment. These eco-cabins, however primitive, provide a benchmark for sustainable building against which more dubious claims of sustainability can be measured, and television publicity for high-profile projects like Ben Law's Woodland House (Law 2006) has popularised the idea of low-impact construction.

Thus, there are many who are trying to extend the idea of sustainable rural development for houses, local schools and facilities, but very few of the pioneering projects proposed so far have gained planning approval. Community Land Trusts have been formed, sometimes with government and official backing from bodies like English Partnerships in an effort to develop alternative models for rural development and to take high land prices out of the equation (Community Land Trust 2007). Most of these projects embrace ecological building ideas.

Low- and zero-carbon development

Environmental organisations like WWF and Bioregional are proposing developments throughout the United Kingdom and in other countries, including South Africa, China and Australia as part of a 'one planet living' campaign. It is intended that they will follow principles of zero carbon, zero waste, sustainable transport, sustainable and local materials, local food, efficient water use, conservation of flora and fauna, respect for cultural heritage, equity and fair trade and happy and healthy lifestyles (WWF 2007). One of the proposed projects is called 'Z-squared' and will involve a development for 2000 homes (5000 people) in the Thames Gateway area. It will be important for these projects to be monitored and assessed to see how close they come to achieving objectives like being 'zero carbon'. Even more ambitious are 'eco-towns' being planned for China. Dongtan, which will be built on an island at the mouth of the Yangtze river, has been designed by Arup. Some claim that it will be a blue print for cities of the future, but already millions of tonnes of concrete have been poured into creating the infrastructure that will make Dongtan accessible. Arup claim that Dongtan will have an ecological footprint of two hectares per person compared with five for Londoners. However:

> A pioneer of the ecological footprint, William Rees of the University of British Columbia in Canada, has mixed views of Dongtan. It is, he says, 'hardly a truly sustainable option' given that it is a new city occupying what is mostly agricultural land near a large ecologically significant wetland. He says that it is being designed to attract wealthy buyers whose way of life will be characterised by 'high levels of personal consumption and large per-capita eco-footprints'. But it could be worse. It is at least less bad, he concedes, than greenfield cities for the rich based on standard urban designs and architecture.
>
> (*The Economist 2006*)

In the United Kingdom, as pressure grows for more housing developments, following the Barker Review of housing (Barker 2006), many new sustainable town extensions and villages are being proposed, though there is some doubt as to whether many of these projects are capable of meeting even the government's own Sustainable Homes Code standards (Building 2007). Most of the current so-called low-carbon housing development that has been proposed uses conventional building methods and has not embraced the alternative eco-building movement. It is not surprising, therefore, that they are struggling to meet even middle-of-the-road green tests.

The development of 'green lifestyles'

There is growing evidence of an increasing number of people from a variety of backgrounds who want to build or live in genuinely ecological buildings in towns as well as the countryside. Many are people with normal jobs who have taken an ethical decision to reduce their impact on the environment. Large numbers of such people are members of environmental organisations, buy organic food, cycle to work and see it as a rational extension of their lifestyle to aspire to living in an environmental-friendly house. As a result, there are enough people wanting to build or live in ecological houses, or include green principles in urban farms, environmental education centres, community buildings and so on, to keep the alternative architecture and building movement reasonably active and busy. There are a number of 'eco-village' type developments, such as The Wintles in Shropshire, under construction or already inhabited, by relatively well-to-do middle-class families (Living Villages 2007). George Monbiot in his *Guardian* column has been quite critical of such middle-class movements, suggesting that they do little to reduce carbon emissions, but, nevertheless, the trend towards greener lifestyles has given a boost to organic food production and a gradual increase in the purchasing of greener building materials. Just as middle-class people will pay more for organic food, so are they also keen to use ecological building materials.

Movement into the mainstream

As the market has grown, mainstream construction has taken an interest, and leading building materials' suppliers such as the Wolseley Group and B&Q are promoting a wide range of ecological products. Wolseley is constructing a demonstration Sustainable Building Centre at its headquarters in Leamington Spa (Wolseley 2006). Designed by ECD architects, the building is being constructed of as many eco-building products as can be packed into one small building and also demonstrates energy-efficiency principles. It will be used to train specifiers about the range of products available. Such investment would not be undertaken purely for a rural self-build market, but is indication that the alternative building movement has been embraced by big business.

There is now a wide range of eco-building products including solar panels, ground-source heating systems, rainwater harvesting products, green roof systems, sheep's wool and hemp insulations and a wide range of organic and ecological paints, finishes and decorative materials. The directory of AECB members shows that there are green architects and builders in most parts of the country, and a growing number of suppliers of ecological building materials and products. Eco-materials companies like Natural Building Technologies (NBT 2007) and Hemcore (2007) are now being backed by government agencies such as the Carbon Trust and the National Non Food Crops Centre.

Sustainable building principles

The alternative sustainable building movement represents a wide range of views and approaches but does share certain common themes. Sometimes these are aspirational rather than fully put into practice. They can be summarised as follows:

- Low energy usage strategy to reduce the amount of fossil-fuel-based energy consumed, through more efficient buildings and technology, particularly higher levels of insulation.
- Use of renewable energy instead of fossil-fuel energy, either integrated into buildings or through renewable energy supply schemes on payment.
- Use of materials obtained from sources that do less damage to the environment. This can be interpreted in many ways, but usually means local sources, using recycled materials and supply from certified sources, e.g. Forest Stewardship Council (FSC 2007) timber. Use of materials and products which do not damage the health of manufacturing workers, builders or building occupants.
- For some, it also means the retention of existing buildings and opposition to wholesale demolition and clearance. There are some overlaps between the green-building and the conservation of historic buildings lobby.

While these aims underpin the objectives of most people concerned with sustainable building, they are often interpreted in different ways. For instance, many organisations feel they are doing their bit for the environment by putting expensive pieces of equipment onto their buildings, such as photovoltaic cells and ground-source heat pumps, but do not consider increasing the insulation in the walls. The energy saved and the cost payback on this kind of high-tech equipment is poor, but it is visible and can be shown off to the neighbours or as a corporate statement, whereas higher levels of insulation are hidden and not a statement of a greener lifestyle, even though they will give a much better payback. Thus, rather than adopt a holistic design strategy, architects and their clients frequently cherry-pick individual items without a full understanding of their environmental impacts or payback. Government-grant schemes to encourage the uptake of renewable energy and 'micro-generation' make building energy-efficiency standards as a condition of the grant, but only to a limited extent. Thus, the renewable energy may go waste for want of higher levels of insulation, which would have had a much better payback. Some larger public buildings will display energy consumption details and government is trying to monitor this in its buildings, but this practice is not yet commonplace, particularly within the private sector. Very little work is being done to check whether buildings actually perform as well as predicted, and thus claims of energy- efficiency may be misleading. Attention paid to evaluating and checking buildings once they are occupied is insufficient, despite campaigning by the Usable Buildings Trust (2007).

Volume housebuilders, housing associations with government encouragement and self-builders are buying expensive housing kits from Canada, Scandinavia and Germany, instead of sourcing their materials and expertise locally, because they believe these are environmentally responsible. Some seem to hold the view that UK technology is not able to achieve good enough standards, and that importing materials and systems from outside the United

Kingdom will be more effective, even though this is not necessarily a sustainable strategy. On the other hand, the 'BEDZED' project in Surrey tried to source all its needs from within 35 miles (and managed to, from within 60 miles) as a sustainable strategy, but this project is and remains the exception, which is why its iconic wind cowls feature on so many reports and publications about sustainable construction (Bioregional/Bedzed 2007).

What are green products?

One of the difficulties revealed in the EPSRC research project (Woolley and Caleyron 2002) was the degree of confusion about the nature of 'green' products. There is a wide range of views and knowledge about the varying performance or composition of certain 'natural' or green products. Even where standards exist, through 'CE marking' for instance, these standards are at such a low level that almost any claims can be made about products being eco or greener than others. While the EU Construction Products Directive was passed in 1988, CEN, the European Committee for Standardisation, has still not finalised environmental criteria for building products. This makes it difficult for companies manufacturing what they believe to be excellent green materials to distinguish themselves from those who may be making bogus claims.

An example of this is in the area of paints. European regulations have forced the reduction or elimination of toxic and carcinogenic materials such as lead and volatile organic solvents in paints, but there is a big difference between many water-based paints and natural or organic paints. This is not only confusing for the ordinary DIY house decorator, but even professionals in the construction industry are ill-informed. Without firm requirements to safeguard indoor air quality (Yu and Crump 2007) and to test buildings once they are completed, the manufacturers of environmental-friendly paints are at a disadvantage. The adoption of a socially responsible policy on the specification and use of paints is very difficult when there are no commonly agreed standards. Environmental Product Declarations (EPDs) are issued by many manufacturers but these are not widely understood or available. EPDs are based on lifecycle analysis (LCA) tools and rely heavily on honest data being provided by manufacturers and the assumptions made in the LCA tools. However, UK Government and EU standardisation may lead to EPDs using common methodologies to be more widely adopted (BERR 2007).

Green-building is aspirational

However, despite the confusion and poor quality of information at a detailed level, in general terms, the aspiration for a more environmentally responsible approach exists. These aspirations are essentially based on a general sense of social responsibility and ethical views about society and protection of the natural environment. Such views have been included in central and local government purchasing policies, which have adopted the concept of sustainable building, and this is changing the nature of the market. Ethical policies are gradually being developed in some detail, which requires designers, specifiers and builders to match relatively tough environmental criteria. These range from certified timber to improved standards of indoor air quality and lifecycle policies affecting reuse and recycling and disposal.

Quite a number of local planning authorities, for instance, have adopted sustainable building policy guidelines for developers and applicants, and although planning law does not necessarily back them up, these policy guidelines are serving to educate the public about the need to embrace environmental principles. Applicants for planning permission are asked to comply with such supplementary planning guidelines, though the guidelines can vary considerably from one authority to another (Woolley 2006b). In areas like south west England, voluntary sector sustainable building or development trusts have been set up with local authority support or encouragement, which facilitate green projects, negotiate with developers, conduct conferences and provide education (Cornwall 2007; Eco Trust 2007; Exeter 2007).

What is significant about these developments is that the adoption of green-building principles and practices does not necessarily have any immediate economic payback or bottom-line advantage. The general view is that green-buildings cost more (even if this is not true), as specifying alternative materials and construction methods is assumed to be more expensive. Thus, those people and organisations that opt for a green-building approach do so in the knowledge that while it might cost them more, at least in the short term, it is a good investment for the future. These decisions are taken because of perceived environmental benefits to society or the planet and are thus motivated by an ethical position on the environment. Green-buildings do not inherently need to cost more, but as the scale of demand is so low, prices tend to be higher. As the market for green-building products increases, so will they be more affordable.

UK Government policies on sustainable building

In 2004, Parliament passed the Secure and Sustainable Building Act that ensures that sustainability criteria can be incorporated into the building regulations. However, it is not clear from the Act what is meant by sustainable building. The Sustainable Building Task Group report in 2003 provided a very limited view of sustainable building, but this was followed by setting up a Sustainable Building Code Steering Group (ODPM 2004) made up of a significant number of property developers!

The steering committee included Richard McCarthy, a civil servant at the ODPM, Peter Fanning, Chief Executive of the Office of Government Commerce, Ian Coull from Slough Estates, Peter Rogers from Stanhope plc and David Pretty of Barratt Homes. This group also included Robert Napier from WWF, Walter Menzies from the Sustainable Development Commission, Michael Ankers from the Construction Products Association and June Barnes from East Thames Housing Association.

This obvious bias towards vested interests led the Parliamentary environmental audit committee to launch a broadside at the Government about the composition of this group. They stated that

> The Government's housing policy is an alarming example of disjointed thinking . . . the principal beneficiary of housing growth will be property developers, with the environment we all depend upon being the principal loser.
>
> (House of Commons 2004: 52)

The report goes on to criticise the composition of the Sustainable Building Code Steering Group, stating that it did not contain experts on sustainable construction.

> [140]. There is ample representation from industry, Government and social housing groups on the Steering Group. Having being told in evidence that organisations such as BRE and the Energy Savings Trust would be invited onto the Group, we were surprised to see that there is no representation from any organisations that are directly involved in how to improve the environmental performance of buildings. It is incredible that the Government has not thought it important to have any representation from the organisations that have the greatest expertise in this area. This omission does not inspire us with confidence that the Code will result in significant and meaningful improvements in how houses are built or how their impacts on the environment are minimised.
>
> (House of Commons 2004: 52)

Michael Ankers from the Construction Products Association could have been seen as someone with expertise in sustainable building, but academic experts and representatives from the Building Research Establishment were not included. Such was the strength of criticism of the government's plans for a sustainable building code that it went through several stages of development and consultation (with a limited number of industry stakeholders). Eventually it re-emerged as the Code for Sustainable Homes, formally launched in December 2006. The attempt to establish a code for the whole of the construction industry had been quietly shelved.

At the time of writing, a further strategy for sustainable construction has been mooted by the Department for Business, Enterprise and Regulatory Reform which has replaced the Department of Trade and Industry under the Gordon Brown Premiership (BERR 2007).

While in some respects this document is closely tied to the usual limited view of government on sustainable construction concerned with energy, waste and water, there are signs of some response to the alternative eco-building perspective with reference to resource depletion, EPDs and renewable materials.

UK government strategies have been driven in the past by a need to make standards as acceptable to mainstream industry as possible, and thus the standards that are usually advocated will not move the industry far from its current practices. Even where more radical ideas are floated in consultation documents, these disappear following pressure from developers. Critics of measures such as the Code for Sustainable Homes, for example, John Broome, writing in *EcoTech*, suggested that the new code was purely intended to greenwash the government's proposals for the Thames Gateway:

> The proposed code is a bitter disappointment to those working in the sector. Some suspect that it has been devised to enable the government to declare the planned waves of new housing in the south east (i.e. the Thames Gateway and other growth areas) to be 'sustainable communities'.
>
> (Broome 2006: 14)

Professional attitudes to sustainable building

Professional bodies in the construction industry appear to be addressing sustainability policies and practices through numerous policy papers, conferences, seminars and CPD events. There is little doubt that raising awareness is on the agenda, but it is hard to see fundamental changes in day-to-day practice. Professionals often complain about over-regulation and see green requirements as just more bureaucracy. Though they may have to profess a commitment to environmental best practices to win tenders, in-depth knowledge as to what is required is often lacking.

Education of future professionals would be expected to address these issues and draw on the knowledge of the alternative eco-building movement, but this is very patchy. Sustainable design is a requirement for all accredited UK architecture courses, but a recent search of accreditation documents for a leading London school of architecture found only two references to 'environmental' and no references to 'sustainability' or 'green' in several hundred pages of documents. For many architecture schools, design comes first and the environment is a grudging bolt-on. Where sustainability is not explicitly referred to, it is justified by saying that sustainability is integrated into all aspects of design and technology. Very little research has been done on the degree to which sustainable building practices have been integrated and whether professional practices and construction companies have adopted ethical and environmental policies, so it is extremely difficult to measure any changes. The industry responds largely to regulation rather than exhortation and voluntary initiatives, and unless tougher rules are introduced to force the adoption of green- and eco-building principles, these will be ignored.

Many professionals, builders and developers will be aware that there are environmental standards that can be used to benchmark their buildings. In the United States of America the 'LEED' (Leadership in Energy and Environmental Design) standard is fairly well-known, and the US Green-Building Council is very successful with attendances of 10,000 at its annual conferences (USBG 2007). LEED is based on BREEAM, which was developed in the United Kingdom, but LEED seems to have been more widely adopted. While a BREEAM rating is now required for most government buildings (BREEAM 2007), there is little evidence of a widespread take-up in the private sector. CEEQUAL, the civil engineering equivalent, seems to have enjoyed wider acceptance, however (CEEQUAL 2007).

Both the lower-level LEED and BREEAM standards are relatively easy to achieve without significant changes from conventional design and construction, and the aim seems to have been to lead the industry gently by the hand towards more sustainable buildings without making standards too radical. However, once professionals step outside these two well-known methods of assessing sustainable buildings, the whole subject becomes very confusing with a very wide range of standards and methods. Internationally there are several hundreds, if not thousands, of sustainable building tools, and this makes it extremely difficult to compare performance or measure the success of designs. A study, funded by the UK Department of Trade and Industry, under the Partners in Innovation Programme, led by Taylor Woodrow Construction, attempted to produce a toolbox and framework which would allow professionals to choose the appropriate environmental method for selecting materials and construction methods (Balanced Value 2007). This proved to be a massive task, given the lack of common criteria, and was probably more successful in drawing attention to the problems rather than coming up with a solution (Woolley 2005).

Many of the environmental and sustainability standards and tools concern themselves with materials and products, and appear to incorporate eco-building principles including health and pollution issues. Others do not, and limit themselves mainly to pragmatic concerns with energy efficiency and waste management on the basis that this, rather than taking a broader principled approach, reduces costs.

The supply of green products and materials: ethical purchasing?

Although, as stated above, there is a perception that green building is more expensive, many green-building methods and materials do not need to be more costly. If they were mass-produced, they could well be cheaper than conventional, synthetic petrochemical-based products. As many alternative materials are renewable, there are real social and economic benefits in producing them. As countries strive to become less dependent on fossil fuels and turn to renewable sources, natural and ecological building products can be seen to have advantages. In Germany, for instance, natural insulation products have been subsidised by the government through grants to household installers. Challenges to this under EU competition rules through the European courts were apparently not successful.

However, at the moment, fibreglass insulation can be sourced for as little as £1.50 a square metre, whereas at the top end of the market some sheep's-wool insulation products cost as much as £10 a square metre. With such a huge price differential, the choice or specification of a natural insulation product is definitely not based on the bottom line but on ethical or environmental principles. Gradually, however, the cost of natural and renewable products is coming down and is becoming competitive with synthetic alternatives.

Not all natural and environmental products are necessarily more expensive, but there is also a perception within the construction industry that they will not be as good as conventional products. A good example of this is the removal of arsenic as a timber treatment chemical through EU legislation to reduce the use of toxic and carcinogenic chemicals. CCA (copper chrome arsenate) gave timber a green colour, so builders knew it had been treated. Less toxic replacement chemicals can be colourless but a green-tinted wood using copper is still made available, perhaps to reassure conventional builders that the timber still has something nasty in it! European legislation is leading to the removal of many toxic and dangerous materials that are confirmed as a danger to health by the World Health Organisation. Creosote, used for external timber treatment, has been outlawed, and the use of formaldehyde, in glues and timber composite products, is being reduced or banned. The EU has not yet been able to ban uPVC (unplasticised polyvinyl chloride), but the plastics industry has been forced to improve its environmental performance and develop ways of recycling plastic products.

There is a great deal of resistance to the use of natural and organic paints because there is a perception that their performance will not be as good as that of synthetic ones. This is partly because ecological building often requires more labour or longer drying times. Toxic synthetic products have been developed with solvents and additives to speed up construction time. Conventional builders do not like having to wait for several days for paints or varnishes to dry.

Environmental builders advocate the use of unfired earth or lime instead of cement and concrete but these are not popular with conventional builders as they involve more labour and are slower to dry or gain their full strength. Despite this, mainstream industry is now developing unfired earth bricks and blocks, some incorporating waste materials. Lime mortars for building are now available in ready-mixed form and are being used, in place of cement mortars, on large buildings.

In the EPSRC study referred to above (Woolley and Caleyron 2002), six companies who were trying to manufacture and/or distribute green products were interviewed. They faced several problems, in particular, the lack of agreed standards that would certify environmental-friendly products and distinguish them from conventional. But, a second problem was that of the perception that eco-products would not be as good as conventional ones.

At that time they were all preoccupied with the need to break out of a 'green ghetto' and into the mainstream so that volumes of production would increase and their products could benefit from economies of scale.

In another unpublished study carried out for a manufacturer of building systems, the company wanted to evaluate the benefits of becoming more environmental-friendly. However they were concerned that it would be a risky marketing strategy to make too many green claims. Not only would it be open to challenge from environmental organisations that what the company was doing was not really green, but the company also felt that many of its customers would see its products as inferior if the products received a green label. Despite this view, the company was surprisingly motivated by a corporate ethical policy and had introduced many environmental reforms long before it was required to adopt any, by legislation. Its products were very energy efficient; it used environmentally certified timber and took recycling very seriously. However, this information was not in its marketing strategy, as many other materials that it used were environmentally less acceptable. It felt it was preferable to say nothing rather than make false green claims.

This responsible position is not shared by many others in the construction industry who can be seen to be making green claims for their products and buildings, knowing that the absence of any widely accepted benchmarks and standards meant that no one can challenge them. As 'greenwash' becomes increasingly common with the growing demand for sustainability, architects, their clients, manufacturers and builders may not be able to distinguish between genuine and bogus green claims, as Environmental Product Declarations (EPDs) are not generally recognised within the industry. Although EPDs can now be found, there are differences in methodology, and little independent scrutiny of claims that are made based on the life-cycle analysis work carried out by the private sector. UK consultants Bousteads are perhaps the best-known and their methods have been widely adopted, but there are many others in the field (Boustead 1999). In Sweden, there does seem to have been some institutionalisation of LCA and EPD work (Environdec 2007).

Timber and forest management is one of the few areas in which independent certification of environmental claims is available through the international Forest Stewardship Council system. A whole range of organisations has come together to ensure that sustainable timber can be sourced (Forest Forum 2007). Standards applied by FSC include not only good environmental forest management practices but also ethical standards about the treatment of timber and involvement of local people (FSC 2007). The timber industry in Europe, however, was slow to adopt FSC standards and instead offered an alternative (PEFC 2007) that is run by the industry. Most timber suppliers tell their customers that their timber is from 'well managed forests', but there remains a great deal of ignorance in the construction industry about the difference between the FSC and other labelling systems. The term 'well managed forests' is about as convincing as 'the cheque is in the post', but architects and specifiers can be surprisingly naive about accepting such assurances without checking them carefully.

Public sector procurement policies may well dictate the use of sustainable timber, and even specifically require FSC certification, but these requirements can often be ignored on site. Specification substitution is a big problem and may well have led to the use of illegally logged hardwood on government projects. It is essential for the architect or building supervisor to ensure that if green and sustainable materials are specified, they are actually used on building projects. The degree of vigour with which such issues are pursued is largely down to the degree of commitment to green- and eco-building principles. Although there is also a great deal of ignorance and confusion about what is really green or ecological, professionals, builders and clients as well as developers who are really determined to protect the environment can manage to find a way of adopting eco-building principles. This will reflect their degree of commitment to social responsibility.

The example of hempcrete

An interesting example of environmental innovation in building construction can be found in the development of 'hempcrete'. This is a method of construction in which a natural concrete can be cast around timber frames or used as infill between concrete or steel structural frames. It provides an insulating wall, floor or roof that is both structure and insulation. Weatherproof and solid and fireproof with very good acoustic properties, it solves many of the problems that have dogged lightweight timber-frame construction. The solid wall can breathe allowing moisture to permeate through the fabric; it can even regulate temperature and moisture creating better indoor air conditions. The material is made simply by mixing a special lime binder with chopped-up stalks of hemp (Lime Technology 2007). As the hemp is an annual renewable material, it can be grown and replaced each year. As the plant absorbs CO_2 whilst it is growing, it can be argued that the building fabric is storing CO_2. Thus, buildings constructed with hemp, timber and other natural crop-based materials can offset the carbon emissions of other less benign aspects of the construction such as concrete, steel or glass that emit large amounts of CO_2 in their manufacture.

So exciting is this innovative sustainable form of construction that it has attracted investment from mainstream industry; Lhoist (2007), the world's largest producer of industrial limes, Castle Cement and other businesses have got involved. A manufacturing plant Lime Technology in Abingdon, near Oxford, has been set up to produce 'hempcrete'. Already, quite a few social housing schemes have been constructed out of hemp, though mostly in Suffolk, owing to the pioneering work of local architect Ralph Carpenter. Also in Suffolk, a huge warehouse to store beer and wine has been built by Adnams the brewers with 15-metre high walls of hemp and lime. This highly innovative building, which includes many other green features, cost the client considerably more than a conventional industrial shed, and considerable risks were taken to try something that had never been done before. The client took the decision to proceed for the good of the environment, not for its own bottom line. However, it is already seeing a payback (Haymills 2006). The natural materials are so effective in providing a stable thermal environment that the installation of an expensive cooling plant was cancelled. Running costs are also therefore reduced. PR for the company has come by way of winning awards and recognition for its courage in embracing the green agenda.

Hemcore (2007), a company that processes hemp from farmers in Essex, is also investing millions in expanding production. Hemp fibre is a valuable product, but buildings are constructed from the left-over straw. The UK Department for Food and Rural Affairs (DEFRA) is strongly backing the use of 'non-food crops' through the establishment of a National Non-Food Crops Centre in York (NNFCC 2007). A great deal of exciting experimentation is going on, producing industrial lubricants from plant oils, pharmaceuticals from crops, and glues, resins and a wide range of bio-composite materials. Although investment in these initiatives is still very small in relation to overall GDP, it is an indication that some sort of corner has been turned in the recognition that alternatives are here to stay.

Conclusion

The main challenge facing the construction industry is to fully embrace the implications of climate change and environmental issues without indulging in greenwash. Business-as-usual will not reduce resource consumption, wastage, pollution, energy wastage and planetary damage; radical changes are required. Leadership for such radical changes has already been provided by the alternative eco-building movement. Much can be learned from these pioneers, though there is still a tendency on the part of mainstream construction to sneer at what has been done. Making buildings out of straw, mud and hemp is seen as a bit of a laugh, but not as something that can be taken on board by real architects and builders who would prefer to stick to steel, acres of glass and concrete. However, once everyone (politicians, professionals and the public) realises that this is no longer acceptable, future steps must be underpinned by standards enforced by government, that will really make a difference.

It is no good if legislation pussyfoots around the issues, scared of upsetting developers and materials' producers. Tough environmental standards qued to be implemented and policed to make non-compliance more transparent. Everyone should be operating on a level playing field, and this will inevitably mean independent scrutiny and open disclosure of environmental impacts. A commitment to this can come only from a 100 per cent acceptance of social and environmental responsibility grounded in ethical principles.

In the long run this will mean that many design and building solutions currently in use will not be environmentally or ethically acceptable. Instead of defending such practices, industry will have to embrace alternatives and ensure that investment and employment is directed to such innovations. This should ensure that our building and infrastructure needs can be met in the future without serious environmental damage.

References

AECB (2007) *The Association for Environment Conscious Building*. http://www.aecb. net/index.php (accessed 24 June 2008).

Alexander, C. (1979) *The Timeless Way of Building*, Oxford: Oxford University Press.

Balanced Value (2007) *Balanced Value Online Toolbox*. Online. Available at: http://www.balancedvalue.com/Index.aspx (accessed 29 October 2007).

Barker, K. (2006) *Barker Review of Land Use Planning*, London: HM Treasury.

Bau-Biologie (2007) http://www.buildingbiology.net/ (accessed 24 June 2008).

Bioregional/Bedzed (2007) *BedZED & Eco-Village Development*. Online. Available at: http://www.bioregional.com/programme_projects/ecohous_prog/bedzed/bedzed_hpg.htm (accessed 29 October 2007).

Boustead, I. (1999) Eco-labels and eco-indices: do they make sense? *Fourth International Ryder Transpak Conference*, Brussels, Belgium.

Boyle, G. and Harper, P. (eds) (1976) *Radical Technology*, London: Wildwood House.

BREEAM (2007) http://www.breeam.org/ (accessed 24 June 2008).

Broome, J. (2006) EcoTech, *Architecture Today*, 13(May): 14.

Building (2007) Brown's eco new towns will fail green test, *Building*, 18 May.

CEEQUAL (2007) www.ceequal.com (accessed 24 June 2008).

Chapter 7 (2007) http://www.tlio.org.uk/chapter7/ (accessed 24 June 2008).

Community Land Trusts (2007) http://www.communitylandtrust.org.uk (accessed 24 June 2008).

Cornwall (2007) *Cornwall Sustainable Building Trust*, www.csbt.org.uk/ (accessed 24 June 2008).

BERR (2007) *Department for Business Enterprise and Regulatory Reform*, http://www.berr.gov.uk/sectors/construction/sustainability/page13691.html (accessed 24 June 2008).

Eco Trust (2007) *Somerset Trust for Sustainable Development*, http://www.ecostrust. org.uk/ (accessed 24 June 2008).

Economist (2006) September 21.

Elizabeth, L. and Adams, C. (2000) *Alternative Construction: Contemporary Natural Building Methods*, New York: John Wiley and Sons.

Environdec (2007) http://www.environdec.com (accessed 24 June 2008).

Exeter (2007) http://www.sustainablebuild.org/exeter_pres.html (accessed 24 June 2008).

Forest Forum (2007) www.forestforum.org.uk (accessed 24 June 2008).

FSC (2007) http://www.fsc-uk.org/ (accessed 24 June 2008).

Greenpeace (2007) *Yet More Illegal Rainforest Timber Found in Westminster*. Greenpeace admin's blog posted by admin on 28 September 2006. Online. Available at: www.greenpeace.org.uk/blog/forests/yet-more-illegal-rainforest-timber-found-in-westminster (accessed 29 October 2007).

Haymills (2006) *Adnams Distribution Centre, Southwold, Suffolk*. Online. Available at: www.haymills.com/project_print.php?id=51 (accessed 29 October 2007).

Heinberg, R. (2003) *The Party's Over Oil War and the Fate of Industrial Societies*, Forest Row Sussex: Clairview Books.

Heinberg, R. (2004) *Power Down Options and Actions for a Post-Carbon World*, Forest Row Sussex: Clairview Books.

Hemcore (2007) www.hemcore.co.uk (accessed 24 June 2008).

House of Commons (2004) *Housing: Building a Sustainable Future*, House of Commons Environmental Audit Committee, HC 135-l, p. 52. Online. Available at: http:// www.publications.parliament.uk/pa/cm200405/cmselect/cmenvaud/135/ 135.pdf (accessed 29 October 2007).

Lammas (2007) http://www.lammas.org.uk/ (accessed 24 June 2008).

Land for People (2007) http://www.landforpeople.co.uk/ (accessed 24 June 2008).

Law, B. (2006) *The Woodland House*, East Meon: Permaculture Publications.

Lhoist (2007) www.lhoist.co.uk (accessed 24 June 2008).

Lime Technology (2007) *Hemcrete*. Online. Available at: http://www.limetechnology. co.uk/pages/hemcrete.php (accessed 29 October 2007).

Living Villages (2007) *The Wintles*. Online. Available at: http://www.living village.com/locations_wintles.html (accessed 29 October 2007).

NBT (2007) www.natural-buildings.co.uk (accessed 24 June 2008).

NNFCC (2007) www.nnfcc.co.uk (accessed 24 June 2008).

ODPM (2004) www.odpm.gov.uk/pns/displaypn.cgi?pn_id=2004_0320.

Papanek, V. (1984) *Design for the Real World*, London: Thames and Hudson.

Pearson, D. (1989) *The Natural House Book*, London: Gaia Books.

PEFC (2007) www.pefc.org/internet/html (accessed 24 June 2008).

Rosen, N. (2007) *How to Live off the Grid: Journeys Outside the System*, London: Doubleday.

SEDA (2007) http://www.seda2.org/ (accessed 24 June 2008).

Usable Buildings Trust (2007) www.usablebuildings.co.uk (accessed 24 June 2008).

USBG (2007) http://www.usgbc.org/ (accessed 24 June 2008).

Wolseley (2006) *Wolseley to Build a National Showcase for Sustainable Building Products*. Press release 28 September. Online. Available at: www.wolseley.com/ MediaCentre/media_169.asp (accessed 29 October 2007).

Woolley, T. (1998) Green architecture: man myth or magic? *Environments by Design*, 2 (2) Autumn Kingston University, 127–138.

Woolley, T. (2005) Balanced Value: a review and critique of sustainability assessment methods, in Yang *et al.* (eds), *Smart and Sustainable Built Environments*, Chapter 18, Oxford: Blackwell.

Woolley, T. (2006a) *Natural Building*, Ramsbury, Marlborough: Crowood Press.

Woolley, T. (2006b) The blind leading the blind, *Public Sector and Local Government Magazine*, October.

Woolley, T. and Kimmins, S. (2000) *Green Building Handbook*, Vol. 2, London: Spon Press.

Woolley, T. and Caleyron, N. (2002) Overcoming the barriers to the greater development and use of environmentally friendly construction materials, *3rd International Conference on Sustainable Building*, Oslo. Also available at: http://greenbuilding.ca/iisbe/gbpn/documents/policies/research/barriers_sust_mat_ use_woolley.pdf (accessed 29 October 2007).

Woolley, T., Kimmins, S., Harrison, P. and Harrison, R. (1997) *Green Building Handbook*, Vol. 1, London: Spon Press.

WWF (2007) www.wwf.org.uk/oneplanet/ophome.asp (accessed 24 June 2008).

Yu, C. and Crump, D. (2007) Indoor air quality criteria for homes for assessing 'health and well being' Wang, B. Song, J. and Hideki, A. (eds), *Advances in Eco-materials: The Proceedings of the Eighth International Conference of Eco-Materials (ICEM8)*, 9–11 July 1977, Brunel University, UK.

Chapter 11

Corporate social responsibility and the UK housebuilding industry

David Adams, Sarah Payne and Craig Watkins

Introduction

One of the most obvious tests of the construction industry's commitment to corporate social responsibility concerns the development of new homes for owner occupation, in which business success is marked by explicit engagement with the varied and changing agendas of government and the home-buying public. The emergence of a specialist housebuilding sector can be seen as a distinctive feature of the UK construction industry that reflects both the rapid growth of owner occupation in the post-war years and the development of specialist financial systems to facilitate wider access to mortgage funding. Although rising real incomes have encouraged consumers to purchase housing increasingly for investment as well as use, few homeowners commission an architect to design a bespoke product and a builder to implement that individual design.

Instead, mass production characterises UK housebuilding in a manner similar to the production of many other consumer durables. As a result, those who own and occupy residential property in the long term find themselves subservient in key decisions on its design and construction to housebuilders, whose short-term interest lies in rapid completion and departure for the next building site. This separation between the use and investment interest on the one hand and the development interest on the other has caused long-standing tension between the housebuilding industry and its customers. In this sense, the emergence of corporate social responsibility represents not a wholly new challenge for housebuilding, but rather a broadening of the older challenge of tackling short-termism in an industry whose products are intended to last significantly longer than most other consumer durables.

A useful insight into the extent to which particular businesses and industries are prepared to think more broadly is apparent in their engagement with Business in the Community (BITC). Since its foundation in 1982, BITC has grown to a movement of over 700 member companies, including 75 of those listed in the FTSE 100. Although its members include the United Kingdom's top two commercial property development companies (British

Land and Land Securities) along with some important names in construction and property consultancy, membership among the leading housebuilders is sparse, with only half of the UK's top ten housebuilders opting for it. In addition, it is conspicuous that one of BITC's current initiatives, that of helping to alleviate the crisis in rural housing, lists ten exemplary case studies involving housing associations, community land trusts or private landowners, but not one that involves a major housebuilder. Is the housebuilding industry simply slow off the mark in recognising its wider responsibilities or is there something about the business of housebuilding that makes it narrowly focused?

At the time when BITC was founded, in response to the urban unrest of the early 1980s, corporate social responsibility (CSR) meant little more than sponsoring sporting events or helping to clean up despoiled beaches. With the rise of the sustainable development agenda from the late 1980s and the emergence of new forms of governance in the 1990s, the concept of CSR has since changed significantly. In a series of reports (DTI 2001, 2002, 2004), the Labour Government redefined and actively promoted the concept of CSR as part of its broader agenda to reconstitute relations between the state and the market. In a whole series of policy arenas, the Government sought to overturn the long-standing adversarial relationship between the state and the market and instead draw the private sector into policy formulation and implementation by means of partnership to achieve shared goals and common purposes. CSR thus became 'essentially about behaviours that go beyond basic legal compliance' (DTI 2002: 7) through which UK businesses take account of their economic, social and environmental impacts to help address key sustainable development challenges (DTI 2004). Although the Government is committed to the creation of fiscal and regulatory frameworks that encourage rather than hinder CSR, it recognises both the voluntary nature of CSR and the crucial need to encourage the corporate sector to 'go the extra mile' beyond that required by regulation, if development is ever to become truly sustainable.

Nevertheless, the Government's appeal for greater CSR, or more responsible business behaviour as it is sometimes termed, is not directed merely to the goodness of executives' hearts but rather to their enlightened self-interest. Cannily, for example, it argues that close attention to CSR is an indispensable component of modern business management along with actions to 'build brand value, foster customer loyalty, motivate their staff and contribute to a good reputation among a wide variety of stakeholders' (DTI 2002: 2). Moreover, as the *Financial Times* (2004) commented:

> Responsible business behaviour has been equated with risk management, placing it high on boardroom agendas. Fear of lawsuits or legislation has hastened the adoption of social, environmental and ethical programmes. However, some companies are seeing not just risk but opportunity in making a positive contribution to society.

Such sentiments, of course, call for a particular type of management and a particular type of business organisation and are grounded in modern management theory and literature. For example, in setting out the business case for CSR, the Government argues: 'The key was to look at CSR as an investment in a strategic asset or distinctive capability, rather than as an expense' (BERR 2008).

Although we take as a starting point for our analysis in this chapter, the UK Government's concerted attempt to define CSR as a link between the implementation of its sustainable development agenda and its particular concept of multi-level and multi-agency governance, it is important to pause and acknowledge other perspectives and initiatives. A simpler definition of CSR, which allows it to stand independently of government policy, suggests that: 'CSR is about how companies manage the business process to produce an overall positive impact on society' (Baker 2005: 1). Moreover, the widespread interest in CSR across the industrialised world by non-governmental organisations as well as governments is shown, for example, by *Business Ethics Online*,[1] the US magazine of corporate responsibility, in the international emphasis of the Corporate Social Responsibility Forum[2] and in the conferences organised by the CSR campaign.[3] The latter once made the rather worrying claim

> like the Olympic Games, a CSR torch is carrying the message from conference to conference, building up to the CSR Business Olympics in 2005. This major event will celebrate positive business activities and best practice throughout Europe and reward achievements in the field of CSR. Like the Olympic movement, it will be an ongoing event recurring every 4 years.
>
> (www.csrcampaign.org)

Now that CSR is therefore big business (and big for government), is it big for one important sector, whose activities and products have a direct impact on the sustainability of development, namely, the United Kingdom's speculative housebuilding industry?

In this chapter, we suggest that the rapid dissemination of CSR as an embodiment of a new business ethic presents the housebuilding industry with a particular challenge to its corporate identity and relationships. We explore how the raised expectations for more responsible business behaviour, which are now evident in government and among the public at large, make particular demands on a housebuilding industry that has experienced an ever tighter regulatory environment in recent years. We trace the immediate efforts of the largest housebuilders to respond to the CSR agenda and question how far the adoption of corporate strategies and policies at head office is likely to make any local impact in an industry where production is notoriously mobile and decentralised. This leads us to ask important questions about the capacity of the industry to adapt to new agendas in its business and policy environments and, specifically, about whether the varied responses to CSR among different housebuilders might reflect their future potential. Since this chapter is built primarily upon analysis of published company reports supplemented by limited interview work with housebuilders and their representative bodies, our findings are necessarily tentative and require to be tested by more detailed research at a local level. It is therefore appropriate that we conclude by outlining how a research agenda might be constructed to test the commitment of the housebuilding industry to CSR at the local level.

The chapter is presented in five sections. In the next section, we outline how UK housebuilding is structured and organised. We draw particular attention to the highly competitive, and indeed turbulent, structure of UK housebuilding and to the nomadic and still rudimentary nature of its craft-based production process, both of which make the industry highly focused on the short term. In the third section, we present an analysis of how the United Kingdom's top ten housebuilders in 2003 had responded to the Government's CSR agenda, recognising of course that the articulation of strategies and policies at the top of any business is meaningless, unless systems are in place to ensure their effective dissemination and implementation throughout the organisation. Our fourth section therefore concentrates on what we perceive to be the main cultural and institutional barriers to the widespread adoption of CSR in the housebuilding industry. To some extent, this section draws on previous work (Adams 2004; Adams and Watkins 2002), but it is also informed by our direct discussions with the industry. In our final section, we suggest that the UK housebuilding industry, with a few notable exceptions, is yet to fully embrace the principles of CSR in the development of housing. This raises important questions about the adaptive capacity of the industry, which are taken forward in our suggested research agenda.

The structure and organisation of UK housebuilding

At first glance, the UK housebuilding industry would seem to be well placed to respond to the CSR agenda. Although there are about 18,000 housebuilders registered with the National House Building Council (NHBC), speculative production of new homes in the United Kingdom is dominated by a small number of major companies. In 2003, the three main builders (Barratt, Wimpey and Persimmon) completed over 12,000 units each, while a further eight companies built over 2,000 each and 12 more over 1,000 each (Wellings 2004). This pattern reflects the long-term trend towards increased concentration of housebuilding capital, at least in terms of unit output (Nicol and Hooper 1999). For example, in 1980, there were 24 housebuilders completing more than 500 units who collectively claimed a 39% share of the market, but by 2000, the number of such companies had risen to 43 and their market share to 71%. Despite the existence of numerous small builders operating at a local level, the concentration of capital in housebuilding means relatively few companies would need to be persuaded of the merits of CSR to achieve a disproportionate impact on the sector's practices.

The sustained economic growth and rapid house price inflation of recent years have made the sector highly profitable, reflecting its financial reliance on the fortunes of the housing market and development cycle. As Fraser-Andrews (2004: 11) comments,

> The housebuilders have been enjoying supernormal returns in the last three years as a result of stock profits with operating margin and return on capital employed not far off levels recorded in the 1980s boom at 18% and 30% respectively in 2003.

Of course, past experience of market collapse and company bankruptcy may produce caution about future prospects, but with the industry's concentrated structure matched by favourable finances, it might appear that speculative housebuilding would be reasonably well placed to give CSR serious consideration. Nevertheless, more detailed scrutiny of the sector's organisation, processes and culture suggests four main reasons why CSR might find it hard to take root in speculative housebuilding.

Potential barriers to corporate social responsibility in housebuilding

First, owing to the very distinctiveness of its processes and products, the industry operates more at the margins than at the centre of business networks and culture where the CSR debate currently appears most influential. Wellings (2006), who has traced the rise of national housebuilders over the past 75 years or so, argues convincingly that companies do best when they focus solely on housebuilding, not when they are combined with general contractors or merged into industrial conglomerates. The emphasis Wellings places on the entrepreneurial nature of housebuilding and the vital role played by dominant individuals in successful housebuilding companies highlights how the sector is largely managed and directed by those whose business experience is entirely within housebuilding, rather than by those who have honed their management skills across different sectors. Although there is an established pattern of middle management moving among housebuilders as their careers develop, there is very little evidence of generic management expertise at senior level bringing modern business attitudes and cultures into housebuilding from, say, high-street retailing or consumer goods production.

If the sector's leaders could demonstrate exposure to generic management concepts through their education, the industry's relative isolation might be less significant. In this context, however, Wellings' (2005: 96) research is particularly revealing.

Although many of the successful housebuilding entrepreneurs use a related trade or professional qualification as a means of entry into the industry, most of these skills are inborn or honed in the real world. Few have been university educated let alone had formal business training.

The very practical emphasis of management training in the sector may well make housebuilders at best slow to adopt and at worst hostile to such modern ideas as adaptive capacity, core competencies and, indeed, corporate social responsibility. Our interview with a representative body of the housebuilding industry revealed the following interesting comment in response to an explanation of the redefinition and broadening of CSR:

> Well I have to say that the definition of corporate social responsibility, which you're using here, is not one that I'm familiar with. I mean when the term is used in relation to industry generally, we do get involved, you know, with organisations like Scottish Business in the Community and so forth. It tends to mean something different; it's all about sponsoring the local football team, building a community centre as a gesture to your local community and companies playing their part within the local community. I mean this may be something that we've missed, but I never really heard it considered in this context.

Second, the industry's corporate structure, although dominated by major housebuilders, is highly turbulent with an almost constant pattern of merger and takeover (Adams and Watkins 2002). Thus, only five of the top housebuilders of the early 1980s retained a top-ten position by the end of the decade and only two of those remained in the top ten by 2001 (Wellings 2005). Most of those that fell by the wayside during these two decades (including such well-known names as Comben, Ideal and William Leech) succumbed to takeover by competitors. Another company, Beazer, which was still a regional operation in the early 1980s, rose rapidly to national prominence to become the third largest UK housebuilder by 2000, only to run into financial difficulties itself and be taken over by its rival, Persimmon, in 2001. Such corporate turbulence remains endemic in housebuilding, with the takeover of David Wilson Homes by Barratt and the merger between Taylor Woodrow and Wimpey, both announced in 2007, merely being the latest examples of a well-embedded feature of the industry.[4] As a result of this unsettled corporate environment, most successful housebuilders are constantly on the lookout for competitors whom they might take over, whereas those who run into even temporary financial difficulties must battle hard to save their respective companies' independence. This is hardly the best starting point from which to prioritise broader social responsibility or to visualise long-run corporate benefits from more ethical behaviour when the short run can be so perilous and uncertain.

Third, speculative housebuilding is an inherently transient process in which the dispersed and localised nature of production bestows a certain independence of mind at the local level and limits the extent to which head office can fashion outlooks, uphold standards, or determine behaviour where they matter most. The three basic skills required to establish development feasibility in housebuilding, namely, controlling ownership through land acquisition, securing planning permission and other public consents, and creating attractive marketing images to entice customers, must be delivered not by head office but at the local level (and, indeed, at the site level). Apart from those locations or times of poor housing demand, the viability and profitability of speculative residential development has thus depended on finding land at the right price, gaining planning permission and marketing the completed product, the first two of which are essentially localised activities (Adams and Watkins 2002).

Since shortages of available building land, particularly at pressured times and locations, most constrain the housing-development process, the industry is inherently 'land-focused' rather than 'customer-focused' or 'community-focused'. As Barker (2004: 106) commented:

When land is in relatively scarce supply, fewer permissioned sites mean that there will be fewer competing housebuilders in any one area. This can reduce consumer choice. In such situations, competition focuses on land. Once land is secured, competitive pressures are reduced: to a large extent housebuilders can 'sell anything'.

In such a business environment, some housebuilders may be reluctant to engage in CSR if they consider it has nothing extra to contribute to short-term profitability or long-term business health. Of course, as we subsequently indicate, far-reaching changes are under way in the policy environment for housebuilding, as a result of which companies that demonstrate a broader outlook may be better placed to flourish in future. Specifically, while the three basic development skills have been polished and sharpened primarily through the construction of mass, standardised greenfield sites, they now need to be refashioned to a brownfield environment, where fresh approaches to land acquisition, planning and marketing are required.

Fourth, housebuilding is an essentially conservative industry in which strong reliance is placed on tried and tested methods and in which new ideas and concepts are treated with a certain caution. Sir John Egan, who has previously headed the Government's construction task force and its enquiry into the skills needed for the development of sustainable communities, commented, when giving evidence to the House of Commons Environmental Audit Committee (2005: para 152), that the housebuilders 'that have survived have pared themselves down to a relatively comfortable life but that is not the way you stimulate innovation. These are comfortable people doing a comfortable job.' Interestingly, an equally derogatory comment was made by a representative of the industry whom we interviewed:

> Housebuilders, if they haven't told you already, will tell you that they're all simple men, simple house builders and they'll abide by whatever is mandatory, anything that is statutory and some of what isn't statutory, but they do need to be helped.

This attitude of 'comfortable survival' might help explain why, according to Ball (1999: 9), 'British housebuilding has an exceptionally poor record of innovation in design and production methods'. Barlow (1999: 23) too chides the industry as being 'notoriously slow to innovate'. Empirical studies of housebuilding practice tend to support such claims. Gibb *et al.* (1995), in their study of Scottish housebuilders, found that the majority questioned could not identify a single technical innovation in the previous five years, while Barlow and Bhatti (1997) undertook a survey of over one hundred firms involved in the UK industry and found that fewer than 10% were developing new designs or trying out new technologies.

Ball (1999) further suggests that consumer conservatism constrains innovation in terms of both output and input markets. The fact that house purchasers are conservative in their tastes in house styles and worried about future resale values reinforces the producers' tendency towards standardisation and deters builders from introducing innovative or exciting house designs. The conservatism of consumers on design is also reflected in the reluctance of lenders to offer mortgages on non-traditional forms of construction, while the uncertainty introduced by cycles in the housing market persuades builders to organise in ways that maximise flexibility. These include widespread subcontracting, traditional building techniques and low amounts of fixed capital on building sites. Such structural constraints encourage firms to adopt low-cost strategies (apart from land acquisition) that require minimum sophistication and forecasting ability, while discouraging innovation.

According to the European Commission (2001: 366), corporate social responsibility represents 'a concept whereby companies integrate social and environmental concerns in their business operations and in their interaction with their stakeholders on voluntary basis'. In summary, however, despite the dominance of the industry by large companies with the resources actively to engage in corporate social responsibility, we believe that the relative isolation of the sector, its corporate turbulence, the essentially localised nature of its production process and its inherent culture of conservatism are together likely to slow down the widespread diffusion of CSR in housebuilding. Indeed, if CSR is to provide the means by which speculative housebuilding begins to engage more fully with the concept of sustainable development, fundamental changes are likely to be required in the industry's organisation, processes and culture.

The policy challenge for housebuilders

Two recent government documents in England, the Barker Review and the Sustainable Communities Plan, highlight the challenges now facing the industry. Kate Barker's review of housing supply proposes that an additional 120,000 dwellings per annum should be constructed, almost doubling completions, rate in the recent past (Barker 2004). This is an ambitious target given that housing production is at a record low post-war level. Although the current low levels of production are undoubtedly partly attributable to the constraints imposed by the planning system, some responsibility must also lie with the strategies of firms and the low production capacity of the industry. For instance, it is clear that builders often control the rate of housing production to avoid deflating the prices paid for their product. They can do this because there is little competition for customers, and they retain a monopolistic control of their local markets (Crook *et al.* 2005).

The Sustainable Communities Plan seeks to enhance housing supply and affordability in the south east of England and to renew markets in parts of the north and midlands (ODPM 2003). Interestingly, it explicitly places the onus on the housebuilding industry to provide three of the key requirements for a 'sustainable community'. These features are development of sufficient size, scale and density and of the right layout to support basic amenities in the neighbourhood and minimise the use of resources; buildings – both individually and collectively – that can meet different needs over time, and that minimise the use of resources; and a well-integrated mix of decent homes of different types and tenures to support a range of household sizes, ages and incomes.

Both Barker and the Sustainable Communities Plan highlight their expectations of a strategic role for the housebuilding industry in contributing to social and economic welfare, although they are largely unconcerned with environmental issues. In this context, the cross-party House of Commons Environmental Audit Committee (2005: para 156) noted that: 'While we are encouraged by the attitude of some housebuilding companies, the majority are nowhere near achieving the kind of record with regard to environmental performance we would consider acceptable.' The Committee called for housing policy, including the outcomes of Barker and the Sustainable Communities Plan, to be explicitly set against environmental limits and suggested a major role for housebuilders in helping to achieve sustainability targets.

The combined effect of these government documents is to identify the housebuilding industry as the key means of delivering several different (social, economic and environmental) policy objectives that are potentially in conflict. As a consequence, housebuilders now find themselves operating in a complex and dynamic regulatory environment in which they are expected to behave in a far more socially responsible manner than in decades past. Despite the structural and institutional barriers to the potential adoption of

CSR identified in this section, there is emerging evidence that some of the UK's largest housebuilders are beginning to develop clear CSR policies and strategies. We therefore review a selection of these in the next section, while identifying some practical difficulties to their implementation in the following section.

The housebuilding industry's public response to CSR

This section reports our detailed analysis of the CSR reports produced by the top ten housebuilders in 2003, listed by Wellings (2004) according to unit completions. These were, in alphabetical order: Barratt, Bellway, Berkeley, Miller, Persimmon, Redrow, Taylor Woodrow, Westbury, Wilson Bowden and Wimpey.

Table 11.1 Key delivery objectives for CSR in the UK housebuilding industry.

Delivery objective	Features
1. 'Mainstreaming' the implementation of CSR	• Environmental Management Strategy • Key Performance Indicators • Transparency • CSR management • FTSE4Good • Business in the Community
2. Altering the construction process	• Modern methods of construction • Efficiency • Recycling • Waste • Innovation • Procurement
3. Changing the housing product	• Affordability • Brownfield development • EcoHomes • SAP ratings • Environmental efficiency
4. Engaging stakeholders and the community	• Effective engagement • Customer relations • Understanding • Initiatives and donations
5. Modifying workforce practice	• Training and development • Health and safety
6. Influencing supplier relations	• Supply chain management • Training and development • Innovation

The study revealed that all ten companies recognised the relevance of CSR in the land development process, with each having dedicated sections on their websites. However, only seven housebuilders published a specific CSR report. Of the remaining three, Berkeley produced a sustainability report, Miller's website accepted the need for 'corporate and social responsibility' without publishing any additional material, while Westbury acknowledged CSR but published only a broad environmental policy.

To structure our analysis, we developed a list of six key delivery objectives for CSR as a means of testing the top ten housebuilders' stated intentions against the UK Government's published approach to CSR (DTI 2001, 2002, 2004). These six objectives are shown in Table 11.1. Subsequent discussion in this chapter concentrates on the first four of these.

'Mainstreaming' the implementation of corporate social responsibility

This first objective is concerned with 'mainstreaming' CSR into companies' modus operandi through the implementation of a CSR management system that includes, for example, the publication of key performance indicators. The typical system adopted by the housebuilders is shown in Figure 11.1.

All ten companies studied had members of the main board sitting on a CSR committee. One company, Berkeley, had named a main board director with specific responsibility for sustainability issues. This company's sustainability working group met on a quarterly basis to review policies, progress and implementation issues. The group was made up of senior executives from each of the main divisions and the minutes were presented to the main board.

Effective communication of the CSR management strategy between the main board and each regional division is fundamental to implementation. Berkeley was again the only builder that explicitly demonstrated this. In Berkeley's case, each division was required to set sustainability objectives and to report to the main board quarterly on its progress, including performance against the company's sustainability key performance indicators (KPIs). Each division was also responsible for ensuring that sustainability issues were integrated into its own management systems (Berkeley Group 2004). No other housebuilder in the top ten in 2003 required each division to set sustainability or CSR targets and measure performance.

Only half of the top ten housebuilders (Barratt, Berkeley, Persimmon, Taylor Woodrow and Wimpey) had developed KPIs and tabulated progress made. The rest were content to demonstrate their 'potential' performance through objectives and targets for the coming years. Eight out of the ten housebuilders produced such targets for 2005–2006. Wilson Bowden Homes produced objectives for the period from 2004 to 2007 together with yearly targets, reporting progress in an annual CSR report.

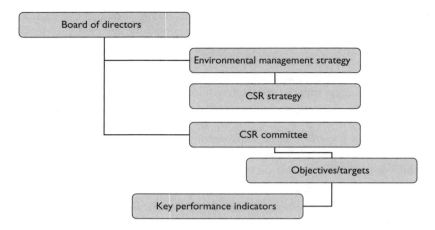

Figure 11.1 A typical CSR management system in housebuilding.

Altering the construction process

The endeavours of housebuilders to minimise the negative impacts of the construction process by taking necessary measures to recycle and reduce waste, increase resource efficiency and improve procurement processes is central to delivering the second objective. Key performance indicators formed a large part of housebuilders' transparent commitment to product and process responsibility. For the construction process, housebuilders had developed KPIs in such areas as brownfield utilisation, safety training, number of employees involved in the SAYE scheme, charitable donations, and reductions in the amount of yard skips removed, CO_2 emissions and water consumption.

Nine out of the ten housebuilders had a waste minimisation policy. Although the majority wished to increase recycling, the only specific type of recycling for which housebuilders were achieving or setting targets in 2003 involved plasterboard, with five companies able to demonstrate action to recycle this material.

Only five housebuilders appeared to recognise supply chain management as a tool for increasing control over specifications and standards in the construction process. As below, Wilson Bowden, for example, had set up a 'Supply Chain Environmental Forum' and developed an 'Environmental Supply Chain Improvement Plan', the aim of which was to improve environmental awareness by the improved communication of various sustainable-construction targets throughout the supply chain (Wilson Bowden plc 2004). However, no KPIs were evident to demonstrate progress made towards the goals in the improvement plan.

Wilson Bowden's 'Environmental Supply Chain Improvement Plan' (Wilson Bowden plc 2004):

- use green specification;
- increase the volume of certified/sustainable timber;
- reduce CO_2 emissions;
- reduce fuel consumption and vehicle emissions;
- reduce the use of volatile organic consumption compound emissions;
- waste/biodiversity issues;
- increase recycling opportunities and volumes;
- reduce packaging;
- reduce gas and electricity consumption;
- use 100% renewable energy;
- reduce water consumption;
- improve material life cycle;
- improve pollution risk management.

Only two of the top ten housebuilders claimed to conduct risk appraisals for their development projects prior to the start of construction. Bellway aimed to test all large sites and every urban development over 0.5 hectares through an environmental assessment or sustainability audit, while Berkley claimed to conduct a risk appraisal for each project. Redrow had produced a 'quality control file' for each property built and conducted a detailed inspection prior to release, while Wimpey had developed a 'construction code of practice' whose four objectives were safety, environment, respect and communication (George Wimpey plc 2004).

Changing the housing product

Comparatively little attention was paid to the sustainability of the *house* when compared to that of the production process. As Table 11.2 shows, only five out of the top ten housebuilders published Standard Assessment Procedure ratings for houses built in 2004, although the majority of those met the Government's 60% brownfield development target. Key performance indicators were less common for product than process; relevant examples included EcoHomes ratings and increases in dual-flush toilets and customer recommendation.

Very little mention was given in the reports to EcoHomes standards, with only Bellway Homes and Taylor Woodrow aspiring to improve their rating 'on selected developments at their discretion' through performance targets and sustainability objectives. Wilson Bowden, as part of its sustainability objectives for 2003–2007, intended to undertake and achieve EcoHomes good/very good rating for 'appropriate affordable housing projects' only (Wilson Bowden plc 2004). Wimpey, having built 1,453 out of 12,232 homes in 2004 to EcoHomes standards, intended to assess the 'practical and

Table 11.2 Top ten housebuilders' SAP ratings and brownfield completions in 2004.

Housebuilder	Average SAP rating	% Brownfield completions
Barratt	Undisclosed	80
Bellway	95.6	75
Berkeley	77.6	95
Miller	Undisclosed	65
Persimmon	92	52
Redrow	Undisclosed	70
Taylor Woodrow	87–105	64
Westbury	Undisclosed	Undisclosed
Wilson Bowden	Undisclosed	44 (2003)
Wimpey	94	68

commercial viability of using the standard more widely' (George Wimpey plc 2004: 15). Although Berkeley did not then employ the EcoHomes assessment on its developments as standard practice, its sustainability targets in 2004–2005 included a requirement to gather data on the performance of all dwellings using the EcoHomes methodology, with a view to setting performance standards.

Although housebuilders were very communicative in their recognition of the *need* to reduce CO_2 emissions and water and energy use, and to build more affordable and environmentally friendly products, little attention was paid to how they intended to *do* this. Wilson Bowden (2004: 18), for example, stated that:

> We continue to embrace the thinking behind Modern Methods of Construction...In designing our properties, we strive to incorporate materials and techniques which...exceed the Building Regulations in order to reduce impact on the environment, (and) improve the quality of our product.

However, the company did not explain what MMC (modern method of construction) it intended to use, or how such methods would reduce environmental impacts.

On a more positive note, Laing Homes (which was subsequently incorporated within Wimpey) received a 'very good' EcoHomes rating for all 525 homes in a planned development on former MOD land in Guildford. The company had incorporated eco-friendly additions such as solar panels, recycling facilities, rainwater butts, and bicycle storage, while 6.8 hectares of the development's 23 hectares have been laid out as public green space, with preserved woodland (George Wimpey plc 2004). In another example, Redrow, in a joint project with Nottingham University, proposed to construct 'a test house', enabling long-term testing and development of building components and systems including thermal and acoustic insulation methods, hydrogen fuel cell energy, combined heat and power units and wireless technology. Whilst this demonstrated willingness to experiment with MMC, it still remained a meagre exception rather than the rule.

Engaging stakeholders and the community

Developing partnerships with stakeholders and communicating more effectively are key principles of CSR. The importance of stakeholder engagement was recognised by most builders. Taylor Woodrow (2004: 6) considered stakeholder engagement as a 'material CSR issue', Barratt (2004: 25) saw it as 'an essential ingredient of our CSR performance', whilst Wimpey (2004: 6) reported that it would 'strive to be a responsive company and enter dialogue with our stakeholders on a wide range of business and CSR issues'. Overall, all but one of the top ten housebuilders (Bellway) made mention of the need to engage effectively with stakeholders and saw this as a core responsibility of housebuilding. However, only seven out of the ten came up with a list of key stakeholders in the housebuilding process, and none developed targets or objectives. Barratt (2004) identified typical key stakeholders diagrammatically to include national and local governments, non-government organisations, investors, communities, customers, suppliers and contractors and employees.

An understanding of sustainable communities and wider community needs is another important principle behind delivering the fourth objective in Table 11.1. Community engagement and involvement appeared to be the most developed aspect of the CSR reports studied, except for Barratt's. Whether it was sponsoring the Scottish World Champion cyclist, Chris Hoy, (Miller) or making sure schoolchildren were aware of the dangers of building sites (Westbury), for example, housebuilders appeared to have some commitment to supporting local communities financially.

However, only half of the ten housebuilders investigated (Bellway, Berkeley, Persimmon, Redrow and Wimpey) explicitly connected issues of community involvement to that of sustainability. The bullet points below show how one housebuilder tried to achieve this connection in one important London development.

- consultation of over 4,000 households;
- newsletter distribution;
- local exhibitions and feedback;
- local school/resident consultation on layout of the new park;
- on-site information centre – plans and models;
- Neighbourhood Liaison Officer;
- regular community newsletter.

The other housebuilders largely focused on charitable events and sponsorship, and entirely missed the point of creating 'sustainable communities'.

Summary of CSR reports analysed

Whilst the ambitious commitment made by the top ten housebuilders to corporate social responsibility was clear through the publication of CSR reports, the quality of those reports and the material discussed varied. Each report was structured differently, and in some cases, confusingly. All the companies demonstrated some commitment to 'mainstreaming' CSR in their management systems, but only Berkeley demonstrated real permeability of this CSR management system at all levels throughout the company.

The housebuilders' approach to the product and the production process was disappointing, with little mention made of the sustainability and efficiency of the house. The perceived importance of stakeholder engagement was evident throughout the CSR reports, although effective engagement seemed limited. Indeed, only a few housebuilders demonstrated commitment to the development of 'sustainable communities', rather focusing on charitable donations and community engagement. In the next section, we therefore try to pinpoint some of the main barriers to the genuine dissemination of CSR in the housebuilding industry.

Prevalent cultures and practices as an impediment to the genuine dissemination of CSR in housebuilding

Despite the structural and organisational constraints to the widespread adoption of CSR in housebuilding, it was apparent that most of the largest companies had at least adopted and published CSR policies or strategies. Those that had not done so explicitly had articulated new operational practices that implicitly pointed to the concept of CSR. Nevertheless, the dispersed and localised nature of production in the sector made it far from inevitable that pronouncements made at the top of companies would have much influence at the bottom, where much of the debate around the sustainability of development takes place. The key research issue is therefore to investigate whether a policy–implementation gap exists in attitudes to CSR across the housebuilding industry and if so, whether it will threaten the delivery of more progressive policies adopted by the boards of the major companies.

Lack of customer care

There is an important connection here with debates about customer care and consumer satisfaction in housebuilding. Several recent surveys have identified high levels of dissatisfaction with the internal quality and layout and external design of new developments (CABE 2004; Leishman and Warren 2005). According to the *Barker Report* (2004: 112),

> Customer satisfaction levels have fallen since 2000, with only 46 per cent of customers saying that they would recommend their housebuilder. The need to improve standards applies right across the industry: of the

nine companies that performed worse than the industry average on this indicator, four – Persimmon, Barratt, Wilson Connolly and Westbury – were among the top ten housebuilders in 2002.

Three BBC TV *Watchdog* reports in 2004 further highlighted the industry's poor record of customer care. The first programme in February 2004 focused on complaints about the quality of drinking water at a newly built Persimmon estate. One month later, the programme turned its attention to over 200 e-mails it had received from dissatisfied customers of Westbury Homes. One viewer had commissioned an independent surveyor to report on the quality of a new Westbury home, who found at least 43 different faults. In November 2004, *Watchdog* reported that it had received over 100 complaints about David Wilson Homes. One buyer who listed 339 separate snagging faults claimed that she could not get the house warm because the walls and ceilings had not been properly insulated and because of gaps behind radiators and around tiles and cracks in the bathroom. Such individual complaints about the quality of newly completed speculative housing are long-standing and led Barker (2004) to call for a strategy from the House Builders Federation to increase the proportion of housebuyers who would recommend their builder to at least 75% by 2007. Failure to deliver such an improvement should, in Barker's view, lead to the Office of Fair Trading conducting a wide-ranging review of whether the market for new housing works well for consumers. Such controversies suggest that the industry might do well to develop a better sense of responsibility to its own customers before it embarks on the greater challenge of demonstrating its new CSR credentials.

Of course, poor customer satisfaction is never deliberately planned by top management in the industry, but just happens and seems to happen with enduring regularity. Indeed, it is likely that shortages of skilled labour, difficulties of subcontracting (which is used extensively in housebuilding) and poor on-site supervision all combine to frustrate the best efforts of the industry to deliver high-quality products of which the producer and consumer might be equally proud. This failure to deliver what is planned and intended by the industry can thus be attributed partly to failures in the design and operation of the relevant systems and partly to ingrained habits, attitudes and cultures at the local level which take poor quality as a normal component of day-to-day experience. Could the same happen to undermine management's best intentions to turn housebuilding into an industry resplendent with CSR awards?

Confrontation with the public sector

If a sense of corporate social responsibility is indeed to pervade all through the housebuilding industry within five or ten years at the most, a fundamental change will be required in its institutional culture at the local level. There has long been evidence, going as far back as the land availability studies of the early 1980s, of the housebuilding industry at a national, and, to a lesser extent, regional level working constructively with governments across a range of mutual interests, including land release, and health and safety. The widespread adoption of CSR throughout the industry will require this sense of partnership to permeate all the way to the lowest levels of operation, thus ensuring that a practical difference is made in the social, economic and environmental impact of housebuilding at the local level.

Unfortunately, however, relations between housebuilders and local authorities are too often characterised by confrontation and lack of trust, especially over land release. Such lack of cooperation comes to a head in the intense planning battles fought out at public inquiries where inspectors have the difficult task of mediating between the demands of builders for a greater supply of land and those of local action groups who wish to preserve valued open countryside. In one sense, it is possible to agree with Barker that the planning system needs to take greater cognisance of the market pressures upon housebuilders and deliver a greater supply of land for development. In another sense, however, as Barker recognises, housebuilders themselves need to take a more rounded view of the impact and potential contribution of their development to local communities, if their proposals are ever to prove more generally acceptable.

Unimaginative and repetitive design

One interesting aspect of this debate concerns the form and design of new speculative housing, which has, until recently, highlighted the tendency among housebuilders to be unimaginative in method and unsophisticated in outlook. Most builders have devised and use standard house types, which contrast with customised one-off designs for specific sites, and they are reproduced in a repetitive unmodified way in a range of different locations (Nicol and Hooper 1999). Such frequent standardisation of product derives from the intense pressures within the industry to maintain profitability by cost minimisation and price competition. Standard house types facilitate construction by a low-skilled workforce, enable central purchasing of components and limit design costs both directly and through blanket building-control approval. Moreover, by using tried and tested products, housebuilders are able to reduce risks by more accurate cost-forecasting when they bid for land and by reliance on designs known to have sold well in the past.

Corporate social responsibility, however, might require housebuilders to take a more flexible approach to development and, instead of standardising, tailor products towards the particular regional or local vernacular. Nonetheless, as we saw in the section, 'The housebuilding industry's public response to CSR', most of the leading housebuilders are yet to take CSR seriously when it comes to their product, although there is evidence that some are beginning to do so in respect of their production processes. Unfortunately, most housebuilders are interested in better design only if it produces higher sales, achieves a higher price, wins planning approval more speedily or contributes to marketing strategies. According to Carmona (1999), this is because house purchasers are primarily driven by cost, value for money, functionality and size of the property, and only then by intrinsic design. In design terms, housebuilders' main incentive is to improve the 'kerb appeal' of their products (Goodchild and Karn 1997) by attaching mock features to façades to give the pretence of individuality, improving internal amenities and facilities and developing detached and semi-detached houses rather than terraces and flats.

Reluctance to create sustainable communities

In this context, evidence from our interviews highlighted 'the lack of engagement between government and industry' and 'the absence of dialogue'. The ever-increasing role of planning-gain in providing for affordable housing and other community benefits came in for particular criticism from one interviewee:

> The only problem we've got is we don't think we should provide it, you know, pay for it. We didn't create the problem, and we're just another industry. Just in the same way you don't ask a bus company to give free tickets to people who can't afford buses, why should the housing industry give free houses or free land to effectively, the country?

In production terms, the industry might well claim that it is already among the most regulated in the United Kingdom. Alongside the recent emphasis on brownfield rather than greenfield housing and the higher design standards now required by many local authorities, the industry faced much stronger energy-efficiency requirements as a result of changes in the Building Regulations in 2002 (Adams 2004). Where an industry already perceives itself to be highly regulated, it requires a particularly imaginative outlook to see the point of going beyond the basic requirements of the current regulatory régime and voluntarily offering society extra benefits which may have no immediate financial return. An industry representative who was interviewed, although recognising that '[h]ouse building tends to be sensitive politically – very sensitive politically at local level and that translates into national level', concluded that: 'We don't see it necessarily as our responsibility to follow government policy. Why would we? That's quite fundamental'.

On the other hand, it was apparent from the opinion of the senior representative of a regional housebuilder that the industry needs to move beyond its past focus on the house plot or even the housing estate and engage with the new discourse of sustainable communities, if it is ever to break out from its confrontations with, and distrust of, locally elected politicians. What was fascinating from this discussion was that our interviewee revealed that he had previously advised the company chairman to alter the focus as well as the language of his public pronouncements. Specifically, he had suggested to the chairman that

> [t]o be successful, to open doors, will you stop talking about money, please start talking about things that are valuable to people, and that subject could be energy efficiency, it could be jobs, it could be education, it could be how we support the hospital, for instance.

Examples of good practice: how housebuilding cultures can change

This sort of progressive attitude does appear to be on the rise, especially among housebuilders outside the top ten. There are some interesting cases where firms have aligned themselves with public-policy objectives, including affordability and environmental and neighbourhood renewal goals, as part of a revised corporate strategy. Crest Nicholson (the thirteenth largest UK housebuilder in 2004), for instance, has explicitly sought to become associated with the delivery of affordable homes and 'ethical' housebuilding, while Berkeley (which entered the 'top ten' only in 2000) has established a subsidiary, Berkeley Community Villages, that is specifically concerned with the creation of sustainable communities planned to the highest design standards. There are many examples of this sort of niche marketing and the evidence suggests that the strategy has been successful in commercial terms. In early 2005, for example, Crest Nicholson reported a 130% increase in the production of affordable homes and estimated that the urban renewal market was worth more than £30 million (Macalister 2005). Such examples provide support for the DTI's contention that CSR practices can contribute to competitiveness and that 'CSR should be good for long term business success' (DTI 2004: 3).

These activities have also delivered wider social and environmental benefits. There are many prominent contributions to the regeneration agenda including Crosby Homes' (a part of the Berkeley Group) redevelopment of Clarence Docks in Leeds, Bellway's work on the new residential quarter in Littlehampton and Urban Splash's high-profile neighbourhood renewal scheme at New Islington/Cardroom in Manchester. In addition, in the sustainable development arena, Countryside Properties (the sixteenth largest UK housebuilder in 2004) have performed commendably in delivering on their commitment to environmentally sensitive development (House of Commons Environmental Audit Committee 2005). Nevertheless, such strategic responses remain the exceptions. As we suggested earlier, the high level of conservatism that pervades the industry continues to act as a significant constraint on the extent to which CSR activities are genuinely embraced.

Conclusion

To summarise our arguments in this chapter, we suggest that the slow participation of the speculative housebuilding industry in the notion of corporate social responsibility can be traced to the distinctive characteristics of housebuilding and, in particular, to its long-established organisation, processes and culture. The industry remains relatively isolated from mainstream business practices and network, is highly turbulent in its corporate structure, is not particularly well-focused on customers, let alone communities, and does not have a strong record of openness to new ideas and concepts.

Nevertheless, it is clear that, at a national level at least, most of the major housebuilders are beginning to take CSR seriously and many have adopted policies or strategies that indicate the desire to take a much broader view of their business activities in future than they might have done in the past. This view is consistent with two surveys of top housebuilders commissioned by the WWF in 2003 and 2004, as part of its 'One Million Sustainable Homes' campaign. When compared, the evidence in the two reports (WWF 2004, 2005) pointed to a significant improvement in the way housebuilders addressed strategy and risk management and assessed their environmental and social impacts. Interestingly, as the second report commented:

> On the whole, the companies now demonstrate a better understanding of the relevance of sustainability issues to their business. This appears to be driven by two important factors: first, at a corporate level, the increased emphasis on non-financial risks and opportunities, and secondly, at a project level, through increasingly demanding planning requirements.
>
> (WWF 2005: v)

Of course, the real test of any commitments made at corporate level will lie in their impact or otherwise at the local level. This will mark out those companies that are truly prepared to engage in community building rather than mere estate construction, and, in the more regulated environment in which housebuilders now do business, may well suggest to investors which companies have greatest potential for asset growth in the future. It can thus be argued that the inherent business logic that many consider responsible for the rapid growth of corporate social responsibility as a whole applies equally to housebuilding, and that the next decade is likely to see a transformation of the structure and organisation of the industry as it moves closer to the construction sector in particular and to encompass modern management strategy in general.

Adams (2004) draws an important distinction between those housebuilders whose behaviour suggests they are over-reliant on the technologies and practices of the past and those prepared to take the risk and invest in what they consider to be the necessary competencies of the future. Unless there is an unexpected and significant policy shift, recognition and engagement with the concept of corporate social responsibility are likely to be important key determinants of the success of housebuilders in the coming decade. The key difference will be between those who merely espouse CSR strategies and policies on their websites and those who take trouble to permeate the whole of their organisations with the concept of CSR.

This is no mere academic test, but is likely to reach to the heart of future business success in housebuilding. Since CSR is part of a wider corporate responsibility agenda that challenges in bred thinking and behaviour in the speculative housebuilding industry, it will require a revolutionary change in approach if it is to be 'mainstreamed' into all that the industry does. Yet, in a sector whose activities are dependent so much on the prevailing regulatory environment, companies who fail to take CSR and, indeed, the whole sustainability agenda seriously are likely to find themselves at a long-term disadvantage when it comes to securing land and obtaining the necessary approvals for development. CSR is therefore another example of the pressing need upon housebuilders to adapt if they wish to survive. Although most of our analysis in this chapter has concentrated on the top ten housebuilders in 2003, there are signs that those companies immediately below the top ten may be more dynamic and have greater adaptive capacity when it comes to 'mainstreaming' the implementation of CSR. If this proves to be the case, it may contribute further to the sector's turbulence, with more innovative companies gaining from their investment in CSR competencies to outwit some of the more conservative but currently larger builders.

Of course, such a theory can be tested only by detailed empirical work to discover whether, for all their published CSR reports and strategies, any substantial change is taking place in thinking and behaviour within the largest housebuilders where it matters most: at the point of local production. We therefore end this chapter by calling for more thorough research on the local implementation of CSR policies by major housebuilders. Specifically, we would identify three research questions that might usefully be addressed:

1. What training programmes, if any, have those housebuilders with CSR strategies and policies instituted to ensure that the policies permeate the whole of the organisation?
2. To what extent does CSR form an important consideration at the time of land purchase, both in evaluating alternative sites and in considering the financial offer to be made?
3. How far do those employed locally by the housebuilding industry consider they have responsibilities to the sustainability of the communities in which they work and to the management and shareholders of the companies by which they are employed?

In conclusion, it is against the answers to such questions at the local level that the real commitment of the major housebuilders to corporate social responsibility will be tested.

Notes

1. http://www.business-ethics.com/
2. http://www.csrforum.com/
3. http://www.csrcampaign.org/
4. It is, as yet, too early to know whether the emergence of 'super-builders' as represented by the new 'Taylor Wimpey' will create a stronger platform upon which to embed corporate social responsibility within business culture or whether CSR will suffer from corporate cost-cutting within such merged companies.

References

Adams, D. (2004) The changing regulatory environment for speculative housebuilding and the construction of core competencies for brownfield development, *Environment and Planning A*, 36: 601–624.

Adams, D. and Watkins, C. (2002) *Greenfields, Brownfields and Housing Development*, Oxford: Blackwell.

Baker, M. (2005) *Corporate Social Responsibility – What Does it Mean?* Online. Available at: http://www.mallenbaker.net/csr/CSRfiles/definition.html (accessed 31 October 2007).

Ball, M. (1999) Chasing a snail: innovation and housebuilding firms' strategies, *Housing Studies*, 14: 9–22.

Barker, K. (2004) *Review of Housing Supply – Delivering Stability: Securing our Future Housing Needs*, London: HMSO.

Barlow, J. (1999) From craft production to mass customisation: innovation requirements for the UK housebuilding industry, *Housing Studies*, 14: 23–42.

Barlow, J. and Bhatti, M. (1997) Environmental performance as a competitive strategy? British speculative housebuilders in the 1990s, *Planning Practice and Research*, 12: 33–44.

Barratt Developments plc (2004) *Corporate Social Responsibility Report*, Newcastle-upon-Tyne: Barratt Developments plc.

Berkeley Group plc (2004) *Sustainability Report*, Cobham, Surrey: Berkeley Group plc.

Business, Enterprise and Regulatory Reform, Department of (2008) *The Business Case for CSR*, http://www.csr.gov.ut/businesscasecsr.shtml (accessed 28 April 2008).

CABE (2004) *Housing Audit: The Design Quality of New Homes*, London: CABE.

Carmona, M. (1999) Innovation in the control of residential design: what lessons for wider practice? *Town Planning Review*, 70: 501–528.

Crook, T., Henneberry, J., Tait, M., Rowley, S. and Watkins, C. (2005) *Planning and Market Signals: A Review*, Report to the Office of the Deputy Prime Minister, Department of Town and Regional Planning, University of Sheffield.

Department of Trade and Industry (2001) *Business and Society, Developing Corporate Social Responsibility in the UK*, London: DTI.

Department of Trade and Industry (2002) *Business and Society, Corporate Social Responsibility Report 2002*, London: DTI.

Department of Trade and Industry (2004) *Corporate Social Responsibility – A Government Update*, London: DTI.

European Commission (2001) *Green Paper on Promoting a European Framework for Corporate Social Responsibility*, Brussels: European Commission.

Financial Times (2004) Social innovation could pay useful dividends, *Financial Times*, 29 November.

Fraser-Andrews, J. (2004) Stock market outlook, in Wellings, F. (ed.), *Private Housebuilding Annual*, London: The Builder Group 10–12.

George Wimpey plc (2004) *Corporate Social Responsibility Report*, London: George Wimpey plc.

Gibb, K., Munro, M. and McGregor, A. (1995) *The Scottish Housebuilding Industry: Opportunity or Constraint*, Edinburgh: Scottish Homes.

Goodchild, B. and Karn, V. (1997) Standards, quality control and housebuilding in the UK, in Williams, P. (ed.), *Directions in Housing Policy: Towards Sustainable Housing Policies for the UK* London: Paul Chapman Publishing, 156–174.

House of Commons Environmental Audit Committee (2005) *Building a Sustainable Future*, London: HMSO.

Leishman, C. and Warren, F. (2005) Planning for consumers' new-build housing choices, in Adams, D., Watkins, C. and White, M. (eds) *Planning, Public Policy and Property Markets* Oxford: Blackwell, 167–184.

Macalister, T. (2005) Crest Nicholson claims lead in ethical building, *The Guardian*, 27 January.

Nicol, C. and Hooper, A. (1999) Contemporary change and the housebuilding industry: concentration and standardisation in production, *Housing Studies*, 14: 57–76.

Office of the Deputy Prime Minister (2003) *Sustainable Communities: Building for the Future*, London: ODPM.

Taylor Woodrow plc (2004) *Corporate Social Responsibility Report*, Solihull, West Midlands: Taylor Woodrow plc.

Wellings, F. (2004) *Private Housebuilding Annual 2004*, London: The Builder Group.

Wellings, F. (2005) *The Rise of the National Housebuilder: A History of British Housebuilders through the Twentieth Century*, Unpublished PhD thesis, University of Liverpool.

Wellings, F. (2006) *British Housebuilders: History and Analysis*, Oxford: Blackwell.

Wilson Bowden plc (2004) *Corporate Social Responsibility Report*, Leicester: Wilson Bowden plc.

WWF (2004) *Building Towards Sustainability: Progress and Performance among the UK Listed House-builders*, Godalming, Surrey: WWF.

WWF (2005) *Investing in Sustainability – Progress and Performance Among the UK Listed House-builders Revisited*, Godalming, Surrey: WWF.

Part V

International perspectives on corporate social responsibility in construction

Chapter 12

Occupational health and safety (OH&S) and corporate social responsibility

John Smallwood and Helen Lingard

Introduction

In this chapter the concept of corporate social responsibility (CSR) is explored, and it is argued that occupational health and safety (OH&S) is an essential component of CSR, as employees, including subcontractors, are key stakeholders whose interests must be accommodated by construction organisations. The moral dimension of OH&S is highlighted, and the provision of a healthy and safe work environment as a basic human right is motivated. It is argued that construction organisations need to develop organisational cultures that are ethical and supportive of OH&S. Furthermore, it is suggested that evolving such cultures requires more than the establishment of formal OH&S management systems, and that OH&S must become an embedded value and part of the organisational culture. Consequently, the need for employees at all levels in construction organisations to demonstrate leadership in OH&S and be held accountable for their OH&S actions is discussed. This is followed by the motivation for the regarding of 'at risk' behaviour as universally unacceptable, and the establishment of new criteria for project performance in the form of 'zero injuries and illnesses'. The practice of construction organisations taking increased responsibility for OH&S in the supply chain, and adopting socially responsible buying and contracting practices is discussed. The responsibility of construction organisations for workers, who are injured or have become ill as a result of their work, and the industry's appalling performance in the important area of occupational rehabilitation, are also discussed. A case study of Australian building products company, James Hardie Industries, highlights the possible consequences arising from public censure of an organisation's failure to take responsibility for workers whose health is affected by work. It is contended that OH&S performance should be reported alongside other social performance indicators by construction organisations in annual reports, and that 'Triple Bottom Line' reporting should be considered. The chapter concludes with a discussion of the role of construction educators in inculcating an 'OH&S ethic' among future built-environment practitioners.

The poor OH&S performance of the construction industry

The construction industry is recognised internationally as one of the most dangerous industries in which to work. The International Labour Organization estimates that there occur at least 60,000 fatal accidents on construction sites around the world each year. This means one construction fatality occurs every ten minutes. Construction accounts for a staggering 17% (one in six) of all fatal workplace accidents (ILO 2005). According to McWilliams *et al.* (2001), 256 people were fatally injured in the Australian construction industry between 1989 and 2002. The fatality rate is 10.4 per 100,000 full-time equivalent employees, which is similar to the road accident fatality rate. The construction industry's rate of occupational injury and disease is 44.7 per 1,000 persons, which is nearly twice the all-industry rate. In Australia, a Royal Commission into the Building and Construction industry reported that there are on average 50 deaths a year on Australian construction sites and that, in 1998–1999, accidents and deaths cost AU\$109 million a year and the loss of almost 50,000 weeks of working time (Lindsay 2003). Similarly, high figures are reported for the United States of America. According to Gillen *et al.* (2002), in 1998 the construction industry accounted for 20% of workplace deaths, more than any other industry.

Apart from the unacceptable number of deaths, construction workers suffer a disproportionately high number of serious, disabling work-related injuries. Guberan and Usel (1998) followed a cohort of 5,137 men in Geneva over 20 years and reported that only 57% of construction workers reached the age of 65 without suffering a permanent impairment. Falls and manual handling are prominent risk factors associated with serious injuries and long-term disability in construction (Gillen *et al.* 1997; Nurminen 1997). In a study of workers' compensation data in Victoria, Australia, Larsson and Field (2002) reported 11 prominent construction trades to have average injury durations longer than the all-industry average. The social costs of long-term disability caused by work-related injury are considerable. Armstrong *et al.* (2000) suggest that many people who suffer work disability spiral into economic hardship, and as many as 30% live below the poverty line.

Occupational fatalities, injuries and disease in construction result in considerable human suffering affecting not only the workers directly involved, but also their families and communities as well as contributing substantially to the cost of national medical care, rehabilitation and social-welfare schemes (Smallwood 1996). However, occupational fatalities, injuries and disease also contribute to the variability of resource in the construction industry, increasing project risk. Such risk can manifest itself in damage to the environment, reduced productivity, non-conformance with quality standards and time overruns, and ultimately in an increase in the cost of construction. Other possible detrimental outcomes of the industry's poor OH&S performance include damage to client property, impaired production processes and a poor client and/or contractor image as a result of negative publicity and public outrage (Smallwood 1996).

Imperatives for change

Despite its traditionally poor performance in the area of OH&S, there is a growing recognition among construction organisations' executives that work-related death, injury and illness are unacceptable. This recognition has manifested itself in statements and OH&S inclusions in annual reports. Greg Clarke, Managing Director of Bovis Lend Lease in Australia, Bovis Lend Lease, recently told the company's annual general meeting that 'nothing is more important than the safety of those who actually work in the [construction] industry'. Clarke's personal vision is of an injury- and incident-free industry reflecting an elevated culture of OH&S in the construction industry (*Australian Financial Review* 2004d). Leighton Holdings, Australia's largest construction organisation, published details of its OH&S performance in its annual report for the first time in 2004. Furthermore, the Leighton Board has formed an ethics and compliance committee with OH&S as a key responsibility. Thus, it appears that certain large construction organisations are actively responding to society's growing belief that work-related death, injury and illness are unacceptable.

The motivation for addressing OH&S

The motivation for addressing OH&S is multifaceted. Some of the primary motivators are: legal considerations; moral/religious beliefs; ethical issues; humanitarian concerns and a respect for people; the desire for sustainability; compliance with national and international standards; the desire to reduce the costs of accidents/incidents; the desire to reduce organisational risk; adherence with total quality management principles; support of local industry OH&S and image initiatives; and the pursuit of better practice.

Ethical business practice includes compliance with legislation. Given that values embrace ethics, the existence of OH&S legislation amplifies the need for the inclusion of OH&S as a value, and as an organisational performance area. The inclusion of OH&S as a value, as opposed to a priority, is important, because organisational priorities are subject to change, whereas values are entrenched and resistant to change. Failure to properly evaluate project OH&S performance creates the impression that OH&S is less important than other objectives and places the focus on the traditional project parameters. Consequently, the potential synergy resulting from the optimisation of OH&S is unlikely to be realised. Therefore, OH&S should be afforded status equal to, or greater than, that afforded to the traditional project parameters of cost, quality, and time.

Flowing from the humanitarian motivation for OH&S is the aspect of 'respect for people', one of the three principles of 'Rethinking construction' initiated by the report of the Construction Task Force chaired by Sir John Egan in the United Kingdom in 1998 (Rethinking Construction Limited 2002). How does 'respect for people' manifest itself? Endeavours and interventions to engender and enhance 'respect for people' cited, include those relative to OH&S, site conditions and welfare.

Also related to the humanitarian aspect is the issue of sustainability. Accidents can result in fatalities, injuries, disease, damage to materials, plant and equipment, all of which result in generation of waste. Waste in solid and other forms impacts on the sustainability of the earth as a result of the use of landfill sites, and the unnecessary consumption of non-renewable resources in the case of rework. Furthermore, fatalities and injuries threaten the sustainability of the victims' families and their communities.

The synergy between OH&S and the other project parameters of cost, environment, productivity, quality and schedule is also cited as a motivator for OH&S. The research conducted among PMs in South Africa investigated the impact of inadequate OH&S on various project parameters – the percentage of responding PMs that indicated such impact is recorded within parentheses: productivity (87.2%) and quality (80.8%) predominated, followed by cost (72.3%), client perception (68.1%), environment (66%) and schedule (57.4%).

The ISO 9000 Quality Management System (QMS) series and ISO 14001 Environmental Management System (EMS) both require the implementation of management systems which can complement and support good OH&S management practice. In addition to many design practitioners and contractors having acquired certification relative to one or both of such systems, there are many clients that have such certification, which in turn requires that their contractors and suppliers are managed in a manner appropriate to the relevant system(s).

OH&S and CSR

Standards Australia (2003) defines corporate social responsibility (CSR) as:

> Mechanism for entities to voluntarily integrate social and environmental concerns into their operations and their interaction with their stakeholders, which are over and above the entity's legal responsibilities.

Shrivastava (1995) identifies different positions on the issue of CSR. First, there are those who believe the only responsibility of business is to make a profit, which view was perhaps most famously stated by Milton Friedman who wrote in the *New York Times Magazine* that business had no responsibility beyond making profits for its shareholders. Proponents of this view believe free markets are the best means by which business conduct can be regulated. According to economic theory, OH&S is a commodity, the value of which is determined by the labour market. Workers are seen as rational actors who assess the level of risk inherent in a job and balance these risks against the wages offered for that job. Thus, the higher the risk, the higher the wage level that needs to be paid. It is argued that the optimum level of OH&S is then determined via the market mechanism. However, this suggestion is problematic because it is assumed that employees have perfect information about risks and are free to move from job to job. These assumptions do not hold true. Workers' ability to switch jobs is usually limited, for example by their skills, and it cannot be assumed that workers have satisfactory means for assessing risks (Clayton 2002).

Second, an opposing position on CSR holds that corporations have significant social side effects, including injury and illness and that these are best handled by reforming corporations and their production systems. This approach forms the basis of the discussion of OH&S in this chapter. Proponents of this view believe that companies, governments and community groups should work in partnership to achieve this. They suggest that a combination of regulations and voluntary corporate actions can be used to achieve the required corporate reform.

Stakeholder theory

The issue of CSR is grounded in stakeholder theory, which holds that managers have a responsibility to all those who have a stake in or claim on the organisation. This is counter to the shareholder-value theory, described above. Stakeholders can be grouped into four main categories, each of which has a potential interest in OH&S. These groups are:

- regulatory stakeholders, including governments and OH&S enforcement agencies;
- community stakeholders, including incident/accident victims' groups and trade unions;
- media stakeholders;
- organisational stakeholders, including employees, suppliers and contractors (Henriques and Sadorsky 1999).

Employees are a key stakeholder group whose interests should be taken into consideration in OH&S decision-making. However, in this chapter the term 'worker' is used rather than 'employee' reflecting the moral and legal responsibilities of organisations to people not directly employed, such as contractors and suppliers. The moral view of workers as stakeholders may necessitate a trading-off of the economic benefit to shareholders against the interests of others. Thus, in certain situations, stakeholder theory requires that profits be sacrificed in the interests of workers' health, safety, and well-being.

Corporate accountability

The power asymmetry between individuals and organisations means that organisations can engage in anti-social activities that would be practically impossible for individuals. Large organisations can produce considerable injustice by accident. In this context, there has been a quest for CSR and accountability. Cropanzano *et al.* (2004) developed a theory of corporate accountability. The basis of this theory is an analysis of what *would* have happened in different circumstances, what *could* an organisation have done differently in a situation and what *should* the organisation have done. Through comparisons of alternative scenarios the answers to these three questions form the basis for evaluating corporate conduct. Organisations will be held accountable for socially irresponsible behaviour if the following conditions are met:

- The current state of affairs is not as favourable as an imaginable alternative;
- The action that produced the state of affairs could have been avoided; and
- The action violated a moral standard.

Thus, relative to OH&S, organisations should be held accountable in the event that they fail to prevent injury and/or illness through the effective management and promotion of the OH&S of workers and others affected by their operations. The issue of criminal accountability of corporations and individuals is addressed later in the chapter.

The moral dimension of OH&S

Eckhardt (2001) says the 'golden rule', which establishes a moral level of care for others that we are responsible to provide, is a common theme in most, if not all, of the world's major religions, and Rowan (2000) suggests that there is something morally significant about persons and that it is morally wrong to treat people in certain ways. Consequently, all people and organisations should be conscious and mindful of the health and well-being of each other, and their workers respectively. The results of managerial actions have extended consequences. These consequences are often experienced by people who have no control over the actions that caused them and, therefore, there is an argument that these consequences should be considered when decisions are made. If decisions can hurt or harm people in ways that are outside their individual control, then the issue is a moral one, which requires some ethical analysis. Consequently, Greenwood (2002) recommends the development of some ethical principles for managing workers.

Rowan (2000) suggests three principles, which are based upon an individual's right to pursue his/her own interests. These are the right to freedom, the right to well-being and the right to equality. If these rights are accepted as being universal human rights, it follows that workers have a moral right to work in a healthy and safe workplace. Requiring workers to work in an environment, which poses a risk to their health and/or safety, prevents workers from pursuing their own interests and therefore violates their basic human rights. Thus, the right to a healthy and safe workplace is a universal right held by workers against employers, who have a concomitant duty to provide a healthy and safe work environment. This view is shared by the World Health Organization, which holds that the right to OH&S is part of basic human rights (*The Lancet* 2003).

Society has recognised this right in the passing of legislation designed to protect workers' OH&S in and arising from work. However, not all moral rights are legal rights. Rowan (2000) cites the example of African slaves in the United States. While there was no law against slavery for many years, the treatment of people as slaves could not be said to be morally right. Thus, ethical conduct in relation to human beings extends beyond the minimum requirements of the law.

OH&S and the law

In many countries, including the United Kingdom and Australia, preventive OH&S legislation is based upon a model proposed by the Robens Committee in 1972. This legislative model is based upon an all-encompassing Act covering every type of workplace hazard and every group of workers. The provisions of these Acts are expressed in general terms: the Acts typically provide for tri-partite decision-making at both governmental and workplace levels. Thus, a key component of the requirements under such Acts is that employers provide consultative processes at the workplace through which workers, as stakeholders, can participate in OH&S decision-making. However, the extent to which the preventive OH&S legislation is enforced has been criticised for being inadequate. For example, in Victoria, Australia, Johnstone (2002) undertook a study of OH&S prosecutions between 1983 and 1999 and reports that, during these years, only a small proportion of visits made by government inspectors to industrial premises resulted in inspectors taking any action. For example, in 1996–1997, 44,703 inspection visits were made, resulting in 3,219 improvement notices and 1,040 prohibition notices. In the same period, only 57 cases prosecuted. It is also noteworthy that the vast majority of prosecutions – 87% between 1983 and 1999 – arose as the result of a fatality or injury (Johnstone 2002). It seems that, in their enforcement of post-Robens OH&S inspectors, inspectorates have continued to see their role, in part at least, as an advisory one. They use education and persuasion to encourage compliance with OH&S legislation, and prosecutions are invoked as a last resort, when persuasion fails.

Johnstone (2002) also examined sentencing outcomes for cases in which charges were proved. He reports that, in 1999, of 120 cases in which charges were proved, 101 defendants were convicted, 17 were fined without conviction, and two received a good behaviour bond. The average fine was 26.7% of the maximum fine available. Given that prosecutions are invoked only as a last resort and usually only when a death or serious injury has occurred, such sentencing outcomes reflect what Carson (1979) has termed the 'conventionalisation' of OH&S crime. Thus, contraventions of OH&S legislation are treated as 'quasi-crimes' and assumed not to have the same degree of gravity as other criminal offences. The hearing of the majority of OH&S offences in Magistrates, Courts, which typically deal with minor offences, such as traffic violations, reinforces this contention.

Possibly, as a consequence of these weaknesses, there has been mounting pressure for the use of the mainstream criminal law in relation to workplace deaths and serious injuries. An argument often cited in support of the use of the mainstream criminal law is that the conviction of a 'real' criminal offence carries greater stigma than a conviction for breaching OH&S legislation. It also captures public attention and, in doing so, is likely to have a greater deterrent effect. It is also argued that the principle of equal treatment means that if people in society can be convicted of manslaughter in the event of a death arising from their negligent conduct, so too should managers who through their negligence cause the death of a worker (Hopkins 1995; Neal 1996).

As the law currently stands, when negligent conduct results in the death or serious injury of a worker, corporations or individual managers may be prosecuted for offences such as manslaughter or the criminal infliction of serious injury. Such charges may be brought in addition to, or as an alternative to, prosecutions for breaches of the preventive OH&S legislation. However, there are significant difficulties in the use of the mainstream criminal law for this purpose. At present, the anthropomorphic treatment of companies requires that, for a corporation to be found guilty of a criminal offence, the criminal conduct must be committed by the board of directors, the managing director, or another person to whom a function of the board has been fully delegated in order for it to be attributable to the company. This principle has been widely criticised because it limits corporate liability to acts and omissions performed at the top of the corporate ladder (Field and Jorg 1991; Wells 1994). Therefore, it is easy, especially in large and complex corporations, for organisational structures to be designed to avoid such liability (Fisse 1994; Johnstone 1999). Consequently, the only corporations that have received convictions under the mainstream criminal law have been very small microbusinesses in which the owner or director had been supervising the dangerous work. Unless prosecutors are able to prove that directors themselves have 'consented or connived' in the failure, resulting in death or serious injury, the corporation cannot be found guilty. Thus, as George Monbiot writes, neglect can be used as a defence against a charge of neglect (*The Guardian* 2004). This situation has led to a 'groundswell' of public sentiment calling for reform of the criminal law. These calls have focused on the creation of a separate criminal offence of industrial manslaughter. The elements of this offence would change the basis of attributing corporate criminal guilt to more adequately reflect the reality of operational decision-making and managerial action within modern organisations.

Sanctions for corporate crime have also been criticised. One difficulty with the imposition of fines is that the real burden may ultimately be borne by people who are entirely innocent. For example, shareholders may bear the cost, or if the company is forced to shut down, innocent workers may suffer. Fines do not necessarily lead companies to take internal disciplinary action against those responsible, or change internal operating procedures (Fisse 1994). A company may simply opt to pay fines and regard them as 'purchasable commodities' or operating costs. Another difficulty with fines as a form of sanction is that they are prone to evasion through techniques such as asset stripping. In some instances, fines that reflect the seriousness of a criminal offence may be too great for a company to pay. This leaves a court with the choice of charging a lesser fine or imposing an appropriate fine and forcing the company into liquidation. Once a company has gone into liquidation, it cannot be held liable for offences committed before dissolution. Fisse and Braithwaite (1993) recommend that a solution would be to combine the firm's capacity with the law's desire to achieve accountability. Thus, corporate offenders would be required to reform themselves and undertake internal disciplinary action, under the threat that if they fail to do this to the satisfaction of the courts, a sanction such as forced liquidation or the withdrawal of a firm's licence or charter to operate may be incurred.

Arguably, there is also a need for *individual* accountability in the maintenance of social control. Company directors make key decisions governing OH&S; for example, they decide how much money will be spent on the prevention of occupational injury and illness, and, arguably, should be held accountable when workers are killed or seriously injured as a result of corporate failings. However, the issue of prosecuting company officers or directors is vexed and successful convictions are relatively rare, because in large, complex organisations, it is hard to identify with certainty the individuals whose conduct satisfies the gross negligence test.

However, just having an OH&S management system is not sufficient. The issue of organisational culture is also of critical importance. Schein (2004) describes organisational culture as a pattern of beliefs and assumptions that are shared by organisational members and that operate unconsciously and define an organisation's sense of self. Thus, organisational cultures act upon organisational members to establish common values and a shared understanding of what behaviours are acceptable and what are not. Kletz refers to this as the 'common law' of an organisation (Kletz 1993). The concept of OH&S culture describes an organisation's norms, beliefs, roles, attitudes and practices concerning OH&S (Turner and Pidgeon 1997). OH&S culture is a subset of organisational culture. A positive OH&S culture seeks to establish the norm that everybody is aware of risks in the workplace and feels responsible for their own OH&S as well as that of others. When a positive OH&S culture prevails, everybody is continually focused on the identification of hazards and the raising of any OH&S concerns with supervisors and management. Schein (1992) suggests that the way that managers conduct themselves will be a particularly important determinant of organisational culture; in particular, perceptions of managers' OH&S attitudes and behaviours is likely to shape OH&S behaviour. Thus, it is vital that managers, from directors to first-line supervisors, demonstrate OH&S leadership and consistently and unequivocally support the OH&S process.

Mounting pressures to be a good corporate citizen

There is a growing recognition that universal accounting standards do not ensure that organisations behave in a way that is responsible to the wider community in which they operate (Saravanamuthu 2004). Consequently, there is increasing pressure on organisations to demonstrate transparency and accountability more than the traditional measures of financial performance. There is an increasing public expectation that organisations will take responsibility for their non-financial impacts.

In Australia, for example, a suite of Australian Standards concerning business governance contains AS 8003 (Standards Australia 2003), dealing with the issue of CSR. The standard provides for the establishment of a CSR policy, the establishment of management responsibilities for CSR, the identification of CSR issues relevant to the organisation, the establishment of operating procedures for CSR, stakeholder engagement, communication, education and training, and monitoring and reviewing CSR performance. OH&S issues explicitly identified in the standard include:

- personal health;
- food and nutrition;
- potable water;
- dormitories;
- factory ventilation;
- emergency evacuation;
- fire safety;
- use of chemicals;
- ergonomics.

In addition, worker-related issues that impact upon OH&S include unreasonable working hours, lack of freedom of association and freedom from discrimination. The Standard also lists supplier ethical issues as relevant to CSR, addressed below.

Socially responsible buying and contracting

Korhonen (2003) argues that measuring the CSR of individual organisations, processes or products is not sufficient in itself because there is the risk of problem displacement between processes and companies. She suggests that the social impacts of a product life cycle are affected by many different social actors crossing product, process, firm, regional and national boundaries. Thus, a network approach to CSR considering a *system* of firms is preferable. This is particularly true in the highly fragmented construction industry, in which it is important to ensure that the social burden of the system as a whole is reduced rather than simply shifted between firms, products or processes. For example, the displacement of an OH&S risk might occur when on-site risks are reduced by opting to prefabricate building components off-site; this simply shifts the OH&S risk of exposure to workers engaged in the prefabrication process rather than reducing the overall OH&S risk inherent in the project. Recently, statutory OH&S responsibilities for clients and designers have attempted to limit the extent to which OH&S risks arising in one stage of a construction project life cycle, such as procurement or design, are displaced to another stage. However, the success of these legislative approaches has been questioned (Entec, 2000; Rigby 2003).

There are growing pressures upon organisations to undertake socially responsible buying, i.e. considering the social impacts of the organisation's purchasing decisions (Maignan *et al.* 2002). For example, in the 1990s considerable pressure was brought to bear upon the manufacturing/retail organisation, Nike, to stop purchasing from suppliers using child labour. Thus far, the push for socially responsible buying has not been strongly felt in the construction industry, but there is evidence that this situation is changing. For example, in Australia, trade unions recently pressured contractors to boycott James Hardie building products in response to that company's alleged failure to meet its obligations to compensate former workers suffering from asbestos-related disease. Thus, socially responsible construction organisations should look beyond the OH&S of their own workers and consider the OH&S performance of their key suppliers and contractors. This can be achieved by designating organisational members responsible for socially responsible purchasing/procurement, ensuring that OH&S criteria are unambiguously stated in purchasing and subcontract documents, educating suppliers and subcontractors in OH&S, monitoring the OH&S performance of suppliers and subcontractors, and, where necessary, sanctioning suppliers and subcontractors for failing to meet stated OH&S criteria.

OH&S management systems

To ensure that OH&S is systematically managed within an organisation, an OH&S management system needs to be implemented.

Defining a corporate OH&S policy is the first step in the OH&S management process. Typically an OH&S policy statement contains a statement of the organisation's commitment to OH&S and identifies the OH&S responsibilities of all workers, including managers in both corporate and project-management roles. The next step in the management of OH&S is structuring the organisation in such a way that roles, responsibilities and relationships support the systematic planning and control of OH&S. The organisation design comprises the way the OH&S is controlled, through allocating responsibility for OH&S functions, identifying OH&S objectives and monitoring progress towards meeting these objectives. It is important that OH&S responsibilities are allocated to managers and that a senior manager, at the top of the organisation, is made responsible for coordinating and monitoring the OH&S management activities.

Responsibilities for OH&S activities must be clearly allocated so that no important tasks 'fall between the cracks'. This is particularly true in construction projects, in which many different organisations are often involved. Holding people accountable for their OH&S responsibilities might involve including performance standards in job descriptions and using OH&S key performance indicators (KPIs) as the basis of measuring how effectively workers are performing their jobs.

Planning is a critical part of OH&S management. At each stage, clear objectives, measurable targets and performance indicators should be set. Measuring and monitoring OH&S performance is essential to making sure that the organisation is implementing its OH&S policy and achieving its OH&S objectives and targets. Without measuring performance, there is no way of knowing whether OH&S is being managed satisfactorily, nor is it possible to hold managers accountable for OH&S. Managers should be given the responsibility for measuring OH&S performance and monitoring the achievement of OH&S objectives in the areas in which they manage. Thus, project managers must ensure that project OH&S performance is measured and that KPIs specified in project OH&S plans are met.

All aspects of the OH&S management process should be subject to regular systematic assessment and review. Reviewing OH&S performance is the key to learning from experience and continuous improvement. There is a very important feedback loop between this stage and all other aspects of OH&S management. The review is concerned with determining the adequacy of performance of the OH&S management system, and ensuring that improvements are made where possible. Organisational reviews of

OH&S performance should be conducted at regular intervals, perhaps every year. However, in the construction industry, post-project OH&S performance reviews are also helpful because they enable lessons to be transferred from one project to the next.

Evaluating project performance

The way that project success is measured needs to be dramatically altered. We suggest that social outcomes, including OH&S, should be measured and reported alongside traditional indicators of project success, such as cost and time performance. At present, the social performance of construction projects is often overlooked. A good example of this is the case of the construction of the Hong Kong International Airport and its related infrastructure. The construction of the HK$158 billion airport at Chek Lap Kok and its related transport infrastructure was lauded as a project management success story, receiving numerous awards for design, construction and operational efficiency (Major Projects Association 2001). The project management team enjoyed these accolades in spite of the fact that 49 workers were killed during the course of building the new airport and its connecting infrastructure (*South China Morning Post* 1997). The construction industry worldwide needs to move to a situation in which OH&S performance is viewed as a critical component of a project's overall success. We contend that a project should not be regarded as being successful when even one life is lost during its construction.

Triple-bottom-line reporting

Triple-bottom-line (TBL) reporting requires organisations to report their performance in accordance with a range of financial, environmental and social indicators. OH&S performance is an important component of these social indicators. At the moment, TBL reporting is voluntary but, if the latest trend continues, clients, shareholders and job seekers will become more socially aware and begin to demand the reporting of non-financial performance. In the United Kingdom, the Health and Safety Commission (HSC) Strategy Statement contains an action point focusing on the public reporting of health and safety by large companies. The HSC have issued guidelines recommending that companies report on their health and safety principles, performance and targets. A recent study of the annual reports of leading British companies revealed that reporting OH&S issues had increased considerably between 1995 and 2001 (Peebles *et al.* 2002). However, the information included in reports was sometimes poor and was often limited to a broad statement of policy. Few companies reported their OH&S targets. There were only two construction companies in the sample, neither of which included any OH&S information in their annual reports. While the reporting of OH&S information is currently voluntary, principles of CSR support transparency and reporting of non-financial performance. Failure to be transparent, or poor performance in these areas, are likely to harm a company's reputation, and may also impact upon financial performance in the future. For example, the notion of ethical investment is becoming more significant, and ethically aware graduates are less likely to be willing to work for organisations whose performance in the social and environmental domains is poor.

Rehabilitation

One aspect of socially responsible business that is often overlooked is the rehabilitation and the return-to-work of workers who are injured at work. The aim of rehabilitation is to 'restore a person who has been injured or suffered an illness to as productive and as independent a lifestyle as possible through the use of medical, functional and vocational interventions' (Association of British Insurers 2002).

The human cost of work disability includes impaired domestic and daily function, strained family relationships, negative psychological and behavioural responses, stress and loss of vocational function (Dembe 2001). Long-duration work-related disability is a serious social problem, and working in the construction industry has been identified as a risk factor for chronic disability (Cheadle *et al.* 1994; Hogg-Johnson *et al.* 1994; McIntosh *et al.* 2000). The construction industry performs poorly in the area of rehabilitation. Proactive rehabilitation and return-to-work programmes have been found to contribute to reduced disability duration measured in lost workdays and enhance rehabilitation outcomes (Amick *et al.* 2000; Hunt and Habeck 1993; James *et al.* 2003). Despite this, recent research by Lingard and Saunders (2004) revealed that formal programmes for the rehabilitation of injured workers have not been implemented by many construction companies. Further, the study provided evidence that some construction employers respond negatively to worker injury, increasing the negative social consequences of injury and reducing the likelihood of a successful return to work.

Workers' compensation

An organisation has a significant moral obligation in ensuring that those who suffer an illness or injury as a result of exposure to its products or processes are appropriately compensated. In most developed countries, workers who suffer a work-related injury or illness can look to two possible sources of financial assistance: they can seek compensation from a statutory workers' compensation system or they can bring a common-law action for damages against the party responsible, usually their employer. Statutory systems are usually 'no-fault' systems requiring employers to pay workers' compensation insurance and ensuring that workers are compensated without necessarily having to demonstrate fault on the part of another party, usually the employer. However, if a worker can demonstrate that his or her injury or illness occurred as a result of negligence on the part of his or her employer, the common law is a second means by which financial recompense may be obtained. In many jurisdictions, tort actions for negligence can be brought against an employer, based on the theory expressed by Lord Atkin in *Donoghue v Stevenson* that people are under a duty not to injure their 'neighbour' through careless behaviour. Workers' rights to sue for common-law damages have been vigorously defended by trade unions and accident victims groups. However, the opportunity for victims to seek compensation via the common law can be severely restricted by corporations that structure themselves so as to avoid liabilities by setting up under-capitalised subsidiaries (*Australian Financial Review* 2004a).

The case study below describes a recent case in which it was alleged that a large supplier of building products restructured its organisation and relocated its business in a deliberate attempt to avoid liability arising from its manufacture of products containing asbestos. This led to a Commission of Inquiry in New South Wales and has been described by trade union leaders as 'the greatest corporate scandal' in Australian history (*The Age* 2004b). The commission outcomes have important implications for the way companies behave as corporate citizens (*Australian Financial Review* 2004b). In particular, the case has raised questions about the wider social responsibilities of business and the extent to which companies seek to maximise shareholder wealth at the expense of moral obligations (*The Sydney Morning Herald* 2004a).

Case study: James Hardie Industries asbestos trust fund

The James Hardie group manufactured asbestos building products in Australia until the mid-1980s. Until 1937, products were manufactured by James Hardie Industries Ltd (JHIL), now called ABN60. Between 1937 and 1986 a subsidiary of JHIL, James Hardy and Co, now called Amaca, manufactured building products containing asbestos. Another subsidiary, Amaba, manufactured brake products containing asbestos.

It is alleged that between 1996 and 2004, the management of the James Hardie Group embarked upon a complicated restructuring to separate the parent company from its subsidiaries and remove assets that could otherwise have been available to pay damages to asbestos victims. In February 2001, ownership of Amaca and Amaba was transferred to a new body, the Medical Research and Compensation Foundation (MRCF), which was left with AU$293 million to fund future claims from thousands of sufferers of asbestos-related diseases, a figure which was reported to be 'woefully inadequate' (*The Age* 2004c).

Soon after the formation of the MRCF, James Hardie's parent company moved to The Netherlands, ostensibly to take advantage of tax benefits. It is alleged that the move was actually made to prevent Australian victims from

(Continued)

seeking further compensation from the parent company because The Netherlands does not have a legal treaty that permits victims to seek compensation in Dutch courts (*The Courier-Mail* 2004). The move, potentially, left thousands of sufferers of asbestos-related diseases without compensation.

In October 2003, the MRCF declared that it faced a shortfall of AU$800 million in funds to compensate sufferers of asbestos-related disease, but James Hardie reaffirmed its position that there 'can be no legal or other legitimate basis on which shareholders' funds could be used to provide additional funds to the foundation and duties of the company's directors would preclude them from doing so' (*The Sydney Morning Herald* 2004b).

In February 2004, the New South Wales Government established a special Commission of Inquiry into James Hardie's asbestos liabilities. The counsel assisting the commission described the company's actions as 'a collective washing of hands' (*The Australian* 2004a). Further, James Hardie has been described as having a culture of 'secrecy and disdain' for genuine victims of diseases caused by its products (*The Age* 2004a).

James Hardie shares fell more than 30% after the company was accused of underfunding the MRCF (*The Daily Telegraph* 2004), and site boycotts of James Hardie products, damaging media coverage and the return of political donations by the Australian Labour Party further tarnished the company's reputation (*Australian Financial Review* 2004c).

On 13 August 2004, just two days before the end of the inquiry, James Hardie admitted to having liability for users of their asbestos products and proposed an offer to compensate all victims of asbestos-related disease arising from use of their product. This proposal was welcomed by unions and victims' groups, but conditions placed upon the scheme were not accepted. In particular, conditions that the scheme be affordable to the company and that it be approved by the board and shareholders were criticised (*The Australian* 2004b).

The Commission of Inquiry sat for nearly six months; its report concluded that James Hardie had misled the public when it claimed it had provided enough money for asbestos compensation (Australian Associated Press 2004). However, the inquiry also found that James Hardie was not *legally* obliged to provide any more funds. Further funding commitments from the company would have to be voluntary.

In the aftermath of the inquiry, James Hardie's Chief Executive Officer and Chief Financial Officer resigned from the company. Finally, on 21 December 2004, after eleven weeks of intense negotiations, it was announced that James Hardie would make up the shortfall, estimated at AU$1.5 billion, to fully fund present and future asbestos claims. This was the biggest financial settlement in Australian history (*The Australian* 2005).

The James Hardie case study illustrates how the social responsibility of companies extends beyond their legal obligations. The abrupt change of direction of James Hardie demonstrates the extent to which public pressure can be brought to bear to force organisations to meet their social and moral obligations, even when they are not legally obliged to do so. As a result of the

James Hardie asbestos issue, Australian state governments have called on the federal government to implement corporate-law reform to ensure that company directors' duties incorporate social responsibilities and, in particular, that companies cannot obtain profit at the expense of human life (Australian Broadcasting Corporation 2004).

Multinational organisations

One issue relevant to the case of the Hong Kong International Airport, described earlier in the chapter, was that of the role of multinational organisations in the construction industry. Much of the airport-related construction work was undertaken by international contractors whose head offices were in Australia, the United Kingdom, France and Japan. Arguably, construction firms involved in international contracting have a responsibility to adopt better practice with regard to OH&S irrespective of local standards and practices of their host country. It is very common, for example, to observe British and Australian contractors using traditional and poorly constructed bamboo scaffolding in Asian construction operations when this would not be accepted in their home countries. These practices constitute double standards and raise ethical questions for the companies concerned. An example of the catastrophic consequences of a lack of organisational consistency in OH&S systems and compliance can be found in the Union Carbide disaster at Bhopal. The accident that occurred in 1984 killed or injured thousands, as well as leaving many families without a primary income earner. Thousands of jobs were lost, and the costs to the Indian Government in providing food, medical treatment and hospital facilities exceeded US$40 million. Union Carbide was held responsible for the disaster and was made to pay US$470 million in damages in a landmark decision, which formally recognised that organisations have a significant responsibility to the society in which they operate. Robertson and Fadil (1998) argue that the communication of standards and procedures from the headquarters to the Indian plant was ineffective, codes and standards were not enforced and Union Carbide's senior management was not held responsible for the management programmes at Bhopal. In construction, the problem of how to apply consistently good OH&S standards and remain competitive in overseas markets remains a vexed one. However, it is incumbent upon socially responsible construction organisations to ensure that OH&S systems and compliance are consistently of a high standard across all of their international operations.

Industry initiatives

In the United Kingdom, the *Rethinking Construction* report, commonly referred to as the Egan report, which resulted from an investigation led by the Deputy Prime Minister, John Prescott, to determine the scope for improving the quality and efficiency of UK construction, identified accidents as a key performance indicator and set a target of 20% reduction in the number of reportable accidents and a 10% increase in productivity and profitability (Department of the Environment, Transport and the Regions 1998). The *Rethinking Construction* report proposed that a 'movement for change' be established in the industry to facilitate improvement. Consequently, the Movement for Innovation (M⁴I) was established, the aim of which is to lead radical improvement in construction in value, money, profitability, reliability and respect for people through demonstration and dissemination of best practice and innovation. Commitment to people is identified as one of five drivers of change. Demonstration projects are intended to benchmark performance, set high standards in OH&S and respect for people, and to disseminate the results of their work. According to Rethinking Construction Limited (2002), demonstration projects are consistently shown to be safer sites, namely, 25% safer than the industry at large. However, it must be noted that this statement is based upon accident statistics, and that the absence of accidents does not mean that sites are necessarily safe. It does, however, suggest that M⁴I sites experience fewer accidents because they are better managed. A further UK discussion document, *Revitalising Health and Safety in Construction*, provides an opportunity to assess where the industry is in terms of OH&S, where the industry wants to be and how to get there (Health and Safety Executive 2002).

Industry collaboration

The persistence of high levels of occupational injury and illness in the construction industry is indicative of the industry's failure to learn from its mistakes. In certain sectors of the industry, particularly among small to medium-sized construction firms that make up the vast majority of construction organisations, there is evidence of a lack of OH&S awareness and knowledge. Within this context, there is enormous scope for the industry to share OH&S information and learn from the experiences of other organisations. The opportunities for industry collaboration with regard to the improvement of OH&S are great. Further, it is argued that the whole industry would benefit from making significant improvements in its OH&S performance, as it would enhance the industry's image. By improving the industry's image and demonstrating that construction firms are professionally managed and socially responsible, the industry would increase its ability to attract talented workers and socially conscious investors.

The Good Neighbour Scheme in the United Kingdom is a good example of inter-organisational collaboration in OH&S. The scheme encourages the sharing of OH&S expertise between organisations. In particular, organisations with good OH&S management systems are encouraged to share practical advice about successful OH&S initiatives within their organisations. Participating firms also provide direct assistance to other firms in the same industry sector. The Good Neighbour Scheme also facilitates the sharing of OH&S information through the supply chain, with the active involvement of subcontractors and suppliers. Many construction organisations have participated in the scheme, including clients, contractors and suppliers.

The holistic role of CSR in OH&S and the role of OH&S in overall performance

Thus far, this chapter has addressed the direct and indirect relationship between CSR and OH&S, and the motivation for addressing OH&S. However, given that this is a chapter in an international CSR text, and that a range of built-environment stakeholders are likely to make reference thereto, it is necessary that an overview of the holistic role of CSR in OH&S and the role of OH&S in overall performance be provided – Figure 12.1 does this in the form of causal loop analysis. The right-hand ellipse indicates the impact of lack of commitment to OH&S on OH&S and overall performance. Conversely, the left-hand ellipse indicates the impact of corporate OH&S awareness/acknowledgement and subsequent commitment to OH&S on OH&S and, ultimately, overall performance and cost.

The right-hand ellipse is addressed first. Lack of commitment to OH&S results in poor ergonomics and contributes to the existence of hazards and risk. Poor ergonomics also contributes to hazards and risk, which results in strains, which in turn results in both absenteeism and ill health, and ill health results in absenteeism. Hazards and risk contribute to the probability of exposures and accidents. Exposures can result in disease, and consequently ill health, and in turn absenteeism. Absenteeism can result in reduced productivity, rework, and falling behind schedule owing to the absence of key crewmembers. Accidents, the outcome of which is largely fortuitous, can result in any, all, or a combination of the following: exposures; fatalities; injuries; reduced productivity as a result of work stoppages; rework as a result of damage to completed work or work in progress; falling behind schedule as a result of work stoppages and damage to the environment. A further consideration is the negative effect that reduced productivity, rework, falling behind schedule, and damage to the environment have on each other. Accidents also constitute a non-conformance, which would not be favourably viewed by many clients. Injuries may require first aid, or medical aid by a medical practitioner, and may result in temporary or permanent disablement. Although hazards and risk need not necessarily result in exposures or accidents, owing to the consequential lack of optimum H&S, they can result in any, all, or a combination of the following: reduced productivity; rework, and falling behind schedule. In essence, hazards and risk contribute to and constitute non-conformances. Reduced productivity also constitutes a non-conformance, and it, along with other non-conformances, contributes to increased cost, and ultimately, poor performance (OH&S/overall).

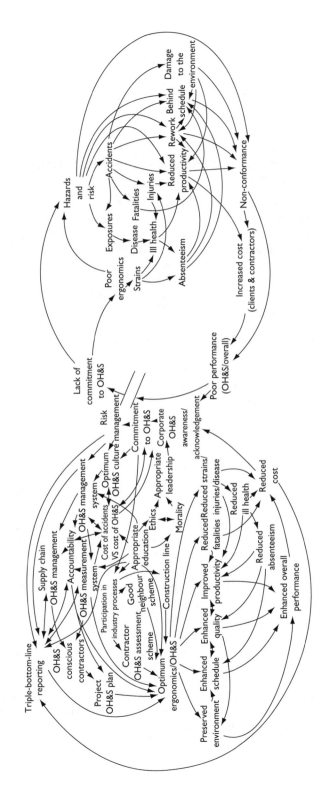

Figure 12.1 The holistic role of CSR in OH&S and the role of OH&S in overall performance.

The left-hand ellipse indicates that the only way to break the cycle, represented by the right-hand ellipse, depicted by the break in the arrow between poor performance (OH&S/overall) and lack of commitment to OH&S, is by way of corporate OH&S awareness/acknowledgement that the lack of such commitment contributes to poor performance (OH&S/overall).

Awareness and acknowledgement of the role and importance of commitment to OH&S is a prerequisite for commitment and change. Such awareness and acknowledgement engenders the realisation of the importance of risk management and provides the catalyst for the development of an optimum OH&S culture. Such a culture is likely to engender a focus on the cost of accidents versus the cost of OH&S, which in turn will enhance accountability and reinforce the OH&S culture, corporate OH&S awareness/acknowledgement, and commitment to OH&S. An optimum OH&S culture is likely to manifest itself in the development and implementation of an OH&S management system, which in turn will result in management of OH&S throughout the supply chain and management being held accountable for OH&S. An OH&S management system would include an OH&S measurement system, which in turn would engender focus on and the quantification of the cost of accidents versus the cost of OH&S. Realisation of the importance of risk management will promote triple bottom line reporting, which in turn will engender optimum ergonomics/OH&S, an OH&S measurement system, accountability and supply chain OH&S management. Supply chain OH&S management and the appointment of OH&S conscious contractors should engender the development and implementation of OH&S plans on projects. OH&S plans in turn engender optimum ergonomics/OH&S, which engenders a preserved environment, enhanced schedule, enhanced quality, improved productivity, reduced fatalities, and reduced strains/injuries/disease. Preserved environment promotes enhanced schedule. Enhanced schedule and enhanced quality, and enhanced quality and improved productivity are mutually reinforcing. Preserved environment, enhanced schedule, enhanced quality, and improved productivity all engender enhanced overall performance.

Appropriate education, one of two 'independent' activities presented in the centre of the ellipse, is a prerequisite for the existence of ethics relative to OH&S and influences morality. Morality in turn includes ethics, and the existence of ethics will promote morality. However, appropriate education, which is the primary medium for the acquisition of knowledge and the development of skills, is a prerequisite for optimum ergonomics/OH&S. Appropriate education also influences the core competencies in the form of self-concept, traits, and motives, which impact substantially on the realisation of optimum ergonomics/OH&S.

Commitment to OH&S and appropriate education increase the likelihood of participation in industry processes such as 'contractor OH&S assessment scheme', 'good neighbour' and 'construction line', which in turn engender optimum ergonomics/OH&S. Furthermore, H&S measurement system, accountability, project OH&S plans, and triple-bottom-line reporting all engender optimum ergonomics/OH&S.

Optimum ergonomics/OH&S in turn engender all, any or a combination of the following: reduced strains/injuries/disease; reduced fatalities; improved productivity; enhanced quality; enhanced schedule and preserved environment, which ultimately result in enhanced overall performance and reduced cost. A further benefit is the synergy between preserved environment, enhanced schedule, enhanced quality and improved productivity. Reduced fatalities result in improved productivity and reduced cost. Reduced strains/injuries/disease result in improved productivity, reduced ill health and reduced cost. Reduced ill health results in improved productivity and reduced absenteeism, which in turn results in improved productivity, enhanced quality and enhanced schedule.

The ellipse cycle should be perpetuated as a result of the reinforcement of corporate OH&S awareness/acknowledgement resulting from enhanced overall performance and reduced cost.

Conclusions

The construction industry worldwide does not have a favourable OH&S record and generates a disproportionate number of fatalities, injuries and disease. In the light of continuing poor performance, OH&S represents a key challenge for the construction industry. Poor OH&S results in considerable human suffering and reduces the quality of life of current and future generations. Construction cannot claim to be a socially responsible or socially sustainable industry until it can improve its OH&S performance. Stakeholder theory, which positions workers as a key stakeholder group, offers significant opportunity for improving the construction industry's OH&S performance and introducing greater equity into the allocation of risk within construction firms through the recognition that people exposed to OH&S risks must have a say in decision-making about the magnitude of these risks and how they might best be controlled. Further, although OH&S legislation requires that organisations address OH&S, there is the issue of accountability, which requires that managers be held accountable in terms of their performance relative to their authority and responsibility, OH&S included. One of the biggest challenges facing the construction industry is the need to move from a culture of finger-pointing, in which clients, designers and contractors blame one another for the existence of hazardous work conditions, to one of collaboration and shared responsibility. Regardless of the role an organisation fulfils in the construction industry, OH&S should be an integral part of the organisation's business or professional conduct; indeed, it should be a core value. While trying to eradicate the culture in which blame-casting is the default response to OH&S issues, the concept of a 'no-blame' culture is equally problematic. Instead, what is needed is a just culture in which responsibility for OH&S is appropriately allocated; organisations and individuals are accountable for their actions; and, where appropriate, persons or organisations are blamed for negligent, wilful or reckless behaviour that results in workplace death, injury or illness.

Traditional financial reporting is being increasingly questioned as it fails to address stakeholders' broader concerns. Consequently, the concept and practice of triple-bottom-line (TBL) reporting has evolved, which entails reporting relative to a range of financial, environmental and social indicators. Given this, organisations need to manage their supply chains in terms of the procurement of materials, plant and equipment and services, and should subscribe to industry schemes and generic systems that promote better practice and CSR ratings. Given the fragmentation and complexity of the construction supply chain, the integration of OH&S presents a challenge to the industry. However, this challenge must be taken up, if the industry is to improve its OH&S performance. Strategies for integrating OH&S into supply-chain management include the development of approved provider lists of suppliers of goods and services and the implementation of mentoring schemes to develop OH&S competency in small-to-medium-sized suppliers and subcontractors.

In conclusion, given that OH&S is concerned with the health, safety and well-being of people, organisations should afford it greater status than that afforded to other performance parameters. Furthermore, a project cannot be regarded as successful if fatalities, injuries and disease have been incurred during its course. The construction industry is presented with many challenges in improving its OH&S performance. Overcoming these challenges will require improved collaboration between industry and project stakeholder groups, an acknowledgement and acceptance of OH&S responsibility by all parties and creative approaches to better integrating OH&S through the construction supply chain.

References

Amick, B.C. III, Habeck, R.V., Hunt, A., Fossel, A.H., Chapin, A., Keller, R.B. and Katz, J.N. (2000) Measuring the impact of organisational behaviours on work disability prevention and management, *Journal of Occupational Rehabilitation*, 10: 21–37.

Armstrong, T.J., Haig, A.J., Franzblau, A., Keyserling, W.M., Levine, S.P., Martin, B.A., Ulin, S.S. and Werner, R.A. (2000) Medical management and rehabilitation in the workplace: emerging issues, *Journal of Occupational Rehabilitation*, 10: 1–6.

Association of British Insurers (2002) *Getting Back to Work: A Rehabilitation Discussion Paper*, London.

Australian Associated Press (2004) Victory brings tears after years of fighting for compensation, Australian Associated Press, 24 December.

Australian Broadcasting Corporation (2004) Call for more socially responsible corporate laws, Australian Broadcasting Corporation, 23 December.

Australian Financial Review (2004a) Law inadequate to protect victims of negligence, *Australian Financial Review*, 31 July, 21.

Australian Financial Review (2004b) Corporate social responsibility moves onto the front page, *Australian Financial Review*, 31 July, 20.

Australian Financial Review (2004c) Hardie backdown leaves many questions, *Australian Financial Review*, 14 August, 6.

Australian Financial Review (2004d) Welcome focus on safety of workers, *Australian Financial Review*, 16 December, 39.

Carson, W.G. (1979) The conventionalisation of early factory crime, *International Journal of the Sociology of Law*, 7: 37–60.

Cheadle, A., Franklin, G., Wolfhagen, C., Savarino, J., Liu, P.Y. and Weaver, M. (1994) Factors influencing the duration of work-related disability: a population-based study of Washington State workers' compensation, *American Journal of Public Health*, 84: 190–196.

Clayton, A. (2002) *The Prevention of Occupational Injuries and Illness: The Role of Economic Incentives*, Working Paper 5, Canberra: National Research Centre for Occupational Health and Safety Regulation, Australian National University.

Cropanzano, R., Chrobot-Mason, D., Rupp, D.E. and Prehar, C.A. (2004) Account-ability for corporate injustice, *Human Resource Management Review*, 14: 107–133.

Dembe, A.E. (2001) The social consequences of occupational injuries and illnesses, *American Journal of Industrial Medicine*, 40: 403–417.

Department of the Environment, Transport and Regions (DETR) (1998) *Rethinking Construction*, London: DETR.

Eckhardt, R.E. (2001) The moral duty to provide for workplace safety, *Professional Safety*, August: 36–38.

Entec (2000) *Construction Health and Safety for the New Millennium*, Health and Safety Executive Contract Research Report 313/2000, Norwich; HMSO.

Field, S. and Jorg, N. (1991) Corporate liability and manslaughter: should we be going Dutch? *Criminal Law Review*: 156–171.

Fisse, B. (1994) Individual and corporate criminal responsibility and sanctions against corporations, in Johnstone R. (ed.), *Occupational Health and Safety Prosecutions in Australia*, Centre for Employment and Relations Law, Melbourne: The University of Melbourne.

Fisse, B. and Braithwaite, J. (1993) *Corporations, Crime and Accountability*, Cambridge: Cambridge University Press.

Gillen, M., Faucett, J.A., Beaumont, J.J. and McLoughlin, E. (1997) Injury sever-ity associated with non-fatal construction falls, *American Journal of Industrial Medicine*, 32: 647–655.

Gillen, M., Baltz, D., Gassel, M., Kirsch, L. and Vaccaro, D. (2002) Perceived safety climate, job demands, and coworker support among union and nonunion injured construction workers, *Journal of Safety Research*, 33: 33–51.

Greenwood, M.R. (2002) Ethics and HRM: a review and conceptual analysis, *Journal of Business Ethics*, 36: 261–278.

Guberan, E. and Usel, M. (1998) Permanent work incapacity, mortality and survival without incapacity among occupations and social classes: a cohort study of aging men in Geneva, *International Journal of Epidemiology*, 27: 1026–1032.

Health and Safety Executive (HSE) (2002) *Revitalising Health and Safety in Construction*, London: HSE.

Henriques, I. and Sadorsky, P. (1999) The relationship between environmental commitment and managerial perceptions of stakeholder importance, *Academy of Management Journal*, 4: 89–99.

Hogg-Johnson, S., Frank, J.W. and Rael, E. (1994) *Prognostic Risk Factor Models for Low Back Pain: Why They Have Failed and a New Hypothesis*, Working Paper 19, Toronto: Ontario Workers' Compensation Institute,

Hopkins, A. (1995) *Making Safety Work*, St Leonards, NSW: Allen and Unwin.

Hunt, A. and Habeck, R. (1993) *The Michigan Disability Prevention Study: Research Highlights*, Kalamazoo: W.E. Upjohn Institute for Employment Research.

International Labour Organization (2005) *Prevention: A Global Strategy: The ILO Report for World Day for Safety and Health at Work 2005*. Online. Available at: http://www.ilo.org/public/english/protection/safework/worldday/products05/report05_en.pdf (accessed 27 November 2006).

James, P., Cunningham, I. and Dibben, P. (2003) *Job Retention and Vocational Rehabilitation: The Development and Evaluation of a Conceptual Framework*, Research Report 106, Health and safety Executive, Norwich: HMSO.

Johnstone, R. (1999) Improving worker safety: Reflections on the legal regulation of OH&S in the 20th century, *Journal of Occupational Health and Safety – Australia and New Zealand*, 15: 521–526.

Johnstone, R. (2002) *Safety, Courts and Crime*, Working Paper 6, National Research Centre for OH&S Regulation, Canberra: The Australian National University.

Kletz, T. (1993) *Lessons from Disasters: How Organisations Have No Memory and Accidents Recur*, Rugby: Institution of Chemical Engineers.

Korhonen, J. (2003) Should we measure corporate social responsibility? *Corporate Social Responsibility and Environmental Management*, 10: 25–39.

Larsson, T. and Field, B. (2002) The distribution of occupational risks in the Victorian construction industry, *Safety Science*, 40: 439–456.

Lindsay, N. (2003) Union steps up worker safety push, *Australian Financial Review*, 10 (February): 7, Sydney: John Fairfax Holdings Ltd.

Lingard, H. and Saunders, A. (2004) Occupational rehabilitation in the construction industry of Victoria, *Construction Management and Economics*, 22: 1091–1101.

McIntosh, G., Frank, J., Hogg-Johnson, S., Bombardier, C. and Hall, H. (2000) Prognostic factors for time receiving workers' compensation benefits in a cohort of patients with low back pain, *Spine*, 25: 147–157.

McWilliams, G., Rechnitzer, G., Deveson, N., Fox, B., Clayton, A., Larsson, T. and Cruickshank, L. (2001) *Reducing Serious Injury Risk in the Construction Industry*, Policy Research Report 9, Melbourne: Monash University Accident Research Centre.

Maignan, I., Hillebrand, B. and McAlister, D. (2002) Managing socially responsible buying: how to integrate non-economic criteria into the purchasing process, *European Management Journal*, 20: 641–648.

Major Projects Association (2001) *Hong Kong International Airport*, MPA Seminar held at the Institution of Civil Engineers, London, 25 April.

Neal, D. (1996) Corporate manslaughter, *Victorian Law Institute Journal*, 30 (10): 39–41.

Nurminen, M. (1997) Reanalysis of the occurrence of back pain among construction workers: modeling for the interdependent effects of heavy physical work, earlier back accidents and aging, *Occupational and Environmental Medicine*, 54 (11): 807–811.

Peebles, L., Kupper, A. and Heasman, T. (2002) *A Study of the Provision of Health and Safety Information in the Annual Reports of the Top UK Companies*, HSE Contract Research Report 446/2002, London: HMSO.

Rethinking Construction Limited (2002) *Rethinking Construction: 2002*, London: Rethinking Construction Limited.

Rigby, N. (2003) *Designer Initiative 17 March 2003: Final Report*, Scotland and Northern England Unit, Construction Division, Health and Safety Executive.

Robertson, C. and Fadil, P.A. (1998) Developing corporate codes of ethics in multinational firms: Bhopal revisited, *Journal of Managerial Issues*, 10: 545–568.

Rowan, J.J. (2000) The moral foundation of employee rights, *Journal of Business Ethics*, 25: 355–361.

Saravanamuthu, K. (2004) What is measured counts: harmonized corporate reporting and sustainable economic development, *Critical Perspectives on Accounting*, 15: 295–302.

Schein, E.H. (1992) *Organisational Culture and Leadership*, Jossey-Bass Inc, San Francisco, CA.

Schein, E.M. (2004) *Organizational Culture and Leadership*, San Francisco: John Wiley & Sons.

Shrivastava, P. (1995) Industrial/environmental crises and corporate social responsibility, *The Journal of Socio-Economics*, 24: 211–227.

Smallwood, J.J. (1996) The role of project managers in occupational health and safety, *Proceedings of the First International Conference of CIB Working Commission W99 Implementation of Safety and Health on Construction Sites*, Lisbon, Portugal: Balkema, Rotterdam, 227–236.

South China Morning Post (1997) Workers dead and buried, 5 July 1998.

Standards Australia (2003) AS 8003-2003 *Corporate Social Responsibility*, Sydney: Standards Australia.

The Age (2004a) Asbestos firm 'lied to save money', 29 July, 1.

The Age (2004b) Hardie's sins stripped bare, 31 July, 1.

The Age (2004c) The road to a corporate inquisition, 31 July, 1.

The Australian (2004a) Inquiry urges jail for asbestos chief, 12 August, 3.

The Australian (2004b) Further doubt on offer by Hardies, 16 August, 4.

The Australian (2005) Combet proved himself a Hardie battler, 4 January, 2.

The Courier-Mail (2004) Latham promises funding for asbestos victim claims, 16 August, 2.

The Daily Telegraph (2004) Hardie to feel the heat after backflip, 16 August, 24.

The Guardian (2004) Far too soft on crime, 5 October.

The Lancet (2003) Who will take responsibility for corporate killing? 7 June, 1921.

The Sydney Morning Herald (2004a) When building wealth turns to shame, 29 July, 21.

The Sydney Morning Herald (2004b) Victims hail first-round win but warn of long fight, 14 August, 4.

Turner, B.A. and Pidgeon, N.F. (1997) *Man-made Disasters*, 2nd edn, Oxford: Butterworth-Heinemann.

Wells, C. (1994) Corporate liability and consumer protection: Tesco v Natrass revisited, *The Modern Law Review*, 57: 817–823.

Chapter 13

Corporate social responsibility in the American continental construction industry

Aguinaldo dos Santos and Fausto Amadigi

Introduction

The notion of 'social responsibility' in construction is not a new theme in the American continent; recently it became a clear source of competitive advantage. International organisations, trade unions, human rights lobbyists and regulators have contributed to bring the attention of construction companies to ethical business behaviour. Nowadays, with an ever-increasing frequency in this continent, construction companies are coming under scrutiny to prove that their activities are conducted in a way that is socially acceptable to those who may be touched by it.

The importance of this change lies in the fact that the construction industry plays one of the most significant economic and social roles when compared to any other industrial sector in this continent. However, large social, economic and cultural differences across all its regions demand different approaches and priorities on the process of implementing more socially responsible business practices in construction. Social responsibility in US construction, for instance, has grown through regulation and industry standards and was initially driven by responsible business operations rather than by community investment. The 'Big Four' regulatory agencies shaped much of the baseline for responsible corporate business operations: OSHA (Occupational Safety and Health Administration), EEOC (Equal Employment Opportunity Commission), CPSC (Consumer Product Safety Commission) and EPA (Environmental Protection Agency).

These agencies created and continue to maintain standards for responsible corporate business practices that have become the benchmark for corporate social responsibility with respect to day-to-day operations of business. More recent examples of industry-specific and sector-wide regulations include the Community Reinvestment Act in the banking sector, the Clean Air Act, the Foreign Corrupt Practices Act and, post-Enron, the Public Company Accounting Reform and Investor Protection Act.

In Latin America, regulations to induce or demand more responsible corporate operations are less prevalent, particularly outside of the Mexican and

Mercosur markets where foreign investment has demanded socially responsible practices. Moreover, workers and social groups who hold less power and influence, such as women or ethnic populations, have been denied the opportunity to progress the case for regulating social responsibility standards in business. Quite often, Latin American corporations possess more wealth and power than the governments themselves in their respective countries. Thus, without strong pressure from civil society, governments are less likely to create standards that imply a cost to such corporations (Gutiérrez 2004).

In the poorest areas of the continent, construction faces two major challenges regarding its social contribution: to reduce the large housing deficit and, also, promote interventions on the large number of houses that do not offer minimum living conditions. In Brazil alone, the deficit of new houses is estimated as 6.5 million and the number of inadequate houses (e.g. lack of toilet, no sewage system) is also estimated to be around the same. The problem here is that most of the people in this market do not have enough income to afford current construction costs.

Valuing diversity

Following ILO conventions 100 (ILO 1951) and 111 (ILO 1958), no discrimination in hiring, compensation, access to training, promotion, termination and retirement is permitted on the basis of race, caste, national origin, religion, disability, gender, sexual orientation, union membership or political affiliation. The employer cannot interfere in the exercise of the rights of personnel to observe tenets or practices, or to meet needs relating to any of the previously listed categories. The occurrence of behaviour that is deemed sexually coercive, threatening, abusive or exploitative shall not be allowed by the company. It should be noted that room is left here for discrimination to occur in terms of age, language, marital status, etc.

Despite the fact that all countries in the continent have signed to ILO conventions, it is still observed that great inequity exists regarding race, particularly with Black people, and it is clearly more evident within the construction industry. Recent statistics in Brazil show that Black people receive lower salaries than White people, and there are no signs of any dramatic changes to this pattern in the medium term. Although poverty has shown a significant reduction since the 1990s (the number of people below the poverty line reduced to five million between 1992 and 2001), the number of poor Black people increased to 500,000 (PNUD 2005).

In general, South and Central America present fewer efforts to value diversity in construction when compared with North America. Gender issues are practically absent in the discussions among construction organisations. In fact, the presence of women in construction is almost absent in the entire American continent. According to OSHA (1999) nearly 60% of women aged 16 and over participate in the workforce in the United States of America. While women have made some gains in occupations traditionally pursued by men, construction trades still remain overwhelmingly male-dominated. In 1970, when OSHA was enacted, women made up less than 1% of workers in the construction trades. By 1995 that percentage had grown only to 2.3% (OSHA 1999), and in 2005 there is no indication that the overall situation has changed.

The small percentage of women working in the construction trades and the serious health and safety problems unique to female construction workers have resulted in a vicious circle. Safety and health problems, along with the macho culture prevalent on construction sites, create a barrier to women entering and remaining in this field. In turn, the small numbers of women workers on construction sites foster an environment in which these problems arise or continue (OSHA 1999).

Despite this situation, new legislation and improvements on education are changing the perspectives of construction regarding diversity. The Department of Labour (DOL) in the United States, for instance, encourages recipients of DOL financial assistance and federal contract awards to comply with laws such as the Civil Rights Act and the Americans with Disabilities Act, and ensures that federal contractors afford minorities, women, individuals with disabilities and veterans an equal opportunity to compete for employment and advancement.

The requirement that US government contractors refrain from discriminating in employment has been an established part of federal contracting policy since 1941, when President Roosevelt signed Executive Order 8802, and it continues today. The Employment Standards Administration (ESA) has the responsibility of ensuring that employers doing business with the Federal Government comply with the equal employment opportunity and affirmative action provisions of their contracts. ESA enforces laws and regulations protecting employees and applicants from discrimination regardless of their gender, race, colour, national origin, religion, disability or veteran status.

Child labour and children's future

Several international legal agreements address the problem of child labour and all countries in the continent have signed these agreements. One such agreement is Article 32 of the UN Convention on the Rights of the Child, which states that children should not be engaged in work that impedes their physical and mental development. ILO's Convention on the Elimination of the Worst Forms of Child Labor, ratified by 174 countries, has inspired a worldwide effort to eliminate degrading child labour.

However, despite this context, child labour is still a reality in the American continent not only in construction but in various other industries. Indeed, in a study carried out, the Associated Press estimated that about 300,000 children work illegally in the United States, some linked to prestigious companies like Heinz, J.C. Penny, Sears, Wal-Mart or Campbell. The list of violations is long; children (mostly White) are to be found working on construction sites in New York or harvesting grapes in California. US industries profit from the work of children in savings of US \$155 million per year (AP and Kruse 1997).

Child labour in construction is not totally banned in the continent by allowing the exceptions included in ILO Convention 138 and Recommendation 146 (ILO 1973). Requirements are established in the Convention on issues such as minimum age (15), work hours, young workers (those under 18), school attendance, workplace conditions and remediation of children. The companies must not use or support the use of corporal punishment, mental or physical coercion as well as verbal abuse. It should be noted that the requirements of ILO on this issue are quite vague and subject to multiple interpretations. Areas such as mental and verbal abuse are not properly detailed, to be auditable. Many other unfair disciplinary practices need to be addressed (e.g. deprivation of work, lowering of standards in working conditions, isolation) (Curado and Santos 1999).

In many cases, workers and their families live on the construction site throughout the building process and everyone in the family participates in the production operations, except for the very young. Because of the strength and skill demanded by construction work, children are engaged mainly in activities such as cement-mixing, fixing of windows and pipes, painting and, in particular, brick-making. In some areas of South and Central America, children as young as five or six years of age still can be found working for an average of 4–6 hours per day throughout the brick-making process.

The manufacture of bricks can be extremely prejudicial to a child's health because working conditions are usually unsanitary, unsafe and have a negative effect on the physiology of the child. Preparing the clay and placing it in moulds require a significant degree of effort, since the clay has a solid, heavy consistency. ILO (1999) calls attention to the fact that this excessive burden retards or deforms the normal physical growth and development of children and this is often further aggravated by undernourishment. Furthermore, lack of potable water and the constant contact with clay contributes to the propagation of infections.

A similar situation is observed in other parts of the construction supply chain in the poorest areas of the American continent. In Brazil, it is still possible to witness children earning money labouring in quarries, breaking stones to be used on concrete or roadworks. Working in a quarry is gruelling for anyone, but especially so for a child. It is not only hard because of the heavy work but, also, the dry air makes breathing difficult.

There is a growing number of initiatives from government, industry and non-government organisations that are contributing to reduce the involvement of child labour in construction in the continent. Examples of such initiatives are the projects run in Ecuador by the National Institute for Children and the Family (INNFA), with the support of the ILO's International Programme on the Elimination of Child Labour (IPEC). The projects are located in the Nueva Aurora area of south Quito and the north-west region of the city of Cuenca. There are 500 families with 300 children in both areas. Half of the men, and most of the women, along with their children, work in brick-making. The aim of the project is to eliminate child labour completely from the brick-making process. Scholarships and apprenticeships are offered to enable children to enter or stay longer in school.

The project provides income-generating opportunities and also runs a centre for technological innovation focused on the increase in productivity and quality of the manufacturing process and, in turn, the family income. Cooperatives have been established to provide further security for the families, helping to strengthen community-based organisation. The project also provides health services, including a community network of health workers who promote awareness of health and safety issues related to the brick-making process.

Child labour is often locked into a vicious cycle from generation to generation. The earlier they start in construction, the less time they spend in school, and the result is a lifetime of poor employability, poor skills, poor productivity and little income. In an effort to break this vicious cycle, Brazil is working closely with the United Nations' International Labor Organization (ILO) and the United Nations Children's Emergency Fund (UNICEF) with the implementation of the Bolsa Escola (School Scholarship), which allows children to attend school full-time. It also provides for special classes in the arts and other enrichment courses that round off their education, including sports and computer lessons. Unfortunately, not every disadvantaged child in Brazil benefits from the Bolsa Escola, and child labour will still occur whilst the vicious cycle of poverty is not broken.

The industry itself is also taking onboard activities that previously were regarded as government responsibilities. Holcim Corporation, for instance, has maintained a partnership with the Vivamos Mejor Foundation – a charitable organisation that initiates and supervises social development projects in Latin American communities. In Maria Auxiliadora, poverty is one of the main causes of high youth unemployment and school drop-out rates among 10- to 17-year-olds. Vivamos Mejor and Holcim try to support children while they receive a solid educational foundation, as this will enable them to sustain themselves both economically and socially throughout their life. Begun in 2003, the project had the goal of reducing the school drop-out rate of 28% by half by 2005, as well as substantially lowering youth unemployment. This initiative is similar to Holcim projects near its plants in Brazil, Nicaragua and Venezuela.

Despite limited government framework and structure and a lack of proper enforcement of responsible business practices, corporations in Latin America interested in social responsible practices have taken initiatives by themselves regarding children's future. For example, ABRINQ Foundation, a non-profit organisation in Brazil, specialising in the toy market, offers a logo to companies that are committed to fight the use of child labour. Corporations are certified through ABRINQ's Child Friendly Company Program after a series of social audits by unions, employees, and NGOs. Companies use ABRINQ's logo to disseminate their corporate value within the community of youth, and this has been shown to be a source of competitive advantage.

Ethics and corporate governance

There is a growing number of efforts in the American continent from government, non-government and industry organisations to embed new values on social responsibility into organisational culture. Instituto Ethos (www.ethos.org.br), for instance, has created key indicators that are in use by the construction industry to assess its own performance in terms of social responsibility. Social responsibility awards based on such indicators are also inducing a more ethical behaviour of construction companies, since clients are gradually becoming more aware of the importance of corporate social responsibility.

Nevertheless, there is a serious need for bringing all dimensions of ethics of construction in the American continent to higher levels. In South and Central America, there is a lack of regulating bodies with instruments to control and support firmly ethical practices. The larger part of the construction industry in this region operates at an informal level, avoiding all paperwork and taxes of the formal market. It is an unfair competition with the formal market since taxes in most countries reach more than 40% even for low-income housing, and the informal market avoids many of these taxes.

Furthermore, most construction clients in South and Central America have at most one experience in their lifetime regarding housing acquisition and, thus, have little knowledge on construction specification. This situation is exacerbated with the lack of institutions that could effectively avoid the commercialisation of sub-standard products. ISO 9000 and regional/national quality programmes dedicated to the construction sector have contributed to improve this situation in the formal market. Notwithstanding these initiatives, construction remains as one of the industrial sectors with the poorest record in terms of business ethics in all dimensions.

The American Council for Construction Education (ACCE) calls for industry associations to take the lead in establishing an ethics code of conduct for the construction industry. Ethics is taught in some construction science and management postgraduate programmes in America. However, few undergraduate programmes in the continent present a formal effort to disseminate social ethics among students, although some argue that it should be part of every learning experience within the academy. Construction companies are lax in including ethics in their company training and decision-making. Hence, ACCE offers ethics training for its partners including such topics as fair contracts, payment and equitable risk sharing, among other initiatives.

An important initiative related to the dissemination of ethics is the strategy adopted by the New York Stock Exchange (NYSE). Listed companies on the NYSE must comply with corporate governance rules and ethics. They must adopt and disclose a code of business conduct and ethics for directors, officers and employees, and promptly disclose any waivers of the code for directors or executive officers. An ethical code can focus the board and management on areas of ethical risk, provide guidance to personnel to help them recognise and deal with ethical issues, provide mechanisms to report unethical conduct, and help to foster a culture of honesty and accountability. However, admittedly, no code of business conduct and ethics can replace the thoughtful behaviour of an ethically inclined director or employee.

Transparency

The most competitive companies in the American continent in terms of market growth and profitability are often the same companies that present a better transparency of their activities and strategies to society in their dialogue with stakeholders, relations with competitors and social reporting. One example of such approach to business is that of Gênesis Group in Brazil where the company made continuous efforts to improve its communication with all stakeholders. The organisation practises a direct communication with its current and former clients as well as potential clients. Every three months the company sends a magazine called *Wellness* (2007) to them with detailed information on its projects along with information on well-being.

Monthly, the CEO of the Gênesis Group sends a signed letter describing the state of the construction work to every client, including actions regarding the protection of the environment. This letter also informs all clients about the company's efforts to improve the workers' quality of life. To further increase the transparency, the company promotes face-to-face meetings among stakeholders during events such as dinners and weekend parties. These events include seminars and photo exhibitions focused on increasing awareness of the environment surrounding the company's existing and former projects.

However, transparency is still lacking is most of the construction industry and that is more evident in the procurement process of government contracts. The lack of transparency is one of the reasons for the widespread corruption in government contracts, particularly in Central and South America. Implementation of e-commerce systems for government procurement is one of the strategies in use to tackle the problem (see Obrasnet in Brazil; http://www.obrasnet.gov.br/). In the Paraná State, Brazil, the state government places all government spending on the Web, and governments in other states are following the same example.

In Guatemala, an agreement between Transparency International and the Ministry of Finance has launched the Transparency Award in Public Procurement – a monthly grant awarded to the institution that best complies with the public procurement law. At the same time, questionable transactions and complaints about institutions that are not using the system properly are made public. In 2003 alone, 37 procurement processes complied with the procurement law, in 2004, over 6,000, and in 2005, at least 12,000. The alliance between Acción Ciudadana, the government, the media, and the private sector has been a key factor for promoting the use of the system.

Transparencia Brazil (2007), a branch of Transparency International, in collaboration with a Rio de Janeiro-based think-tank and the Santa Catarina Court of Audit, launched an innovative public contracting project in the Santa Catarina State municipalities. The project is based on the idea of improving access to, and processing of, information for citizens and authorities, with the goal of enhancing the watchdog capacity of civil society. The tool uses public data on purchasing from 293 municipalities and produces comparisons between costs of commodities and procurement processes.

The process of achieving transparency, particularly regarding contributions to political campaigns, is also under great scrutiny. The scale of corruption is magnified by the size and scope of the construction sector, particularly in the developing countries within the American continent. The lack of transparency in contracting processes for large-scale infrastructure projects can have profound negative consequences for economic and social development. The negative impact becomes especially obvious in the American continent in the context of post-conflict reconstruction and relief projects after human catastrophes.

Corruption in the construction sector not only plunders economies, but it actually shapes them (Transparency International 2004). Corrupt government officials steer social and economic development towards large capital-intensive infrastructure projects that provide fertile ground for corruption and often neglect small projects on housing, health and education programmes which are of so much importance in the poorest areas of the American continent (Transparency International 2004). Hence, there is no doubt that implementing transparency in government procurement, in conjunction with better education of the population, is a necessary step to inhibit corruption.

Formal and informal job market

The law in most countries in the American Continent provides safeguards for workers and companies on dismissals. To guarantee these safeguards, in most countries an official contract is required to formalise the relationship between the worker and the company. In Brazil, for instance, if an employer decides to fire a worker, it has to pay compensation in proportion to the worker's time in the company. Once dismissed, the worker can receive unemployment salary which is set in proportion to his/her previous salary.

The huge burden on the state governments regarding retirement funds has motivated most countries to privatise such funds or allow alternative funds in parallel to state funds. However, in countries like Brazil, it is compulsory for the company to contribute to the federal retirement fund regardless of the option for a private retirement fund. If a construction worker has worked all his/her life without making such contribution, he/she can still retire, but then he/she will receive solely the minimum national salary which is barely enough to survive.

Construction workers in general are badly paid in the continent and often operate in the informal market, which means little protection from social security programmes. In the United States, statistics shows that the industry relies very much on 'illegal immigrants' or 'unauthorised workers'. A similar problem occurs in South and Central America with the migration of unskilled and illiterate people from the countryside to larger towns.

'Unauthorised workers' in the United States make up 4.3% of the nation's workforce. This may seem a small percentage, but the pressure of its presence helps keep wages for unskilled jobs low. And many of the jobs are 'off the books', meaning the government may be forgoing billions of dollars a year in income tax collections. That figure, however, is partially offset by employers' withholding taxes for illegal workers who never file returns or seek benefits. The service sector employs the highest number of illegal immigrants at 33%, followed by the construction industry, production and food processing, and farming (Pew Hispanic Center 2002).

Every day, thousands of illegal workers head to the sidewalks to wait for a job. There, alongside other men, they are picked up by trucks. It is a morning ritual played out regularly in cities and towns in the United States where illegal immigrants scramble for work in a country that comfortably accepts their work while disavowing their right to be there. The work is steady and offers a reasonable payment, and no one asks for documents or identification. On paper, many of these workers do not exist. Fake social security numbers and birth certificates make sure of that, and construction is one of the main destinations for these illegal workers.

Illegal immigrants earn far less than the rest of the US population. Their average family income of $27,400 is more than 40% below the legal immigrant's or native family's income of about $47,700. The reason for such discrepancy is that illegal immigrants work cheaply and do not complain about that; those who do complain are easily replaced. They have little bargaining power, and employers take advantage of that. Enforcement is lax, especially in a post-9/11 world. The government does not have the time or resources to devote to rounding up illegal construction workers; instead, it focuses on national security and critical infrastructure sites (Pew Hispanic Center 2002). Within this environment, and also due to the lack of proper training, Mexican workers are more likely to die on the job than are native-born workers.

Workforce dialogue and participation

The right of the personnel to create and become members of trade unions of their choice must be respected by the employer, according to ILO Convention 87 (ILO 1948). The company must also allow for collective bargaining. In those countries where local laws affect these rights, the company should attempt to find 'parallel means' to enforce them. Following ILO Convention

135, the Standard protects the personnel representatives, by allowing them access to fellow workers and preventing their discrimination.

North America is far more unionised than South and Central America in terms of not only the number of unions but also their diversity. In North America, there are unions for plumbers, electricians, bricklayers, and so on. In South and Central America, construction workers usually get associated under umbrella unions, such as the Construction Workers' Union of São Paulo, which has thousands of associates covering all areas of construction expertise.

Health and safety programmes have also been a channel for participation of workers in construction management decisions in the continent. In countries like Brazil, the law demands each construction site have a health and safety committee that has the power to implement changes and improvements on site.

Efforts to implement quality programmes in construction have contributed to an increase in the participation of workers in the decision-making process in leading companies in the continent. Nevertheless, the significant difference in educational level between construction workers and site managers often creates a burden on promoting more participatory management. Construction companies' hierarchical structures usually adopt a multi-layer configuration where managers often adopt an autocratic approach for dealing with the decision-making process. The reality in most companies is a hierarchical system where holding information means power, and it is used to limit the progress of new workers.

Quality of life at work

Over the years, construction has ranked among industries with the highest rates of both fatal and non-fatal occupational injuries, and this is no different in the American continent. The number of fatal occupational injuries in US construction, which is often pointed out as the most advanced in the continent, reached 1,225 in 2001, with an incidence rate of 13.3 per 100,000 workers employed. For the same year, the construction industry in that country experienced 481,400 non-fatal injuries and illnesses at a rate of 7.9 per 100 full-time workers in the industry (BLS 2002).

The National Traumatic Occupational Fatalities (NTOF) surveillance system, maintained by the National Institute for Occupational Safety and Health (NIOSH), confirms the bad image of the industry regarding working conditions. Analysis of death certificates of workers who died from traumatic injuries in the workplace, at the age of 16 years or more, shows that falls from elevations were the fourth leading cause of occupational fatalities between 1980 and 1994. The 8,102 deaths due to falls from elevations accounted for 10% of all fatalities at an average of 540 deaths per year (NIOSH 2000).

A similar bad record in terms of safety is seen in other regions of the American continent, although with peculiarities in each region. Workers' health issues in the Caribbean, for instance, are reflective of the economic, social and environmental characteristics of the region. General environmental considerations affecting working conditions in construction include high temperatures and humidity, which bring a variety of health problems. Technology and site management are also factors that differentiate the various regions across the continent in terms of health and safety. In South and Central America, for instance, the construction industry is generally more labour-intensive, and the components and materials often demand a substantial human effort to move and assemble.

In this context, it is no surprise that construction has one of the worst public images among the industrial sectors in the American continent. As in other continents, the sector is often perceived as being dirty, dangerous and dull (3Ds), with adversarial relationships at all levels and great environmental insensitivity. This image has a serious negative impact on the recruitment of new young talents. A *Wall Street Journal* Almanac Poll of high-school-aged vocational technology students ranked 'construction worker' 248th out of 250 possible occupation choices, ahead of 'dancer' and 'lumberjack' and just edged out by 'cowboy'.

Relationship with the community

Latin America is a region where inequality has created enormous gaps between social groups, and the construction site is the place where this social gap is most evident. At the same time, there is a growing perception among society, enhanced by the media, that socially responsible activities are not any more confined to elected politicians and public administrators. Thus, multilateral agencies have been turning to the construction sector to support the implementation of more socially responsible practices.

After the various economic crises in Latin America, many more businessmen and businesswomen have come to understand that 'there is no healthy business in a sick society' (Gutiérrez 2004). They have embraced philanthropic traditions that in the past were rooted in religious beliefs and today respond to civic demand for social responsibility. A special issue of *Harvard Review of Latin America* traced such practices from ancient to modern days and presented it as the evolution 'from charity to solidarity' (Erlick 2002).

Altruism and solidarity have been significant drivers in the Latin American construction sector, resulting to a large extent from the tradition of charity derived from the region's Catholic background. The ethical imperative has been the driving force for business leaders who, for example, target their socially responsible efforts on education. Indeed, many construction leaders have led their businesses and those of their colleagues to collaborate with solving public-education problems. Their contributions often came along regardless of any potential advantage for their businesses, since there is a high worker turnover level in construction, and more educated workers might not be there in the next construction project.

Two other kinds of utilitarian motives can be attributed to the altruism and solidarity of construction companies in South America. First, involvement of construction companies with the social sector can be used to minimise the occurrence of an identified risk or to be prepared to face its consequences, particularly when the firm's operations might face opposition from the community. Roitter and Berger (2003) present the example of Ausol, a multinational corporation that constructs and manages highways in Argentina. Since highways have many characteristics that are disruptive, community goodwill became an effective 'operating licence' for Ausol. A second kind of utilitarian motive is the search for competitive advantages through the improvement of a construction firm's image in the matter of its social commitments.

Solidarity of construction through community involvement in Latin America is often seen with projects targeting the provision of low-income houses. An example of such projects is the cooperation between government, construction companies, non-government organisations, and suppliers on the PSH – Programa de Subsídio à Habitação de Interesse Social, a programme that subsidies low-income housing in Brazil. In this programme, one of the initiatives includes the provision of construction materials and training to families in order to allow them to build their own house. Two social issues are tackled at the same time in this programme: the need for reducing the housing deficit in the country and the improvement of employability of the low-income families.

The observed experiences in Latin America show that altruistic and utilitarian motivations can and do coexist and blend in practice when construction companies attempt to get closer to community needs and aspirations. Gutiérrez (2004) argues that altruistic motivations alone might not withstand economic downturns; utilitarian drives alone might distance social partners.

Social responsibility and its implication on design

The design of products and components used in construction can play an important role in enabling better social responsibility not only within construction companies but throughout the entire construction supply chain. For instance, it can enable more equity, allowing a fair distribution of resources at the local level or increase social cohesion by respecting fundamental rights and cultural diversity, helping to combat discrimination in all its forms (e.g. gender).

One particular design approach that has profound social impact in construction is the development of ready-to-assemble or do-it-yourself products. In Brazil, this engagement by users in assembling sub-systems and products is clearly understandable, given the fact that 85% of all Brazilians have an income lower than the minimum wage. Very frequently it is possible to find housing projects with social focus in which prospective residents partly or fully engage themselves in building their home. As a matter of fact, housebuilding in Brazil is mainly 'build-it-yourself', which means that the owner/resident is engaged in building the home or managing it. Figure 13.1

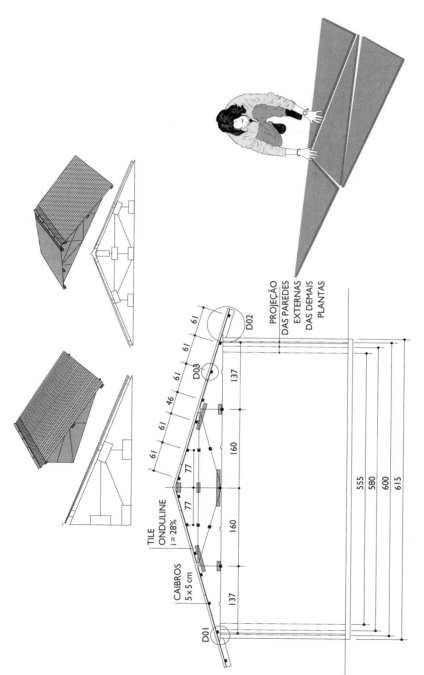

Figure 13.1 A ready-to-assemble product for roof structures of low-income houses.

shows an example of a ready-to-assemble product for roof structures, developed at the Federal University of Paraná and targeting low-income housing projects.

However, the existing solutions such as do-it-yourself, ready-to-assemble, 'build-it-yourself', or even 'joint-building' are quite limited regarding product design. Mostly, and following a pattern, plant and materials have not been developed or designed so as to allow users to truly assemble a product themselves in a safe and efficient way. Product ergonomics, information design, product sustainability, modular building, maintenance, expandability, and other key aspects in product design still need to be properly developed to reach a fully fledged DIY product.

DIY products already have a very well-structured market in developed countries around the world. The reason for this is the high cost of labour and also that people in those countries wish to assemble their own products owing to cultural issues. However, in the poorer areas in the American continent, DIY products have a strong economic appeal, given the possibility of reducing costs as the user himself/herself participates in assembling his/her own product.

Perspectives for the future

There is a clear, growing awareness in the construction sector in South America of new practices of CSR. It has evolved from a philanthropy perspective towards more complex CRS approaches, aligning CSR more directly with competitive corporate strategies. Companies in this continent have begun to consider more systematically the social context in which they operate. Endemic problems of some parts of the continent – lack of houses, insufficient sewerage systems, water supply, and energy systems, amongst others – have started to be seen as business opportunities.

New inclusive business-oriented models have emerged, in which companies not only look at the economic return of their investment but also the social benefit of enabling the access of poor people to employment, for instance. The World Business Council for Sustainable Development calls this approach 'sustainable livelihoods business', defined as 'doing business with the poor in ways that benefit the poor and benefit the company' (WBCSD 2006). The basic difference between this new model of CRS and the traditional model is the approach of the business:

> The focus on business implies a focus on profitability, which is important for several reasons. It means that a company's sustainable livelihood business projects become part of corporate mainstream thinking and activities. It also means – if the projects do indeed realize the goal of profitability – that they have no limited, fixed budget. The new business

can thus become immensely 'growable' and replicable, and thus lead to a much greater impact than corporate philanthropy.

(WBCSD 2006: 16)

One example of corporate initiative in line with this view is reported by WBCSD (2006), based on a case study at Cemex, a global company that produces cement, ready-mix concrete, and aggregates. Through a project called 'Patrimonio Hoy', Cemex developed in Mexico a pioneering experience of buying construction materials for the low-income population, that constitutes the informal pursuing of constructors of the do-it-yourself type.

The project had its beginning in 1998 with the objective to develop a new line of competitive business with poor people. Motivated by the opportunities of the low-income segment, Cemex implemented an arrangement based on groups to save money (they are the partners or customers of the Patrimonio Hoy), and promoters and local agencies of attendance, rendering services and technical assistance to the customer.

The innovation in this initiative is in the method to obtain the money to buy the products. It is based on a system to save money and obtain credit. Groups of low-income people are formed with three members with the objective to save money. With weekly personal contributions, they become capable of acquiring the products from the fifth week. Provided that they are able to save, the deliveries can be anticipated as 'on credit.' Besides counting on a facilitated system of payment, the customer gets the support of a logistic technician and a team of local agencies; the added benefit is the average reduction of 30% in the costs and 60–70% in the construction time. In three years of operation, the Patrimonio Hoy has already established 49 local agencies in 23 Mexican cities with 36,000 customers, to which it adds another 1,500 monthly.

It is inevitable to perceive that this concept of social responsibility is based in a deep sense of business ethics, considering the unexplored social, economic and environmental implications of working at the base of the social pyramid. As Prahalad (2005) argues:

A total new world of chances will confide from the moment where the companies start to consider the poor persons no more victims or social pack and start to recognize them as enterprising untiring and creative and consuming conscientious of value.

This is clearly a business approach feasible in South, Central and North America and, foremost, it is likely to have a more profound and broader social impact than conventional approaches.

References

AP –Associated Press and Kruse, D. L. (1997) *Illegal Child Labor in the United States*, New York: The Associated Press.

BLS – Bureau of Labor Statistics (2002) *Survey of Occupational Injuries and Illnesses. Nonfatal (OSHA Recordable) Injuries and Illnesses. Supplemental Table S08. Nonfatal Occupational Illnesses by Category of Illness, Private Industry, 1997–2001.* Washington, DC: U.S. Department of Labor, Bureau of Labor Statistics, Safety and Health Statistics Program. Online Available at: http://www.bls.gov/iif/oshwc/osh/os/ostb1118.pdf (accessed 1 November 2007).

Curado, M. and Santos, A.D. (1999) Managing for the 3rd millennium: the social accountability challenge, in *Proceedings of the International Conference on Entrepreneurship*, Riga: Turiba School of Business.

Erlick, J.C. (ed.) (2002) Giving and volunteering in the Americas: from charity to solidarity, *Revista: Harvard Review of Latin America*, Spring: 1–87.

Gutiérrez, R. (2004) *Corporate Social Responsibility in Latin America: An Overview of its Characteristics and Effects on Local Communities, Workshop Draft*, Inter-American Development Bank, April.

ILO (1948) *ILO Convention 87: Convention Concerning Freedom of Association and Protection of the Right to Organize*, Geneva: International Labour Organization.

ILO (1951) *ILO Convention 100: Convention Concerning Equal Remuneration*, Geneva: International Labour Organization.

ILO (1958) *ILO Convention 111: Convention Concerning Discrimination (Employment and Occupation)*, Geneva: International Labour Organization.

ILO (1973) *ILO Convention 138: Convention Concerning Minimum Age*, Geneva: International Labour Organization.

ILO (1999) *Child Labour in Construction and Brick-Making*, International Labour Conference, Geneva: International Labour Organization.

NIOSH – National Institute for Occupational Safety and Health (2000) *Worker Deaths by Falls: A Summary of Surveillance Findings and Investigative Case Reports*. U.S. Department of Health and Human Services Public Health Service Centers for Disease Control and Prevention – NIOSH, September.

OSHAS – Occupational Safety and Health Administration (1999) *Report: Women in the Construction Workplace: Providing Equitable Safety and Health Protection*, Advisory Committee on Occupational Safety and Health, June.

Pew Hispanic Center (2002) *How Many Undocumented*, National Survey of Latinos.

PNUD – Programa Das Nações Unidas Para O Desenvolvimento (2005) *Relatório De Desenvolvimento Humano Brasil 2005 – Racismo, Pobreza E Violência*, Nações Unidas.

Prahalad, C.K. (2005) *A Riqueza Na Base Da Pirâmide: Como Erradicar A Pobreza Com O Lucro*, Porto Alegre: Bookman.

Roitter, M. and Berger, G. (2003) *Desarrollando Aliados Y Alianzas: Autopistas Del Sol Y La Fundación SES*, Boston: Harvard Business School Publishing SKS001.

Transparencia Brazil (2007) *Desempenho em Licitacoes nos Municipios de Santa Catarina*. Online. Available at: http://www.licitassist.org.br/Desempenho (accessed 1 November 2007).

Transparency International (2004) *Annual Report 2004*. Online. Available at: http://www.transparency.org/content/download/2274/14262/file/TI%20Annual%20Report%202004.pdf (accessed 1 November 2007).

WBCSD – World Business Council for Sustainable Development (2006) *Doing Business with the Poor: A Field Guide*, Switzerland: WBCSD.

Wellness, Genesis Group (2007) Online. Available at: http://www.revistawellness.com.br (accessed 1 November 2007).

Chapter 14

Corporate social responsibility and public sector procurement in the South African construction industry

Paul Bowen, Peter Edwards and David Root

Introduction

In this chapter corporate social responsibility (CSR) is discussed within the context of the South African construction industry. The chapter commences with an appreciation of the socio-economic legacy of apartheid, comments on the prevailing moral business culture and stumbling blocks to the development of an ethical business culture in South Africa, and provides a brief overview of the South African construction industry, its role in the economy and the challenges it faces.

Thereafter, the concept of CSR and the African principle of *ubuntu* are introduced, and the role of the government in public sector procurement reform as an enabling environment for CSR is discussed. The chapter concludes by exploring CSR practice within the South African construction industry and commenting on the extent to which CSR is actually contributing to transformation and the social upliftment of historically disadvantaged members of society through public sector procurement.

The scope of the discussion is limited to a consideration of CSR in the South African construction industry, in terms of both the public and private sectors. It does not examine the related issues of local government reform, housing and service delivery, municipal service provision, municipal finance reform, urban development strategies, human settlements policy, etc. Insight into these transformation initiatives may be gained from, for example, Tomlinson (2002). However, the chapter does seek to explore the challenges the construction industry faces in adapting to the new socio-economic and political realities of post-apartheid South Africa.

The South African context

By traditional economic measures, South Africa is a middle-income, developing country with an economy marked by substantial natural resources, a sophisticated industrial base and modern telecommunications and transport infrastructure (Western Cape Department of Economic Affairs, Agriculture and Tourism 2000). South Africa has inexpensive electrical power and raw materials, as well as lower labour costs than Westernised, industrialised countries (Western Cape Department of Economic Affairs, Agriculture and Tourism 2000). However, this belies a social reality of two distinct economies operating side by side within the same geographic space. In essence, South Africa is both a First World and Third World country, where affluence and poverty coexist. Despite significant changes within the country resulting from the dismantling of apartheid, since the late 1980s there has been little progress towards improving the low overall income levels of the majority of the population and addressing the highly skewed income distribution between the different race groups. Other serious shortcomings include poor quality schooling, inadequate housing for a significant proportion of the population, the lack of quality social services for all and insufficient growth rates to address the unemployment problem. Poverty, unemployment, lack of skilled labour, corruption, high levels of crime and the accelerating incidence of HIV/AIDS infection all remain significant problems (Whiteside and Sunter 2000).

In a 2004 Labour Force Survey (StatsSA 2004), unemployment in South Africa is officially estimated at being approximately 28% although this excludes informal employment which constitutes just under 20% of the economically active population. Once this fact is taken into account, unemployment (i.e. no formal job opportunity) rises to 41% of the economically active population (StatsSA 2004). Thus, only 59% of the economically active population are employed in the formal economy and are able to enjoy the benefit from the protection, rights and security that this entails.

Income levels in South Africa still show extreme racial disparities, although the situation is improving slowly. It is estimated that 40% of the South African population live below the minimum living level ('the breadline'). In a research project undertaken by the Bureau for Market Research (2000), it was established that 18 million people out of South Africa's population fall below the minimum level and that between eight and nine million Black people are 'completely destitute'.

Economically, the country is enjoying a combination of above-average growth, stable interest rates and growing business and consumer confidence, but the real concern is that the existing economic performance is not sufficient to make a significant impact on unemployment or poverty. Therefore, proactive policies are required to address both the growth rate and the inequalities within society. The South African government has sought to do this through direct government measures in its dual roles as construction client and as legislator.

Government as construction client

The government's first attempt to improve the quality of life of the previously disadvantaged segments of South African society was through the Reconstruction and Development Programme (RDP) (ANC 1994). At the core of this is the issue of housing. Currently, between 2.5 and 3.7 million people are estimated as being either unhoused or under-housed (Porteous and Naicker 2003). The numerous townships scattered around the majority of South African cities and towns are a legacy of apartheid; they are far from where the populace work, with limited public transport and poor service provision. In response to this enforced separation, squatter settlements have sprung up on vacant land closer to the cities; these lack even the most rudimentary municipal services such as sanitation, water supplies and electricity (Dewar and Todeschini 1999).

The RDP was superseded in 1996 by the Growth, Employment and Redistribution strategy (GEAR) (National Treasury 1996) that adopted the RDP's principles (van Wyk 2003). This set out the future macroeconomic strategy of the country and has informed the current growth-orientated, counter-cyclical fiscal framework (van Wyk 2003). GEAR has focused on increased government investment in the country's social and economic infrastructure through the Expanded Public Works Programme (EPWP) (see www.cidb.co.za), which seeks to provide temporary employment and training for the unemployed, together with more mainstream direct investment in construction projects through central, provincial and local government expenditure. Finally, the parastatals (state-owned companies) such as Eskom, Telkom and Transnet have also increased investment in response to the demands of their political masters.

Government as regulator

Since 1994, a deluge of legislation has been passed by government as it has sought to redefine the social contract between government and the governed (and through this, the business climate). Much of this is not sector-specific, although a number of Acts and regulations are specifically aimed at the construction industry, most notably The Construction Industry Development Board Act (No. 38 of 2000) (Republic of South Africa 2000b), which established a Board with a mandate to reconstruct, grow and develop and transform the construction industry in line with the government's GEAR strategy. At a more general level, a large number of Acts and regulations have been gazetted of which the two most important for the construction industry are:

1. The Broad Based Black Economic Empowerment Act (2003) (No. 53 of 2003) (Republic of South Africa 2003) which enables government to adopt practices that promote the empowerment of previously disadvantaged groups in society;
2. The Preferential Procurement Policy Framework Act (2000) (No. 5 of 2000) (Republic of South Africa 2000a) which allows for preference in the allocation of public sector contracts to protect and advance the interests of previously disadvantaged groups in society.

Unemployment remains the biggest economic challenge for South Africa (StatsSA 2004). Addressing this challenge would have a considerable effect on poverty reduction and enhance the quality of life of the previously disadvantaged and discriminated South African communities. However, the problem of unemployment requires a complex solution. It is not enough simply to attempt to create the required number of jobs. To enjoy sustainable employment, it is necessary to provide equal employment opportunities to all. The position taken by government is that this requires the empowerment of disadvantaged individuals (i.e. equipping them with adequate knowledge and skills) and further investment in the development of human capital. These can be achieved, albeit partially, through affirmative action in the South African construction industry and other industries.

Affirmative action, a deliberate intervention on the part of government in its role of client, aims at facilitating the provision, directly and indirectly, of socio-economic opportunities (e.g. skills development or employment) to individuals who, either historically or otherwise, have been denied those opportunities, and at preventing such discrimination from occurring in the future.

The South African construction industry is seen as a key to furthering the government's policies (CIDB 2004). This is summed up in the government's White Paper on Creating an Enabling Environment for Reconstruction, Growth and Development in the Construction Industry (DPW 1999) which states government approach towards the industry as a policy and strategy that

> promotes stability, fosters economic growth and international competitiveness, creates sustainable employment, and addresses historic imbalances as it generates new industry capacity for industry development.

To understand how this new agenda is impacting on the industry and its ways of working, it is necessary to appreciate the role construction plays in the national economy.

The South African construction industry and its role in the national economy

The construction industry is made up of a wide range of stakeholders (including the academic and training sector) which shapes the human capital of the industry: the client bodies who own the activities and projects through which the industry procures its work; and the professionals whose capabilities facilitate the conception, development, implementation and long-term operation of the infrastructure and services. It also includes public and private sector investors, contractors, manufacturers, labourers, specialist subcontractors and material suppliers.

The demand side of the industry is characterised by both public and private sector clients. The former, functioning at local, provincial and national level, accounts for 30% of the output of the construction industry with a further 13% from public corporations (parastatals). In contrast, the private sector is responsible for 57% (CIDB 2004; van Wyk 2003).

Investment in infrastructure is seen as a key driver of economic growth in that government spending improves infrastructure and, in doing so, enables the efficient delivery of other services, reduces business costs and so acts as a catalyst for a higher economic growth and employment creation (National Treasury 2004). Thus, the construction industry is seen as the principal means by which much of this infrastructure is provided and a prime target where the preferred new equity and redistribution policies can be implemented.

More specifically, the industry accounts for some 5.1% of gross domestic product (GDP) (CIDB 2004) and contributes about 30% to gross fixed capital formation (GFCF). The industry currently employs approximately one million people (520,486 formally and 470,514 informally), of which the formally-employed constitute 5.1% of the total formally-employed population (van Wyk 2003). The construction industry's contribution to capital formation is set to increase drastically if projections of future infrastructure provision are realised (DPW 1997), and this is expected to impact dramatically on employment. The government's own expectations are that 65% of the one million jobs that the government has committed itself to creating over the next five years will be generated through labour-intensive infrastructure development (van Wyk 2003). This development activity will be almost wholly construction-driven.

The labour-intensive characteristic of construction also matches the South African situation where the poor quality of schooling and consequent low levels of skills mean that there is a mismatch between the skills offered by the Black majority and those required by the developed formal industry (van Wyk 2003). Internationally, construction industries are seen as a point of entry into the formal labour market, providing preliminary training (Ganesan 2000) and socialisation into the formal work environment. Studies indicate that, in South Africa, the construction industry ranks fourth highest behind agriculture, domestic work and mining, in terms of the number of people employed who have no education (van Wyk 2003).

One achievement of GEAR has been to build a consensus that the country needs to lift its trend of GDP growth rate to 6% to make significant inroads into tackling unemployment and poverty. The implication of this for the construction industry is that it will have to double in size over the next ten years (CIDB 2004; van Wyk 2003). However, this increase in capacity has to take place against a backdrop where employment levels in the industry have dropped over the past 15 years (van Wyk 2003). In addition, the lack of middle-management skills and the transformation of the personnel and procedures in the public sector, which will be the source of the bulk of the growth in new demand for infrastructure, mean that there are severe capacity constraints that will have to be addressed. The high numbers of mismatched skilled and uneducated participants in the industry are a serious impediment to increasing industry capacity, with the industry's skills and knowledge being concentrated in a relatively small and ageing cadre of professionals and skilled artisans.

Despite all this, the formal construction sector is recognised as being a national asset (CIDB 2004). Uniquely within sub-Saharan Africa, it is technologically advanced, operating internationally to developed-world standards and capable of providing a full range of construction and engineering services in South Africa and abroad. However, much of its strength is a product of its development under the former apartheid regime and the self-reliance that the country's then isolation fostered. This has had an inevitable impact as the established industry has sought to reconcile itself and adapt to the rapid changes in the socio-political environment and, in doing so, to develop more ethically-orientated practices.

Establishing an ethical business culture in South Africa

The issue of ethics is often confused with the issue of corporate social responsibility (CSR). However, one legacy of apartheid has been a skewed business ethic that was permeated by the National Party's racist policies. The issue of the prevailing (often immoral) business culture in South Africa must be viewed against this backdrop, where international sanctions resulted in a culture of valuing unethical but effective means of doing business. In some instances, this could result in the prevailing ideology of the apartheid government running contrary to the interests of business (e.g. the restriction of certain skilled occupations to particular racial groups, thus impacting negatively on business efficiency) (Rossouw 1998 citing Massie 1993). More often than not, these issues were driven by a corporation's economic self-interest rather than business morality. This meant that the prevailing business ethic became one of pragmatism driven by economic self-interest, rather than the internalisation of any values shared across society. This culture of questionable business morality was exacerbated by the immoral nature of the apartheid ideology, which deliberately sought to fragment South African society and prevent the emergence of truly shared national values (Rossouw 1998 citing Massie 1993). The result of this was that the majority of the (disadvantaged) population considered the apartheid socio-economic dispensation to be illegitimate and thus felt no moral obligation towards it (Rossouw 1998 citing Massie 1993). At the same time, many of the established economic enterprises that prospered under its shadow had their consciences lulled by their success.

Rossouw (1998) identifies five stumbling blocks that impede the development of a moral business culture in South Africa, namely: lack of commitment to the new dispensation; the prevailing business culture; lawlessness; the role of civil society; and scarcity of goods. A lack of commitment to the new dispensation is characterised by citizens not identifying strongly with the newly formed society, possibly adopting a wait-and-see attitude or even passively resisting as a result of feeling resentment at their hitherto privileged position being under threat (Massie 1993) – in essence, not identifying with the changes taking place, nor accepting responsibility for the survival of the new democracy.

Insofar as economic lawlessness is concerned, whilst South Africa does not suffer from a lack of appropriate laws, it largely lacks the capacity to enforce these laws effectively and the expertise to identify and prosecute the perpetrators of serious economic offences. Quite simply, the police have more pressing issues to attend to. The role of South African civil society in promoting and entrenching ethical, social and business values has been eroded by virtue of the previous regime systematically undermining the very development of civil society; for example, restrictions on public gatherings and the banning of many organisations critical of the apartheid ideology which have resulted in a society underdeveloped in its capacity to play a constructive role in democracy and in the development of the economy (Rossouw 1998).

Rossouw (1998: 1569) suggests a range of measures to overcome the stumbling blocks identified above. His suggestions centre on the notion that 'one should take responsibility for one's business actions in order to ensure that they do not harm the society in which one lives'. More specifically, Rossouw points to the use of externally and internally induced responsibility. Externally induced responsibility, involving persuasion or even forced compliance, itself consists of three components, namely: the state, civil society, and the business community. The state has two instruments at its disposal for inducing responsibility in its citizenry – the education system (inculcating an ethos of responsible citizenship) and the criminal justice system (appropriate laws and sufficient law-enforcement capacity). Civil society has an important role to play in influencing responsible business behaviour, by virtue of the participation of (for example) the media, consumer groups, conservationists, churches and non-governmental organisations (NGOs). The media can exert influence by educating the public and exposing inappropriate business behaviour. The business community itself needs to adopt a proactive stance through commitment to ethical codes or value statements for different industries and organisations.

Internally induced responsibility refers to the situation where persons/organisations themselves take on the desired responsibility for ethical behaviour. Given the prevailing unethical business culture in South Africa, Rossouw (1998) favours the external option, claiming that there are currently insufficient signs of spontaneous self-regulation and that the internal approach is unlikely to succeed in the short run. It will take time for a strong sense of responsibility to develop, and external assistance is required during this period of transition.

Whether or not these arguments are persuasive, this view of the formation of an ethical business culture in South Africa does provide clues about the way forward in fostering CSR in the South African construction industry.

Corporate social responsibility

According to Hemingway and Maclagan (2004), there has been a growing interest in CSR across a range of disciplines. The Confederation of British Industry (CBI 2005) defines CSR as requiring an acknowledgement on the part of companies that they should be publicly accountable not only for their financial performance but also for their social and environmental contribution. The CBI (2005), by asserting that businesses should be accountable for the impact of their activities on *society* and/or the environment, sees CSR as encompassing the extent to which companies should promote human rights, democracy, community improvement and sustainable development objectives throughout the world. Maclagan (1998: 147) provides a useful definition of CSR: 'Corporate social responsibility may be viewed as a process in which managers take responsibility for identifying and accommodating the interests of those affected by the organization's actions'.

The notion of CSR is complex. For example, Hemingway and Maclagan (2004: 33) debate CSR in terms of a *proactively responsible view* (i.e. taking responsibility for CSR regardless of public opinion) and a contrasting *reactive management responsive level of commitment*. Carroll (1996), cited by Hemingway and Maclagan (2004), questions whether complying with the law comprises CSR, or whether CSR should be defined in terms of voluntary actions that *exceed* legal standards or public opinion. Carroll (1996) states that four components need to be present for an organisation to be able to claim it is socially responsible, namely, economic, legal, ethical and philanthropic responsibilities. The philanthropic (voluntary) dimension of CSR is noteworthy, as exceeding the requirements of the law is reflected in the CSR literature as a dominant feature.

Maclagan's definition of CSR, given above, draws attention to the role of *managers* in formulating and implementing an organisation's CSR policy, and introduces the importance of individuals' values and motives. Moreover, it brings into question the corporate, as opposed to individual, status of any CSR initiative. Flowing from this, Hemingway and Maclagan (2004) identify two important dimensions for the analysis of CSR in practice: the motivational basis (is it commercial, idealistic or even altruistic?); and the *locus* of responsibility (is it corporate – as portrayed in the CBI definition of CSR – or individual?).

It is argued here that the concept of *ubuntu* provides a way forward in respect of CSR. *Ubuntu* epitomises the African philosophy of respect and human dignity. According to Mbigi and Maree (1995: 9):

> *Ubuntu* can help organisations to develop corporate citizenship. By building the solidarity spirit of ubuntu it is possible to build co-operation and competitive strategies by allowing teamwork to permeate the whole organisation.

Taken further, the ideology of *ubuntu*, incorporating respect, dignity, togetherness and mutual support, can be extended to the entire construction industry (Rwelamila *et al.* 1999). Indeed, efforts by the public sector in developing a strategy for the construction industry to facilitate empowerment and socio-economic upliftment by incorporating affirmative procurement can be said to epitomise the very essence of *ubuntu*. These procurement reforms are discussed later in this chapter.

The CSR agenda is evolving and is driven by a global shift in the way the role of business is perceived. In the context of sustainable development, business is increasingly seen as a crucial element in the process of social transformation – for the benefit of society and the business itself (Hamann 2003). According to Hamann (2003), the shift is characterised by a number of developments, namely: greater partnering between the public and private sectors (see www.societyandbusiness.gov.uk); greater acceptance by business of the benefits of CSR (see www.bsr.org); the development of CSR principles and reporting guidelines (see www.globalreporting.org; and AA1000: www.accountability.org.uk); and an increasing emphasis on the importance of CSR amongst employees, consumers, investors (i.e. socially responsible investing) and social and environmental activists.

Hamann (2003) goes further, pointing to three key elements of CSR or corporate citizenship, namely:

- an acceptance by organisations that CSR goes beyond philanthropic investment – not simply some percentage of after-tax profit invested in social development; but also a consideration of how profits are actually generated;
- integrating CSR into the core activities and decision-making of the organisation by embracing economic, social and environmental aspects of sustainability in a holistic manner;
- embracing constructive engagement rather than confrontation – on the part of organisations and lobbyists alike.

Maignan (2001) examined, in a cross-cultural context, consumers' perceptions of corporate social responsibilities. She concluded that a positive correlation exists between consumer behaviour and responsible organisations, but points to the need to better understand how consumers in diverse countries and cultures define socially responsible corporate activities. Al-Khater and Naser (2003) surveyed attitudes in Qatar regarding CSR disclosure and accountability, concluding that participants favoured CSR information disclosure in organisations' annual reports, in both monetary and non-monetary forms.

Notwithstanding De George's (1996) apparent cynicism towards CSR, Wilson (2000) offers a number of guiding rules for adopting CSR, namely: corporate rules and stakeholder relations; corporate governance; increased equity and diversity; environmental preservation; employment as a social contract; public–private sector relationships; and ethical performance.

Given this view of CSR, it is worthwhile noting the relationship between CSR and sustainable development within the context of the South African construction industry. The South African government has committed itself on many levels to the principles of sustainability both in the natural and built environments. Indeed, section 24 of the Bill of Rights in the 1996 Constitution states that:

> Everyone has the right...to have the environment protected, for the benefit of present and future generations, through reasonable legislative and other measures that...secure ecologically sustainable development and use of natural resources while promoting justifiable economic and social development.
>
> (Republic of South Africa 1996b: 10)

In many ways, South Africa represents a microcosm of the developmental and environmental issues facing the world economy in the transition to sustainability (Hill *et al.* 2002). South Africa manifests both the environmental problems of the industrialised world, such as acid rain, and those of the under-developed rural areas, such as soil erosion. At the same time, it exhibits the socio-economic issues of both worlds, with its urgent need for economic development contrasting with the desire to protect the choices of future generations.

Given that CSR is inherently interlinked with sustainable development in South Africa, public sector reform would seem apposite in addressing these dilemmas within a new ethical approach to construction practice.

Public sector procurement reform: an enabling environment for CSR in the construction industry

Since 1994, the South African government has enacted legislation aimed at fulfilling its commitment to the objectives of development and growth emanating from the Reconstruction and Development Programme (DPW 1999). The vision is for a construction industry policy and strategy that promotes economic growth, creates sustainable employment and addresses historic imbalances as it generates new industry capacity for sustainable economic development (DPW 1999). Rwelamila (2002) provides a comprehensive overview of the strategy development process for creating an enabling environment for reconstruction, growth and development in the construction industry. Some of the salient points of the strategy include: developing a stable delivery environment; enhancing industry performance through programmes such as partnering, participative management and environmental protection; introducing contractor accreditation; facilitating new industry capacity via the promotion of small businesses; and developing the capacity and role of the public sector to maximise employment opportunities through labour-intensive construction, the empowerment of communities via participation and training, the establishment of a statutory Construction Industry Development Board (CIDB), and the establishment of an Emerging Contractor Development Programme (ECDP) (DPW 1998). These initiatives culminated in the promulgation of the CIDB Act (No. 38 of 2000).

Redressing the inequalities of the past

Preferential procurement has been adopted in South Africa as an instrument to effect socio-economic change through the promotion of employment and business opportunities to marginalised sectors of the population. The use of preferential procurement is one of the central strategies in the Reconstruction and Development Programme (RDP) approach adopted by the government in its White Paper (Republic of South Africa 1994) as a social reengineering policy aimed at redressing the legacy of apartheid. Section 217(2) of the Constitution provides for preferencing policies in the allocation of contracts, while section 217(3) requires that legislation be enacted to provide a framework within which such procurement policy must operate. The Preferential Procurement Policy Framework Act (No. 5 of 2000) requires organs of government, particularly those involved in infrastructure delivery, to determine their preferential procurement policy and to integrate preferential procurement into their procurement of goods, services and works, through an innovative tool entitled targeted procurement (TP). *The Implementation Manual on the Use of Targeted Procurement to Implement an Affirmative Procurement Policy (APP)* defines TP as

> a system of procurement which provides employment and business opportunities for marginalised individuals and communities, enables procurement to be used as an instrument of social policy in a fair, equitable, competitive, transparent and cost effective manner and permits social objectives to be quantified, measured, verified and audited.
>
> (DPW 2001: 5)

An enabling strategy for the construction industry

To achieve the government's stated goals of social and economic reform, the Department of Transport and Public Works (DTPW 2002) has implemented a preferential procurement implementation plan. A basic tenet of the department's preferential procurement policy is to improve the quality of life for all through socially just, developmental and empowering processes. The emphasis here is not restricted to the end product, but incorporates the wider considerations of the delivery mechanisms used to achieve the desired goals (DTPW 2002). The preferential procurement policy aims to promote public sector procurement reform, to introduce a more uniform public sector procurement system, and to provide guidelines for procurement policy. The general procurement policy guidelines formalise the standards of behaviour, ethics and accountability required of the public sector. The guidelines also act as a statement of government's commitment to a procurement system that enables the emergence of sustainable small, medium and micro private sector enterprises, which will add to the common wealth of the country and the achievement of enhanced economic and social well-being (DTPW 2002).

Proper and successful public sector procurement rests upon certain core principles of behaviour – the Five Pillars of Procurement (DTPW 2002). These pillars are commonly known as: value for money; open and effective competition; ethics and fair dealing; accountability and reporting; and equity. The fifth pillar, equity, is vital to socio-economic sustainability as it ensures that government is committed to economic growth by implementing measures to support industry generally, but in a socially and economically fair and just manner.

Objectives of the preferential procurement plan for the construction industry

The fundamental objectives of the preferential procurement policy (DTPW 2002) include:

- a participation target for historically disadvantaged individual (HDI) ownership (in the supply chain) of at least 40% for all goods and services procured;
- a participation target (in the supply chain) for women/disabled persons of at least 20% within five years for all goods and services procured;
- the increase in HDI equity ownership within each company;
- the increased participation of HDIs in the management and control of companies;
- the promotion of South African–owned enterprises and products;
- the promotion of enterprises located in designated areas (e.g. locally based businesses);
- the promotion of export-orientated production to facilitate job creation;
- the promotion of small, medium and micro enterprises (SMMEs);
- the creation of new jobs or the intensification of labour absorption;
- empowerment of the workforce by standardising the level of skill and knowledge of workers.

The *raison d'être* of preferential procurement in the construction industry is to promote an enabling environment for greater HDI participation in the private sector. Central to this policy is the promotion of targeted joint ventures and other *affirmable business enterprises* (ABEs). In this context, an affirmable business enterprise (ABE) is one adhering to statutory labour practices; a legal entity registered with the South African Revenue Service; and one providing a commercially useful function (as opposed to 'window-dressing'). In addition, ABEs need to be at least 51% owned by one or more previously disadvantaged persons (PDIs), and the management and daily business operations must be under the control of the PDIs who effectively own it.

In terms of the Plan, HDIs refer to persons not entitled to political franchise prior to 1994, females or disabled persons, whilst (PDIs) are restricted to persons not entitled to political franchise prior to 1994. Small, medium and micro enterprises (SMMEs) are seen as an important conduit for the generation of employment and more equitable income distribution, activation of competition, exploitation of niche markets and, ultimately, stimulation of economic development (DTPW 2002).

In a sense, the public sector preferential procurement policy and procedures can be viewed as having a coercive influence on the development of CSR in the South African construction industry. The development of the policy sets out the society's expectations (mediated through the executive) of procurement and business practice formalising 'the standards of behaviour, accountability and reporting, ethics, and equity considerations expected of the public sector in procurement' (Kajimo-Shakantu and Root 2004: 52). In essence, the policy defines the CSR qualities expected of firms who act as suppliers to the government, such 'affirmative action' being justified on the basis of legality, morality and the wider social responsibility (Adele 1996). Questions then arise as to how effective these have been in practice, and to what extent the policy objectives have flowed through to private sector procurement.

CSR in the South African construction industry: comparing theory and practice

The construction industry was one of the first sectors of the economy where preferential or targeted procurement policies were introduced. Therefore, it presents one of the best opportunities to assess the extent to which the policy has succeeded or failed in changing the business climate, and, indeed, the ethics of business practice by organisations.

Currently, there are over 1,500 construction firms registered with the Emerging Contractor Development Programme, of which a significant proportion are enterprises owned by women (www.publicworks.gov.za). By 2003, the proportion of goods and services supplied by HDI-owned enterprises had increased to 43% (beating the DPW's target of 40%) compared to 4% in 1994 (van Wyk 2003). In the Limpopo province, it has been reported that

80% of public sector projects are undertaken by Black- or women-owned enterprises (CIDB 2004), whilst individual projects such as the Cape Town International Convention Centre (CTICC) (Shakantu and Bowen 2002) and the Coega harbour development near Port Elisabeth (www.coega.co.za) have also provided evidence of major changes taking place in the way packages of work have been directed towards HDI-owned enterprises.

These experiences cover a range of ways in which the client has encouraged the targeted enterprises to be involved in the government's supply chain. These align with Madi's (1997) classification of affirmative action as being under three broad themes:

1. *Ownership*: where pressure has been exerted on established firms to increase ownership of the company's equity by previously disadvantaged groups. Typically, this has been around 25%–30% and has normally involved employees buying into a firm they work for, or external parties (both individuals or Black-owned organisations) taking a stake in a firm;
2. *Corporate advancement*: where the policy has encouraged organisations to increase the representation of historically disadvantaged groups in management within established firms by advancing Black people or women within the management hierarchy or through direct appointment at board level as non-executive directors;
3. *SMME (Small Medium and Micro Enterprises) outsourcing*: whereby established firms actively seek out small emerging contractors who are then employed as subcontractors or suppliers. This typically involves splitting up the work packages into smaller sized sections to lower the technical and financial barriers for these emerging enterprises. For example, the CTICC involved some 200 HDI-owned firms (Shakantu and Bowen 2002).

In addition, on larger projects, including those under the Public Private Partnership (National Treasury 2004), established firms (both domestic and overseas enterprises who contribute the bulk of the technical and management knowledge) are frequently required to enter into strategic alliances with HDI-owned firms such as joint ventures or partnership agreements with the intention that technical and managerial knowledge will be transferred to the less-established party. Thus, the policy aims to transform existing capacity, whilst at the same time fosters and develops new capacity within the sector.

In practice, along with the successes, there have inevitably been a number of problems associated with the preferential procurement policy in each of the following themes:

Ownership

Within the domain of large, established organisations, the issue of ownership dominates debate in respect of Black Economic Empowerment (BEE) in South Africa, because of the perception that it has generated a narrow Black elite of politically connected individuals symbolised by prominent Black entrepreneurs who, it is alleged, have built conglomerates by taking significant minority stakes in a range of businesses that have needed to meet various ownership criteria, at below-market prices. However, of more concern is the problem of 'fronting', where shell companies or HDI-owned firms act as 'fronts' for winning orders or contracts, which are then performed by an established firm. In these situations, a Black director or partner acts a sleeping partner providing no risk capital, expertise or other talent other than his/her gender and ethnicity in return for financial payment. On occasions, these can be very exploitative, where, for example, domestic workers have found themselves in business with their employers without being aware of it! Both situations fail to address the notion of building capacity within historically disadvantaged groups.

Corporate advancement

One unforeseen outcome of preferential procurement has been the trend that employees appointed through such affirmative action have been predominantly appointed at board level, where their actions and decision-making are constrained by the need of the board to act as a collective. Thus, such non-executive directors are put in 'safe' locations (Adele 1996), and despite the trappings of executive power, have limited scope to influence the development of their organisation, including its culture and business ethics. This results in what is known as the 'revolving door syndrome' (Adele 1996) where the organisational environment does not change, and HDIs, who enter the business with high expectations of promoting change, rapidly become disillusioned and leave.

Even those organisations consciously seeking to bring on and develop Black management talent find themselves with high staff turnovers as they, in response to the demands of the business environment, try to accelerate the development of their young, Black professionals and managers. This may result in the individuals not having the time and space to develop the maturity and experience necessary to perform in the roles in which they find themselves. They subsequently lose confidence and leave the organisations in much the same way as promising young sportsmen, put into national teams too early, rapidly burn out because they are mentally unprepared for the pressures of playing at international level.

SMME outsourcing

The small-business sector in South Africa has the potential to contribute to job creation and economic growth. However, the policy environment faced by SMMEs determines their capacity to contribute to the process of development (Luiz 2002). The National Small Business Act (No. 102 of

1996) (Republic of South Africa 1996a) classifies small businesses into four categories: micro (including survivalist enterprises); very small; small; and medium. The Ntsika Enterprise Promotion Agency (1997), cited by Luiz (2002), estimates that in 1995 the overall contribution to total GDP was 20.8% for small enterprises, 11.9% for medium enterprises and 67.3% for large enterprises. The contribution to formal employment was estimated at 29.5% for small enterprises, 15.3% for medium enterprises and 55.2% for large enterprises.

Reasons why governments worldwide have assisted SMMEs include the capacity of SMMEs to absorb labour; strengthen cultural traditions and social systems; facilitate the participation in the economy of indigenous peoples; employ local technology and local raw materials; employ more labour per unit of capital and require less capital per unit of output than do large enterprises; display resilience to economic depression; employ women; cater for local differences in taste; and provide a nursery for entrepreneurship and innovation (Harper 1984; Rwigema and Karungu 1999).

When applied to established firms, outsourcing or subcontracting work packages, the preferential procurement policy makes no distinction on the type of work subcontracted. Understandably, established firms have focused on retaining control of those activities with a high added value and a high impact on their core business. This has meant that the work packages that are outsourced to the BEE firms mainly tend to be low-added-value commoditised work where price is more important than service or quality. This is not an ethical issue, but is sound business practice. However, such an approach does not take into account the impact of activities on the local economy even though, in this case, both interests may coincide. A further example is shown in the activities identified under the Expanded Public Works Programme (EPWP), which is aimed at BEE firms with an explicit intent to have the greatest impact on the local economy. Typical projects include school cleaning and renovation, community gardens, removal of alien vegetation, tree planting, etc. (van Wyk 2003). All of these represent relatively low-skilled, highly labour-intensive projects which, whilst beneficial, do not provide opportunities for HDI-owned firms to advance their technical and managerial expertise and move higher up the value chain.

The consequence of this is that these firms can remain 'ghettoised' in a highly competitive business environment where labour is effectively commoditised. Price is the only form of competition and firms are locked into a dependency on tendering for public sector projects without the ability to diversify into other market segments because of a lack of technical expertise, access to capital and low levels of profitability.

Strategic alliances

A response to the limitation of outsourcing is that of developing strategic alliances, although the distinction is not always clear-cut. For example, an established firm may 'spin-off' a subsidiary (in effect, a form of equity transfer), outsourcing a previous activity, but that enterprise remains closely connected to the original parent company through the workload and/or a trade investment by the parent company in that HDI-owned firm. Other forms of strategic alliance may include joint ventures or partnering (project or business partnering).

The relevant literature reveals the existence of various forms of joint ventures (JV). Equity JVs, involving the creation of a separate corporate entity and involving two parent firms, are amongst the most common. Although JV relationships are inter-firm, they are usually controlled by managers assigned to the venture by the parent firms or by headquarters, executives with JV responsibilities (Peng 2002). JVs represent a strategic and economic arrangement, but also constitute a social, psychological and emotional phenomenon, as such requiring nurturing and commitment (Peng 2002). In an international context, JVs exist both as a mode of foreign investment and as a means of technology transfer (Gale and Luo 2004). In developing countries, they are a means of stimulating market development, acquiring advanced technology and developing managerial skills necessary to promote further economic growth (Li et al. 2000). Problems associated with JVs include: attempts by one of the partners to obtain dominant control; disagreements over management and operational strategies of the JV; disagreements over the contribution of each party and the distribution of the profits (Gale and Luo 2004; Holton 1987).

In the context of the South African construction industry, JVs are typically strategic alliances between established, 'White-owned' construction firms and historically disadvantaged, 'Black-owned' construction firms. Such JVs are entered into by the 'White' Firms as a price for gaining access to public sector work. Thus, they are a means to an end, not an end in itself. Such an example is that of Public–Private Partnerships, where the National Treasury has established that all PPP tenders should incorporate a 25% BEE component in the ownership of the Special Purpose Vehicle (a form of JV) (National Treasury 2004). However, difficulties arise here in the sheer cost of tendering, where BEE firms have difficulty in raising their share of the risk equity.

Partnering is a long-term commitment between two or more organisations for the purpose of achieving specific business objectives by maximising the effectiveness of each participant's resources (Cox and Townsend 1998). The relationship is based on trust (competency trust, ethical trust, and intuitive trust), pursuance of common goals and 'an appreciation of each other's expectations and values' (Zaghloul and Hartman 2000). Partnering is increasingly being used on construction projects (Black et al. 2000). It is claimed that partnering can result in greater efficiency on site, reduced construction time, reduced contractual conflict and increased managerial effectiveness (Packham et al. 2003; Wong and Cheung 2004). In other countries, the adoption of partnering has been a reaction to the destructive tendencies of the adversarial relationships that characterise the construction industry, hence its promotion by Latham (1994) and others. It requires a fundamentally different mindset in terms of business practice: a shift from the traditional arm's-length contractual relationships (ACR), that in the past dominated the Anglo-Saxon business world, towards that of an obligational contractual relationship (OCR) that is found in Asia (Sako 1992). Traditional South African business culture has, as a result of the economy being historically dominated by British norms and practices, understandably been described as one that fosters adversarial relationships (Price Waterhouse 1996).

Viewed in this way, the themes of ownership, corporate advancement and outsourcing can all be accommodated to a lesser or greater extent within the traditional 'Anglo-Saxon' business culture with a corresponding lack of imperative for establishing a new business ethic. Outsourcing and corporate advancement can both be rationalised in terms of economic efficiency: outsourcing as the concentrating of an organisation on its core competencies, and corporate advancement as a desire to gain access to a wider labour market. The issue of ownership can also be rationalised under the existing business ethos: as a necessary price for the licence to operate, as disinvestment of subsidiaries or as the opportunity to rationalise holdings and reshape themselves as global and international businesses. Thus, managements can put off any fundamental value-driven change in their practice whilst appearing to acquire the trappings of CSR.

In contrast, partnerships and other forms of strategic alliance are potential avenues for change and developing CSR within the industry. The need to move towards OCRs between organisations resonates with the concept of *ubuntu* and its infusion into a new South African business culture. *Ubuntu* is a form of social solidarity, communicating the idea of interdependency and mutual support. It is a middle ground between individualism and extreme collectivism: recognising heterogeneity as opposed to seeking to impose homogeneity on the collective.

This is most readily apparent within the PPP sector within South Africa. The long-term nature of these contractual arrangements requires deeper relationships with and between employees, customers, suppliers, investors and the community (Root 2005). The government requirement that all projects be 'sustainable', with intra- and inter-generational equity and fairness to be embodied in project objectives (Kaatz *et al.* 2004), demands that CSR is internalised within the organisations involved, whether these are established enterprises or not. Such approaches broaden the boundaries of the construction project to incorporate those viewed traditionally by the industry as 'people outside the project' (Newcombe 2003: 843). Thus, the industry moves away from the isolated delivery of building towards the provision of services that the building delivers, be it a prison, hospital, or school (Kaatz *et al.* 2004; Root 2005). Examples of such PPP initiatives include the Malmesbury and Bloemfontein prisons. The effectiveness of

PPP procurement is yet to be tested over the medium to long term, since no South African projects have been operating over a sufficiently long period.

Summary and conclusions

This chapter has explored CSR in the South African construction industry. Central government initiatives to overcome the unfairness, imbalances and disadvantages fostered under the former apartheid regime have been examined for their potential to encourage and enhance CSR and ethical business practice. These initiatives are largely driven through the implementation of the Preferential Procurement Policy Framework Act, and the targeting of the construction industry as potentially one of the most promising means of alleviating economic and social ills. The effectiveness of this policy implementation still has to be assessed. Indeed, even the tools and techniques for such assessment are not yet fully developed. Sustainability objectives, and the use of Public–Private Partnerships for essential public service delivery, have a role to play in engendering CSR, but their outcomes also await appropriate evaluation. The social reengineering of South African society in general, and the construction industry in particular, will be no easy task. It is clear that the road ahead is long and arduous.

The South African government is playing its part in encouraging CSR in the construction industry. For its part, the industry, particularly with regard to its private sector participants, may need to undergo substantial shifts in organisational culture – arguably a difficult task at the best of times. *Ubuntu*, the African philosophy of respect and human dignity, must be an integral part of this if change is not to be seen eventually as mere 'window-dressing'. Fortunately, the combination of government legislation, public sector procurement, social and economic development and the consensus forged in the past ten years of democracy suggests that the industry will be able to adopt *ubuntu* as an African conceptualisation of CSR as the industry struggles to redefine its business practices.

Websites accessed

AccountAbility (UK): http://www.accountability21.net
Business for Social Responsibility (UK): www.bsr.org
Coega harbour development: www.coega.co.za
Confederation of British Industry: www.cbi.org.uk
Construction Industry Development Board: www.cidb.co.za
Department of Public Works (SA): www.publicworks.gov.za
Global Reporting Initiative: www.globalreporting.org
Statistics South Africa (StatsSA): www.stassa.gov.za

References

Adele, T. (1996) *Beyond Affirmative Action*, Randburg: Knowledge Resources.

African National Congress (ANC) (1994) *Reconstruction and Development Programme*, Johannesburg: ANC.

Al-Khater, K. and Naser, K. (2003) Users' perceptions of corporate social responsibility and accountability: evidence from an emerging economy, *Managerial Auditing Journal*, 18/6(7): 538–548.

Black, C., Akintoye, A. and Fitzgerald, E. (2000) An analysis of success factors and benefits of partnering in construction, *International Journal of Project Management*, 18: 423–434.

Bureau for Market Research (2000) *The South African Provinces: Population and Economic Welfare Levels*, Report No. 276, Pretoria: Bureau for Market Research.

Carroll, A.B. (1996) *Business and Society: Ethics and Stakeholder Management*, Cincinatti: Southwestern Publishing.

Confederation of British Industry (CBI) (2005) *Policy Work: Corporate Social Responsibility*. Online. Available at: http://www.cbi.org.uk (accessed 22 April 2005).

Construction Industry Development Board (CIDB) (2004) *SA Construction Industry Status Report – 2004: Synthesis Review on the South African Construction Industry and its Development*, Discussion Document, Pretoria: CIDB.

Cox, A. and Townsend, M. (1998) *Strategic Procurement in Construction*, London: Thomas Telford Publishing.

De George, R.T. (1996) *The Myth of Corporate Social Responsibility: Ethics and International Business*, London: Rowman and Littlefield Publishers.

Department of Public Works (DPW) (1997) *Green Paper: Creating and Enabling Environment for Reconstruction, Growth and Development in the Construction Industry*, Policy Green Paper, Pretoria: Department of Public Works.

Department of Public Works (DPW) (1998) *Emerging Contractor Development Programme*, Pretoria: Department of Public Works.

Department of Public Works (DPW) (1999) *White Paper: Creating and Enabling Environment for Reconstruction, Growth and Development in the Construction Industry*, Policy White Paper, Pretoria: Department of Public Works with Departments of Transport, Water Affairs and Forestry, Housing, and Constitutional Development. Government Gazette, No. 20095, May.

Department of Public Works (DPW) (2001) *The Use of Targeted Procurement to Implement an Affirmative Procurement Policy: Implementation Manual*, Pretoria: Department of Public Works.

Department of Transport and Public Works (DTPW) (2002) *Preferential Procurement Implementation Plan*, Pretoria: Department of Transport and Public Works.

Dewar, D. and Todeschini, F. (1999) *Urban Management and Economic Integration in South Africa*, Cape Town: Francolin.

Gale, A. and Luo, J. (2004) Factors affecting construction joint ventures in China, *International Journal of Project Management*, 22: 33–42.

Ganesan, S. (2000) *Employment, Technology and Construction Development*, Aldershot: Ashgate.

Hamann, R. (2003) Mining companies' role in sustainable development: the 'why' and 'how' of corporate social responsibility from a business perspective, *Development Southern Africa*, 20 (2): 237–254.

Harper, M. (1984) *Small Businesses in the Third World*, Chichester, UK: John Wiley & Sons.

Hemingway, C.A. and Maclagan, P.W. (2004) Managers' personal values as drivers of corporate social responsibility, *Journal of Business Ethics*, 50: 33–44.

Hill, R.C., Pienaar, J., Bowen, P.A., Kusel, K. and Kuiper, S. (2002) The transition to sustainability in the planning, construction and management of the built environment in South Africa, *Special Issue of the International Journal of Environmental Technology and Management*, 2 (1–3): 200–224.

Holton, R. (1987) Making international joint ventures work, in Otterbeck, L. (ed.), *The Management of Headquarter-Subsidiary Relationships in Multinational Corporations*, Aldershot: Gower.

Kaatz, E., Root, D. and Bowen, P.A. (2004) Implementing a participatory approach in a building sustainability assessment model, *Proceedings of SB'04 International Conference 'Sustainable Building for Africa'*, Stellenbosch (CD-Rom), 14–15 September.

Kajimo-Shakantu, K. and Root, D. (2004) The preferential procurement policy: implications for development of the construction industry, *Proceedings of the 2nd Postgraduate Student Conference*, Cape Town: Construction Industry Development Board, 10–12 October: 47–57.

Latham, M. (1994) *Constructing the Team; Joint Review of Procurement and Contractual Arrangements in the United Kingdom Construction Industry*, London: HMSO.

Li, J., Qian, G., Lam, K. and Wang, D. (2000) Breaking into China – strategic considerations for multinational corporations, *Long Range Planning*, 33 (5): 673–687.

Luiz, J. (2002) Small business development, entrepreneurship and expanding the business sector in a developing economy: the case of South Africa, *Journal of Applied Business Research*, 18 (2): 53–69.

Maclagan, P.W. (1998) *Corporate Social Responsibility as a Participative Process*, London: Sage Publications.

Madi, P.M. (1997) *Black Economic Empowerment in the New South Africa: The Rights and the Wrongs*, Randburg: Knowledge Resources.

Maignan, I. (2001) Consumers' perceptions of corporate social responsibilities: a cross cultural comparison, *Journal of Business Ethics*, 30: 57–72.

Massie, R.K. (1993) Understanding corruption, *Die Suid-Afrikaan* (August/September): 38–41.

Mbigi, L. and Maree, J. (1995) *Ubuntu – The Spirit of African Transformation*, Randburg: Knowledge Resources.

National Treasury (1996) *Growth, Employment and Redistribution: A Macroeconomic Strategy*, Pretoria: National Treasury.

National Treasury (2004) *National Treasury PPP Manual*, Pretoria: National Treasury.

Newcombe, R. (2003) From client to project stakeholders: a stakeholder mapping approach, *Construction Management and Economics*, 21: 841–848.

Ntsika Enterprise Promotion Agency (1997) *The State of Small Business in South Africa*, Pretoria: Ntsika Enterprise Promotion Agency.

Packham, G., Thomas, B. and Miller, C. (2003) Partnering in the house building sector: a subcontractor's viewpoint, *International Journal of Project Management*, 21: 327–332.

Peng, M.W. (2002) Joint venture dissolution as corporate divorce, *Academy of Management Executive*, 16 (2): 92–106.

Porteous, D. and Naicker, K. (2003) South African housing finance: the old is dead – is the new ready to be born? in Khan, F. and Thring, P. (eds), *Housing Policy and Practice in Post-Apartheid South Africa* Sandown: Heinemann: 192–227.

Price Waterhouse (1996) *Affirmative Action in South Africa* (Special Report), Price Waterhouse.

Republic of South Africa: Government Gazette (1994). *White Paper on Reconstruction and Development* (Government Gazette No. 16085), 23 November, Pretoria: Government Printer.

Republic of South Africa: Government Gazette (1996a) *The National Small Business Act* (Act No. 102 of 1996), Pretoria: Government Printer.

Republic of South Africa: Government Gazette (1996b) *Constitution of the Republic of South Africa* (Act No. 108 of 1996), Pretoria: Government Printer.

Republic of South Africa: Government Gazette (2000a) *The Preferential Procurement Policy Framework Act* (Act No. 5 of 2000), Pretoria: Government Printer.

Republic of South Africa: Government Gazette (2000b) *The Construction Industry Development Board Act* (Act No. 38 of 2000), Pretoria: Government Printer.

Republic of South Africa: Government Gazette (2003) *The Broad Based Black Economic Empowerment Act* (Act No. 53 of 2003), Pretoria: Government Printer.

Root, D. (2005) Introducing sustainability into public private partnerships in South Africa, *Proceedings of CIB W92/T23/W107 International Symposium on Procurement Systems 'The Impact of Cultural Differences and Systems on Construction Performance'*, Las Vegas, Nevada, USA, 7–10 February, 1: 243–251.

Rossouw, G.J. (1998) Establishing moral business culture in newly formed democracies, *Journal of Business Ethics*, 17: 1571–1653.

Rwelamila, P.D. (2002) Creating an effective construction industry strategy in South Africa, *Building Research and Information*, 30 (6): 435–445.

Rwelamila, P.D., Talukhaba, A.A. and Ngowi, A.B. (1999) Tracing the African project failure syndrome: the significance of 'ubuntu', *Engineering, Construction and Architectural Management*, 6 (4): 335–347.

Rwigema, H. and Karungu, P. (1999) SMME development in Johannesburg's southern metropolitan local council: an assessment, *Development Southern Africa*, 16 (1): 107–124.

Sako, M. (1992) *Prices, Quality and Trust: Inter-firm Relations in Britain and Japan*, Cambridge Studies in Management, Cambridge: Cambridge University Press.

Shakantu, W. and Bowen, P.A. (2002) Tracking targeted procurement policy: a success story at the Cape Town International Convention Centre (CTICC) project, *Proceedings of the 1st Postgraduate Student Conference, Construction Industry Development Board*, Port Elizabeth: 12–14 October, 166–176.

Statistics South Africa (StatsSA) (2004) *Labour Force Survey September 2004* (P0210). Online. Available at: http://www.statssa.gov.za/publications/ statsdown load.asp?PPN=P0210&SCH=3358 (accessed 4 November 2007).

Tomlinson, R. (2002) International best practice, enabling frameworks and the policy process: a South African case study, *International Journal of Urban and Regional Research*, 26 (2): 377–388.

van Wyk, L. (2003) A *Review of the South African Construction Industry*, Pretoria: CSIR (Boutek).

Western Cape Department of Economic Affairs, Agriculture and Tourism (2000) *White Paper on Preparing the Western Cape for the Knowledge Economy of the 21st Century*, Cape Town: Western Cape Department of Economic Affairs, Agriculture and Tourism.

Whiteside, A. and Sunter, C. (2000) *AIDS: The Challenge for South Africa*, Cape Town: Human and Rousseau Publishers.

Wilson, I. (2000) The new rules: ethics, social responsibility and strategy, *Strategy and Leadership*, 20 (3): 12–16.

Wong, P.S. and Cheung, S. (2004) Trust in construction partnering: views from parties of the partnering dance, *International Journal of Project Management*, 22: 437–446.

Zaghloul, R. and Hartman, F. (2000) Construction contracts: the cost of mistrust, *International Journal of Project Management*, 21: 419–424.

Chapter 15

Corporate social responsibility in the Hong Kong and Asia Pacific construction industry

Steve Rowlinson

> Everything that can be counted does not necessarily count; everything that counts cannot necessarily be counted.
>
> (Albert Einstein 1879–1955)

CSR in perspective

The growth of interest in corporate social responsibility (CSR) has been stimulated by a series of scandals, not least of which was the Enron debacle. Professor Ghoshal of the London Business School, in the *Financial Times* (18 July, 2003) under the headline 'Business Schools share the blame for Enronitis,' pertinently opined that the blame lay squarely on the shoulders of the professors of business and that business schools 'need to own up to their own role in creating Enronitis'. He went on to blame the 'misinterpretation' of Michael Jensen's agency theory, Oliver Williamson's transaction cost economics and Michael Porter's competitive strategies as a prime cause of this unbalanced view which was bad for society. Ghoshal concluded that

> by incorporating negative and highly pessimistic assumptions about people and institutions, pseudo-scientific theories of management have done much to reinforce, if not create, pathological behavior on the part of managers and companies.

So, who is to be blamed for this lack of corporate social responsibility? Has CSR evolved in the same way in Asia Pacific? Are there different CSR issues in Asia Pacific? This chapter takes examples from China, Hong Kong, Singapore, Macau and Malaysia – two countries and three city states with very different cultures with all but one in the top twenty globally in terms of (GDP) gross domestic product – to investigate and illustrate these issues. The issues will be put into perspective by a brief review of the economies of these countries and then specific examples will be cited and issues raised. But first, a little vignette which reflects, perhaps, one of the drivers for the changes currently taking place in the region.

CSR catches on in China

This was the headline in Air China's in-flight magazine (Anon. 2005a) that leapt out of the page at me on a flight between Shanghai and Beijing. The drive for this appeared to be coming from the European Chamber of Commerce in China:

> Chinese businesses are awakening to the virtues of corporate social responsibility (CSR). Over the past decade, CSR has blossomed as an idea, if not as a coherent practical programme, in large corporations. Multinationals in China are now coaxing local counterparts to join the cause of responsible business behaviour. A new CSR working group at the European Chamber of Commerce in China is encouraging international and Chinese companies operating in China to uphold best labour and environmental practices and to demand the same from local suppliers. European and Hong Kong firms, both well practiced in the arts of putting forward a best face, say helping out local communities is good for image – and the bottom line.

However, a major issue in terms of corporate social responsibility and governments' driving initiatives in this area is the commonly held belief in Asia that the WTO is using this as a bargaining chip to reduce the competitive advantage which Asian countries currently hold. Welford (2004) opines:

> However, many of these issues have not been welcomed by many governments of developing countries who see the potential inclusion of environmental, social, ethical, and human rights issues in the current round of WTO negotiations as a ploy to reduce their competitive advantage and a way in which developing countries can retain a degree of protectionism. The Havana Declaration of the Group of 77 that followed the South Summit held in April 2000 was unequivocal in this respect, stating that it rejected all attempts to use these issues for resisting market access or aid and technology flows to developing countries.

So, is there any truth in these assertions? Are Asian countries being singled out because of the state of development or is the issue a much wider one that affects all developing nations. Perhaps the truth of the matter lies in the pace of development. China, Vietnam and Thailand, for example, have developed at a far greater pace than many other nations, both in the past and presently. Hence, their ability to adapt and change to the international expectations in terms of trade has led to very agile organisations and flexible approaches to trade. In another report on corporate and environmental governance, dealing with business and human rights, Welford and Hills (2004) indicated many countries in which Hong Kong companies are at risk in terms of exposure to human-rights violations. These countries were as far afield as Russia,

Mexico, Brazil, and Turkey as well as many countries within the Asian region. The types of violation identified ranged from bonded labour, including child labour and forced labour; denial of women's rights; denial of freedom of association and of expression; and hostage-taking and disappearances. Many of these issues are intimately linked to culture, ethics and political systems that allow such 'corrupt' practices to prosper and continue.

So, is CSR a Western stick with which to beat the East in order to ensure 'fair competition'? Are Western exploitative practices condoned by the East, or exported from the East? Let us examine these issues by way of examples from Hong Kong, China, and other Asia Pacific countries.

The Singapore construction industry

Economic overview

Singapore has a highly developed and successful free-market economy. It enjoys a remarkably open and apparently corruption-free environment, stable prices and a per capita GDP equal to that of the four largest West European countries. The economy depends heavily on exports, particularly of consumer electronics and information technology products. It was hard hit during 2001–2003 by the global recession, by the slump in the technology sector and by the knock-on effect of the outbreak of SARS (Severe Acute Respiratory Syndrome) in 2003. Fiscal stimulus, low interest rates, a surge in exports and internal flexibility led to vigorous growth in 2004–2006 with real GDP growth averaging 7% annually (see Table 15.1). The government has plans to establish a new growth path that will be less vulnerable to the global demand cycle for information technology products – it has attracted major investments in pharmaceuticals and medical technology production – and plans to continue efforts to establish Singapore as south-east Asia's financial and high-tech hub (CIA 2007).

In the second quarter of 2007, all sectors saw healthy growth, and expansion was particularly strong in the financial services and construction sectors, which saw double-digit growth. Manufacturing posted gains and, driven by the strong rebound in building activities, construction

Table 15.1 GDP growth rate in Singapore.

Year	Growth rate (%)
2003	2.20
2004	1.10
2005	8.10
2006	6.40
2007	7.90

Source: CIA (2007).

increased its workforce substantially at the start of 2007 (Ministry of Trade and Industry 2007a).

Construction growth

The city state's economy is growing strongly, partly because of major construction works for the massive Marina Bay integrated resort project which started in 2007 and partly because of planning and investment activities for a second integrated resort. Singapore's GDP increased 8.2% on a year-on-year (YoY) basis in the second quarter of 2007, up from 6.4% in the previous quarter. On a quarter-on-quarter seasonally adjusted annualised basis, real GDP grew by 12.8%, up from 8.5% in the first quarter (BCI Asia 2007).

Language and culture

There are a multitude of languages spoken in Singapore that reflect its multi-racial society. The Singapore government recognises four official languages: English, Malay, Mandarin and Tamil. The national language is Mandarin, while English is mainly used as the business and working language (Wikipedia 2007a). As Singapore is a small and relatively modern amalgam of Chinese, Malay, Indian and European immigrants, the culture of Singapore expresses the diversity of the population as the various ethnic groups continue to celebrate their own cultures though they intermingle with one another.

(Wikipedia 2007b)

CSR in Singapore

As a holistic and sustainable approach to bring forward the CSR movement in Singapore, the National Tripartite Initiative (NTI) on CSR was launched in May 2004. The NTI on CSR serves as a steering committee to review and formulate broad CSR strategies, taking a tripartite approach to include the key stakeholders including business, unions and the government. The national steering committee has since founded a society known as Singapore Compact for CSR as a platform for fostering dialogue and collaboration among various CSR stakeholders. It will play a pivotal role in defining the direction and landscape of CSR in Singapore, and will help Singapore embrace CSR as a coordinated national initiative (CSR Singapore 2007).

The Hong Kong construction industry

Economic overview

Hong Kong returned to China from Britain in 1997; since then it has been a Special Administrative Region of China. Hong Kong's population was 6,921,700 at mid-2007, representing an increase of 64,600 or 0.9% over

Table 15.2 GDP in Hong Kong.

Year (HK$million)	GDP (nominal 2000 base)	GDP per capita (nominal* 2000 base)
2004	1,509,915	222,586
2005	1,623,479	238,284
2006	1,735,882	253,151
2007	438,498	

Source: National Income Section, Hong Kong Census and Statistics Department, 2007.

mid-2006. Of this population, about 95% are Chinese, and English and Chinese are Hong Kong's two official languages. The Cantonese dialect is the most commonly spoken language in the territory, though English is the language of the business and service industries; hotel employees, many urban Hong Kong residents, most young people and shop and service personnel understand and speak it to some degree. Other Chinese dialects such as Mandarin (Putonghua), Shanghainese and Chiu-Chow can be heard as well. The Hong Kong economy suffered a downturn following the Asian economic crisis and the SARS pandemic but since 2003, the economy has begun to pick up slowly. In the second quarter of 2007, the GDP increased by 6.9% in real terms over the previous year.

Key characteristics of the Hong Kong construction industry

Hong Kong's construction industry is characterised by a small number of large local contractors; a high level of subcontracting; the presence of many overseas contractors; and a substantial number of companies being both developers and contractors. Most of Hong Kong's construction companies are small in size; about 97% of them performed less than HK$10 million gross value of construction work in 2004. Hong Kong construction companies have earned a reputation for rapid construction of quality high-rise apartment blocks and office towers. Specialised construction techniques, such as reclamation and design-and-build methods, have made Hong Kong a regional leader. There is no formal restriction for entry to the contracting business in Hong Kong. Foreign and local contractors are treated alike, and they can tender for public sector projects so long as they have a good track record and sufficient financial capacity.

Hong Kong construction companies have become leaders in the export of services to Asia, with the Chinese mainland and Macau being the main markets. Major service categories include project management, contracting and engineering consulting.

The local construction sector accounted for 3.2% of GDP in 2004. As of June 2006, the sector employed over 52,000 site workers. The employment level for the broader building, construction and real-estate sectors is around 250,000 including such professionals as architects, surveyors, structural engineers, building services engineers and civil engineers.

Within the housing sector, the actual completion of private residential units in 2005 was 17,300, down from 26,000 a year before. On the public housing front, annual production is expected to drop from 19,600 for 2007–2008 to 19,000 for 2010–2011.

In the second quarter of 2006, the overall gross value of construction work completed at construction sites was HK$10.2 billion, down 17.6% YoY. Of that figure, construction work completed in public sector construction sites accounted for 39.5% or HK$4 billion (down 32% YoY). Construction work completed in private sites accounted for the remainder, which was valued at HK$6.1 billion, down 4.3% YoY.

On the non-site construction front, the gross value of related activities reached HK$12.9 billion in the second quarter of 2006, up 24.1% YoY. For the same period, general trade, which excludes carpentry, electrical and mechanical fitting, plumbing and gas work, was valued at HK$9.2 billion up 39.2% YoY, (Hong Kong Trade Development Council 2006).

CSR in Hong Kong

As the world's eleventh largest trading economy, Hong Kong cannot afford to overlook the sweeping influence of CSR over business decisions. As a matter of risk management, companies strive to ensure that practices of other companies, including international suppliers, do not tarnish their reputation, with CSR initiatives undertaken to show that their operations meet or exceed legal requirements and societal norms in their home country or abroad. Failing that, acts of 'punishment' by consumers and investors (including shareholder activism via pension funds) could lead to very negative repercussions. Conceptually, CSR practices are largely voluntary acts, though the translation of CSR principles into supplier code requirements leaves little leeway for Hong Kong's manufacturers and exporters – full compliance is expected or business relationships will be affected.

According to a local survey conducted in 2003, Hong Kong SMEs, including those in the trading and manufacturing sectors, focused their attention on compliance with mandatory government rules and environmental management, with CSR considered better suited to larger companies and such practices considered to be adopted when the economy was more robust. Nevertheless, there is a growing interest over the past couple of years among Hong Kong manufacturing, construction, real-estate and trading companies in CSR practices, which has taken a variety of forms ranging from community investment and corporate philanthropy to better codes of conduct and business ethics (Hong Kong Trade Development Council 2005).

The Macau construction industry

Economic overview

Macau gained its status as a Special Administrative Region of China in 1999 on its return from Portugal. It has an estimated population of about 520,400 of which about 95% are Chinese and about 3%, Portuguese. As a creation of the Portuguese, Macau represents a peculiar blend of Oriental and Western influences. Indigenous languages spoken are Chinese-Cantonese (Yue dialect and Min dialects, about 96% of the population) and Portuguese (about 4%). Beijing-Chinese (Putonghua dialect) is a second language and growing in influence (e.g. it is used in education). English is also expanding as a language in commerce and tourism. The old Macanese language (Patuá, or Makista) was a typical Creole language, based on Portuguese but heavily influenced by various Chinese dialects and by Malay. It has now virtually died out.

As a casino-centred city, its economy is primarily based around the gaming industry, and the associated service industry. Tycoon Stanley Ho monopolised the gaming industry from the 1960s until it was liberalised in 2003. A phenomenal growth in the gaming industry then ensued. Indeed, Macau overtook Las Vegas as the world's No. 1 gambling market with spending in slots and table games estimated at about US$6.5billion in 2006. Revenue from the gambling sector alone accounted for about 70% of all yearly total revenue of the government between 2004 and the first quarter of 2007 (Macau S.A.R. Statistics and Census Service 2007). An overall growth in the economy has also followed with a reported nominal GDP growth rate of 32.9% and real growth rate of 25.6% for the first quarter of 2007 (see Table 15.3). Observers, however, caution against the single industry focus of the economy.

Table 15.3 GDP in Macau.

	2004	2005	2006	2007 (1st Quarter)
GDP (10⁶ MOP)	82,966	92,951	11,4364	33,248
GDP per capita (10³ MOP)	181.6	195.2	227.5	—
Growth rate (Nominal)	30.5	12	23	32.9
Gross gaming receipts (Gratuities excluded, 10⁶ MOP)	43,511	47,134	57,521	18,598
Revenue from gambling sector	15,236.6	17,318.6	20,747.6	7022.7
% of total revenue	63.85	61.41	75.83	75.79
Population	462,637	484,277	513,427	520,400 (est.)

Source: Macau S.A.R. Statistics and Census Service, August 2007

Table 15.4 Macau construction expenditure.

	2004	*2005*	*2006*	*2007* *(1st Quarter)*
Expenditure on public works	2814.2	3652.1	3682.5	1.9
Private sector construction ('000m2)				
Buildings started (gross floor area)	715	2133	1054	841
Buildings completed (gross floor area)	215	391	1276	167
Building units sold	27,823	33,644	26,400	10,324

Source: Macau S.A.R. Statistics and Census Service, August 2007)

CSR in Macau

In terms of CSR, the government of Macau has set up a Centre for Sustainable Development Strategies (CEEDS), which aims to assist in defining development policies and strategies. The Chief Executive of the Macau government, Edmund Ho, said that the centre would support

> the Government in formulating public strategies and policies for sustainable development, which will contribute to ensure harmonious compatibility between economic, social and environmental objectives and a progressive improvement of the quality of life of the Macau population.

He said the centre has been created to ensure the 'sustainability of the development process' and the 'progressive elevation of quality of life of the resident population within the framework of boosted regional integration and cooperation with the Great Pearl River Delta Region and the Asia-Pacific Region'. The internationalisation of the Macau economy, particularly as a platform for promoting cooperation between China and Portuguese-speaking countries, is also cited as a reason for setting up the new centre to support the government.

The People's Republic of China construction industry

Economic overview

China has been developing at high speed since the reform of its economic system in 1978. The system is still in transition from a planned economy system to a developed-market economy. The government is making efforts to remove provincial and sectoral barriers to free the movement of resources.

To be integrated into the world market, China joined the WTO in 2001 and is pursuing free trade agreements (FTA) with more Asian countries. Lim *et al.* (2006: xvi) in their report of China's economics predicted that:

> in 25 years, China, with another two to three doublings of its GDP, will be a major economic power and a leader, along with the USA, the EU and Japan, in setting international economic policies.

China had a population of 1.31 billion by the end of 2006 with around 40% of the population living in rural areas. The Chinese culture emphasises relationship and ethics in management, which is perceived to blur the effect and roles of contracts and regulations.

The domestic Gross Output Value of the construction industry has increased by more than 20 times, accounting for 3.8% of GDP in 1978 and 7.0% in 2006 (see Table 15.5). By the end of 2006 it employed a workforce of 2.84 million, but the construction industry suffers from low capital investment, a preponderance of small-scale units, low technology and over-competition. The construction industry is highly dependent on the economic policy of China. For example, in 1987–1990, when China conducted a macro-economic control policy, the construction industry output decreased 1.6% every year but during 1992–1995, in an overheated economic environment, it increased 27% every year.

Table 15.5 GDP and construction statistics in China.

	2000	2001	2002	2003	2004	2005	2006
GDP (billion yuan)	992.15	1,096.55	1,203.33	1,358.23	1,598.78	1,838.68	2,108.71
Per-capita GDP(yuan/person)	7,858	8,622	9,398	10,542	12,336	14,040	15,973
Construction(billion yuan)	55.22	59.32	64.65	74.91	86.94	101.34	116.53
Gross output value of construction (billion yuan)	124.98	153.62	185.27	230.84	290.21	345.52	409.76
Floor space under construction (million sq.m)	1,601.41	1,883.28	2,156.09	2,593.77	3,109.86	3,527.45	3,996.06
Floor space completed (million sq.m)	807.15	976.99	1,102.17	1,228.28	1,473.64	1,594.06	1,641.22
Number of persons engaged in construction (million)	1.99	2.11	2.25	2.41	2.50	2.70	2.84
Population (million)	126.74	127.63	128.45	129.23	129.99	130.76	131.45

Source: China Yearly Macro-Economics Statistics (National) chinadata@umich.edu.

CSR in China

Under the previous planned economic system, enterprises were not independent profit-making units. Aside from production, they assumed responsibility for health care, education housing, pension and employment of family members of their workers. Financially, the heavy social responsibilities made it impossible for enterprises to survive the move to market economics. After redefining the boundary of enterprises, the functions of pension, hospital, education and accommodation were separated from corporations and shifted to the newly established social insurance system. However, 'driven by profit' enterprises went to a further extreme by showing no accountability towards society, except tax payment (Zhu 2003). Thus, one might conclude that under the 'old' regime CSR was built into the system, but under the 'liberalisation' CSR was effectively put on the back burner.

The issue of CSR came back into public attention again from 1985 in a report of a chemical factory which attempted to change attitudes by making energy-saving products (Hua 1985). The issue of CSR was discussed in terms of the boundary between corporation and society, a distinction which is seen differently in Asia than it is in the West (Zhang 1994). Since 2000, Western theories of CSR were introduced into China (Ning 2000; Chen and Jia 2003). Reflecting on the situation in China, Ning (2000) pointed out that enterprises' lack of social responsibility was commonly manifested by speculatory investments, loan defaults, fake commodities, low employee well-being, tax evasion, etc. The business community criticised voluntary CSR activities such as charity donation, offering employment opportunities and preserving the environment as being much less effective than they were under the previous planned economic system.

Since 2003, CSR issues have been introduced systematically through government dictat. Chen and Jia (2003) reviewed the Western CSR theories and concluded that CSR under planned economic systems is essentially different from the Western view of CSR, which is based on an integrated social contract. CSR under China's transitionary planned economic system was an expansion of a production unit's boundary, whereas the Western view of CSR is a rational decision of a profit-making corporation. This is an important cultural difference that should be kept in mind. Shi (2002) categorised the content of CSR into five aspects: production-related responsibilities, human resource–related responsibilities, law-related responsibilities, responsibilities to outside stakeholders, and voluntary activities. Zhai and Liu (2003) operationalised the CSR concept into seven guidelines for enterprises: keep the law, pay tax, create employment opportunities, maintain quality, preserve the environment, pay for charity and make technological innovation. However, there has always been some form of CSR evident in Chinese industry. For example, the bridge collapse in Sichuan province in 2007 was followed immediately by the arrest of company directors associated with the construction project, both design and construction organisations. Indeed, it is not uncommon for company directors to be executed in China following terrible events such as mine explosions and collapses. Thus, the concept of corporate manslaughter is already ingrained into the Chinese system but perhaps in a manner that the West finds at least draconian, if not unacceptable.

The globalisation of economics challenged China's CSR with the criteria of Social Accountability 8000 (SA 8000). More and more international companies require their partners to follow the SA 8000 standards of CSR (Yin 2005). Yu (2005) compared the SA 8000 with the Labor Law of China and she reported that SA 8000 is more detailed and operation oriented, but some of the issues in the Labor Law are stricter than the corresponding ones in SA 8000. She suggested government customise the international criteria SA 8000 for China's market system.

CSR issues in Asia in general

Employment practices

In their report on employment practices and corporate social responsibility in Hong Kong, Welford and Mahtani (2004) made the following observations. Whilst focusing on internal employment practices, they indicated that policy in the areas of gender, pregnancy, marital status, sexual harassment, disability, family status, sexual orientation, race and age were key issues which companies should be addressing. In fact, Hong Kong currently has only three equal-opportunity ordinances, and these deal with sex discrimination, disabilities and family status. However, there are no laws dealing with race, sexual orientation and age at present. The overall impression given in their report is that the best Hong Kong companies have addressed most of the discrimination issues in their policies, but that the majority of Hong Kong companies still lags a long way behind in dealing with these issues.

Another issue raised is working hours. Welford and Mahtani (2004) discuss in detail some of the issues which make Hong Kong peculiar, if not unique, in its attitude to working hours and flexible working systems. The issue of lack of work–life balance, and its symptom, burnout, are poorly dealt with by many Hong Kong companies. The ethos of long working hours and staying at work until the boss leaves is deeply ingrained in Hong Kong society. In recent research, Lingard *et al.* (2007) indicate that burnout, which is typified by emotional exhaustion, depersonalisation and reduced personal accomplishment, is prevalent in Hong Kong, both in industry and in tertiary education students. The levels of burnout were at least comparable to those registered in Australian construction professionals, and the burnout was brought about by high workloads compounded by insufficient resources, dissatisfaction with pay and fringe benefits, and the belief that favouritism and unfairness were prevalent in supervisory practices. Tertiary education students were found to suffer from high levels of emotional exhaustion and a low level of self-esteem. This, in part, stemmed from the almost 'indentures-like' practices of the teaching professionals who also ran their own businesses, which were nurtured by students' efforts. In discussions and workshops held in China's five big cities (Beijing, Shanghai, Tian Jin, Guangzhou, ChunChing), a similar pattern emerged. However, unlike Hong Kong society, the Chinese professionals accepted this as a normal way of life and rationalised this view on the basis that rapid economic development was necessary and expected, so one had to sacrifice oneself for the greater good of society, a view stemming from Confucian values and the collectivist concept so typical of Chinese society. The Hong Kong attitude, perhaps following the property crash of the late 1990s, was more focused on personal accomplishment and well-being than it is in China.

Health and safety

In her discussion of CSR in the Australian construction industry, Lingard and Rowlinson (2005) state:

> There has been an increasing public expectation that organisations will take responsibility for their non-financial impacts, including their environmental performance and impacts upon the community. This has led to the notions of social responsibility and Triple Bottom Line (TBL) reporting.

A new suite of Australian Standards has just been published concerning business governance. AS 8003 (Standards Australia 2003) deals with the issue of corporate social responsibility. Here CSR is defined as:

> A mechanism for entities to voluntarily integrate social and environmental concerns into their operations and their interaction with their stakeholders, which are over and above the entity's legal responsibilities.
> (SA 2003: 4)

Such standards are not in general use in Hong Kong nor in China, but things are starting to change. For example, following the boom in casino construction in Macau, a new labour law has laid down minimum standards for workers, most of whom come from north of the border in China. These standards include provisions for personal health, food, nutrition, potable water and canteen facilities, and dormitories. In addition, the standard of performance in terms of safety on the sites is surprisingly good in Macau: Gammon Construction Company Limited managed to complete 2.7 million man hours of work without a single lost-time accident. This contrasts sharply with the previous accident record in Macau in the construction of major infrastructure some years earlier. Indeed, during interviews with Hong Kong-based management and supervisory staff in Macau, the respondents indicated that they found working in Macau far less stressful. Why was this? The answer related to the fact that the workers have more free time and more time off; they lived close to the work sites in well-appointed dormitories provided by the employer and they found time and had the inclination to socialise with colleagues from other organisations. This is something they were unable to do and were unused to in Hong Kong. As a consequence, they believed that their employment conditions and their enthusiasm for work were far greater now in Macau than in Hong Kong. Similar views were reported by Lingard *et al.* (2007) in an infrastructure project in Queensland, Australia.

Another example of a successful approach to CSR comes from China Light and Power (CLP), a Hong Kong power provider, but heavily involved in construction and development. The company regularly organises open days when many of its employees' families come along to enjoy both entertainment and education. As a supplier of electricity and electrical products, the company focuses its open days around safety issues both at home and at work. The prime idea is that the family cannot survive if the breadwinner is injured or maimed at work, and also that the worker needs to be satisfied as to the well-being of the family in order to be motivated to work effectively and efficiently. As a developer, CLP has an exceptional OH&S record.

Environment and development

When discussing CSR in the real-estate industry of Malaysia, Tay Kay Luang in the *New Straits Times* in November 2004 indicates that few of Malaysia's 2,000 property developers actually subscribed to CSR. He opines that although some companies see CSR as a means to a competitive advantage, most see it purely as an additional cost. He goes on to discuss the Malaysian government's efforts to promote CSR. Six property and construction companies were awarded CSR certificates during a government home-ownership campaign. These companies saw CSR as a means to integrate practices into the company functions, which allowed them to flourish and to meet and direct market demand.

This clearly reflects the acknowledgement of the importance of the link between the product and the company's marketing strategy and also highlights the difficulty of the first mover in such circumstances. Tay Kay Luang reports that Putrajaya Holdings Sdn Bhd was issued with the CSR certificate for its work in coming out with the first 'Garden City, Intelligent City' concept in Malaysia through the development of the Federal administrative capital of Putrajaya where 40% of the township has been designated as 'green areas'. This clearly reflects a risk in moving away from conventional designs and processes and attempting to forge market opinion by promoting responsible development. Again, a long-term view is important, but this cannot always offset the competitive disadvantage owing to competitors' lower prices or the development of cartels. In the same article Tay stated that many 'non-CSR' companies found CSR unattractive because it was perceived to be an added cost. However, he states:

> The irony is that those which have integrated CSR practices with their business have flourished. For example, Gamuda Land's policy to set aside 40 per cent of each development parcel to green lungs, parks, lakes and landscaping has made its branding synonymous with eco-friendliness.
>
> (Luang 2004)

He quotes the example of Gamuda which, at its flagship Kota Kemuning project, rehabilitated swampland into a wetlands park, which also functions as a natural drainage system. The same company went further at the Bandar Botanic project where over 20% of the development was set aside for greenery and landscaping at a cost of RM40 million, although the local authority required only 14%. This translated into higher prices, but sales hit a record RM750 million; thus, the CSR approach led to improved financial performance.

Hence, taking a long-term view of CSR can lead to competitive advantage. To stay ahead of the competition, property and construction companies must strive to build an image of being more caring and thoughtful as part of their profile and maintain this throughout the project life cycle. Luang commented that in Malaysia while many recognised the need to generate excellent publicity and PR, they have not yet demonstrated sensitivity towards the environment and CSR. As far as the Malaysian property and construction industry goes, CSR covers such areas as health and safety at work, fair pay and a commitment to community development. However, Luang points out that there are several areas which property and construction companies must address to become more socially responsible. These include product stewardship, community, environment and work conditions. As in many developing Asian countries, housebuyers are no longer tolerant of shoddy workmanship and late delivery. In Asia, most households spend 40% or more of their disposable income on mortgages, so property and construction companies can no longer ignore consumer demands, particularly as a tide of democracy and empowerment sweeps the region.

Elsewhere, Hong Kong Land has taken the lead in socially responsible development in Hong Kong by addressing the issue of the maintaining of wetlands and other areas of natural beauty and interest as part of their development process. The project which has received most publicity thus far is the proposal from a consortium comprising Swire, Sun Hung Kai and Hong Kong Land, which aims to turn the 225-hectare Tai Ho site into an Ecology Park and private housing. In an article entitled 'Firms issue proposals to develop green sites', Cheung (2005) noted that urn niches, herb gardens, spa hotels, elderly homes and private housing are among proposals submitted by developers for some of the city's most sensitive sites.

Cheung reports that six development proposals and four management-agreement schemes were submitted to the government by consortia, individual developers and green groups. Under the planned public–private partnership scheme, 12 priority ecological sites would be chosen for limited, responsible development. The government encouraged collaboration between green groups and landowners by developing the management-agreement initiative. The sites that had been identified included flora- and fauna-rich areas, very special interest sites such as a particular dragonfly-rich site and areas containing native woodland. These developments are in marked contrast to the typical high-rise and high-density mass concrete structures of recent years.

Business practices

In the *Sunday Morning Post* of 8 May 2005, Kenneth Ko reported on the demand for more transparency in property sales. He was specifically raising concern over developers' tactics and the way they had concealed pricing information from buyers in the property market and maintained levels of sales through an internal sales mechanism.

In the boardroom, the Chief Executive and Chairman of the Bank of East Asia took a 44% increase in remuneration in 2004, despite the fact that the Hong Kong economy was relatively weak; many workers had actually taken pay cuts, and the bank's profit growth was a mere 26%. These issues are a matter of corporate governance and social responsibility. However, critics have suggested that the government's view that CSR is a good starting point for a private company may not always be appropriate.

The formation of cartels and monopolies has long been a feature of business in Hong Kong and Asia. Sometimes, these practices are defended as a process of vertical integration, but they are more commonly a means for controlling supply and reducing competition. The ready-mixed-concrete supply industry is a prime example of this in Hong Kong, where a small number of developers control the vast majority of the market. This leads to a situation where one or two organisations can control the way in which the majority of the industry works, and this situation militates against the development of a young and vital supply chain which is sustainable.

Freedom of association and human rights

Another area where Hong Kong and south-east Asia lag behind is in terms of freedom of association and supply chain management. Other countries in the Asia Pacific region, such as Australia, have much more open policies on these issues and make a serious attempt to engage their supply chain. The construction industry in Hong Kong has been repeatedly vilified for its use of non-value-adding subcontracting and for the quality and safety problems that this approach brings on.

In an article in the *South China Morning Post* of 30 April, 2005, Dicky Sin reported on the abhorrent practice of spying on injured workers who are claiming compensation from insurance. It appears that telephone taps and hidden cameras are often used during investigations into claims lodged by workers. The aim of the insurers is to prove that the injury is not as serious as reported; yet, the fact that the injury has taken place is evidence of poor safety management within the industry and poor direction by the insurers to their clients.

Investment

In his study of awareness and attitudes to environmentally and socially responsible investment, Welford (2004) reports that his respondents indicated that companies should focus on profitability and take responsibility for their environmental and social impacts and have high standards of corporate governance. Despite opinions to the contrary, Welford found that many of his respondents in Hong Kong felt that profit should not be the primary focus for a business and that social and environmental issues were important. The point seemed to be that the respondents required a balanced position to be taken by the companies that they invested in.

Philanthropy

However, Hong Kong and China score very highly in terms of community investment and philanthropy. The most notable example of this recently was the donation of HK$1 billion to the University of Hong Kong by Li Ka Shing, one of the world's top fifty billionaires, and CEO of, amongst others, property development company Cheung Kong Holdings. This donation had no strings attached, and the university was at liberty to make use of the donation as it felt fit. However, the vice-chancellor fell foul of the sensibilities of graduates of the university's medical school when he outlined plans to rename the medical faculty as the Li Ka Shing medical faculty. This was an act of appreciation that was not appreciated by many of the vice-chancellor's colleagues. The Li Ka Shing Foundation has a long history of philanthropic donations, both in Hong Kong and in his native Shantou province. The donation to Hong Kong University amounted to only 30% of all of the Foundation's donations over the past ten years. Other organisations, such as Arup and Associates Hong Kong Ltd, have a long tradition of sending, or allowing to be sent, their top staff to assist in ventures such as *Médecins Sans Frontières* (MSF), and Arup's staff provided logistical assistance during the aftermath of the tsunami in Aceh province in Indonesia. The engineering and logistics skills provided by Arup's staff are invaluable and their absence is certainly a drain on the company's resources; these are given willingly and in a spirit of philanthropy.

Culture

Culture, the values and beliefs held by a group or a nation, has an impact on the way CSR is viewed and implemented in different locations. Hofstede defined five dimensions of culture including a power-distance and uncertainty avoidance. In his research, Hofstede found that the Philippines exhibited a very high power-distance score; there is a deference to people in power or who have social status or economic power. In such circumstances the checks and balances that maintain a degree of CSR break down, as the vast majority is unwilling and unable to question the deeds and actions of the mighty few. Thus, in such societies, the development and maintenance of a culture of CSR is difficult. In other countries, the dimension of uncertainty avoidance can be very high. In such countries there is a dependence on rules and procedures and protocols to maintain a social structure and organisation. China has a high uncertainty avoidance score, but this does not mean that CSR will be implemented any more effectively in China. An intervening factor is the pace of development of the country. When development takes place rapidly, there is a need to circumvent and then bypass these rules and procedures as the necessity for the rapid take-up of opportunities is seen as an overridingly important factor. As a consequence, in the rapid development of the Chinese economy over the past ten years, the concept of *guang xi* has been an important moderator in terms of CSR. *Guang xi*, in its simplest form, refers to relationship. This relationship may be through family connection, but is often through business connection. As such, the building up of a relationship takes time, but once established, there is an obligation to perform to the benefit of those with whom the relationship has been built and to the exclusion of others. It will be obvious to the reader that such a favouritist-type approach can run counter to the concepts of CSR.

In many societies, such as China, Thailand and Hong Kong, trust needs to be built up gradually and be well-established before an appropriate business relationship can develop. There is a heavy reliance upon this relationship and trust, and less of a reliance on legal contracts and detailed agreements. This then leads to a situation where the details of a contract may not be seen to be as appropriate as the nature of the relationship between the parties. In such a society, this is a commonly accepted and understood way of doing business. But for, say, an American company undertaking business in Hong Kong or China, this might not seem to be a fair and appropriate way of going about business and may be seen to be as an outworking of the favouritism often attributed to *guang xi*. However, this is really a reflection of a different set of cultural values rather than a disregard for the principles of CSR. Hence, when one looks at CSR in this global context, it is very important to include the cultural dimension, both in terms of cultural values and beliefs and the stage and pace of economic development, before making judgements as to the applicability of CSR.

There is a tendency in Chinese societies for a predominance of family-based businesses. Such businesses work on the principles of harmony and respect for elders, loyalty and a collectivist view of the good of the business and exhibit a long-term orientation. As such, these businesses may appear to build a barrier around themselves which precludes non-family members from participation and may be seen to be, in some eyes, a restraint on trade. This again, perhaps, is a reflection of the different cultural values and beliefs rather than a disregard for the Western concepts of CSR.

In fact, it is quite reasonable to expect that a high level of CSR can exist within such family businesses exhibiting these Confucian characteristics. However, such businesses exist in a situation where their stakeholders are minimal and in the main come from the family network. As such, there is a particularly strong common mindset which drives the organisation to achieve its goals to the exclusion of consideration of environmental and other forces acting upon the organisation. Again, such a mindset can lead to the underlying feeling that there is no concept of CSR in such organisations. However, that is not true; but the direction and drive of the organisation is often dominated by a senior family member, and there may well be a lack of recognition at this level of the issues which are at stake. Hence, judgements about such organisations need to be carefully assessed and circumstances and culture properly understood.

When looking at the concept of CSR and the issues that arise from it, there are a number of themes which commonly recur. Four of these are corporate manslaughter, insider trading, ethics and professionalism. These are issues that directors are expected to address and provide examples of good practice in. Take, for instance, the issue of corporate manslaughter. This specifically addresses issues relating to occupational health and safety and has been a matter of much debate in the Australian State of Victoria. In discussing this issue, Lingard (2005) stated that in 2002, the Crimes (Workplace Deaths and Serious Injuries) Bill was defeated. The Bill would have created criminal offences of corporate manslaughter and negligently causing serious injury by a body corporate and imposed criminal liability on officers of a body corporate. Although welcomed by trade unions, it was fiercely opposed by employers' groups. Lingard opines that although the Bill was defeated, the general public and trades unions support the move towards the institutionalisation of industrial manslaughter charges. Indeed, in China the law currently allows for such charges, and there have been many instances of company officers being tried, found guilty and even executed for such acts.

Thus, this issue indicates the need for a different view as to how companies and their directors should be operating. The idea of setting a good personal and corporate example is paramount. There is a recognition, globally, that companies and their directors should not be focusing in the direction of profit maximisation, but should be balancing this view with their social, health and safety responsibilities to be contributors to society. This is ably argued in both Ghoshal's and Lingard's commentaries, and it reflects a move away from traditional financial expectations of the company and indicates the need for shareholders, who are major stakeholders in companies, to adjust and adapt their expectations. These expectations can be rationalised in the ideas expressed and put forward in the concepts of value-based management and intangible-assets monitoring (www.valuebasedmanagement.net). These ideas and directions indicate to individuals, investors and companies how a balance can be attained between profitability and social responsibility and how such concepts in the long term lead to sustainable policies and strategies.

Ethics

The issue of ethical behaviour and the expectations of professionals are two interrelated and important issues. Fellows *et al.* (2004) discuss ethics in the process of bid preparation for contractors and indicate how ethical issues need to be addressed in the bidding process. Indeed, the case study, presented below, of the situation surrounding a Hong Kong project is an important example of the confluence of ethical and professional behaviour. Professionals are expected to abide by codes of practice which their respective professional bodies and institutions lay down as the mechanism by which professionals are to be judged. The regulation of professional behaviour by the professional bodies is a central tenet of the mechanism for maintaining business ethics. The case, presented below, goes into these issues and shows how a lack of professionalism and an unethical viewpoint can lead to serious problems in the real-estate and construction industries. These problems, which beset the real-estate and construction industry, are not the fault of one or two individuals, but actually come from the attitudes which corporations, real-estate investors and construction companies have ingrained in their values. By having a non-responsible attitude to the industry as a whole, these companies set up a situation in which various individuals are tempted to behave unethically and unprofessionally and thus bring a poor image to the industry.

A large property development company had a long-established practice of inviting bids for new projects from a number of well-known, highly competent and respected contractors. When these bids are submitted, the directors of the development company would compare the bids for their planned methods of construction and alternatives presented and, after lengthy discussions and interviews with the bidders, would put together what could be described as a hybrid proposal based on these bids and would then invite the participating organisations to rebid on this hybrid version of the project which had been developed from a combination of all of the companies' 'intellectual property'. This had the effect of driving down tender prices as far as the developer was concerned, but also of reducing potential profit margins for the participating contractors. As this situation continued, there was an increased pressure in a declining market for directors of construction companies to gain inside knowledge of competitors' bids in order to pare their prices to a competitive level and also to improve their construction project plans. When the opportunity arose to elicit this information from a consultant to the developer, a number of the participating contractors were tempted to access such information by mechanisms which could be considered unethical and unprofessional. In so doing, the contracting company may have been able to secure a competitive advantage in the bidding process, but the professional and ethical values of the consultant to the developer and the director of the construction company were compromised by the need to make some form of offer for the information to be provided. Although this information may have been obtained in a legal manner, the mechanism by which it was revealed to only a small number or a limited number of bidders reflected a lack of responsible management of the bidding process and provided the opportunity for more serious corrupt practices to start to take root in the industry.

Hence, this is an example where individuals were tempted to sail very close to the wind in terms of gaining access to potentially confidential and important information, and where this process was in a sense encouraged by the practices adopted by the development company. The whole issue of professional ethics and corporate social responsibility can come together in such a tempting situation where competition is intense and fierce. Such issues, echoing the views of VanderKamp, can be addressed by governments. In a recent edition of *CSR Asia Weekly* (14 September, 2005), it was reported that the Malaysian government was imposing a code of ethics on its contractors.

Code of ethics for the construction industry

The estimated 63,000 contractors in Malaysia will have to abide by a special code of ethics which the government introduced in 2006 to improve their efficiency. Government's view was that there was a need for such a code to provide guidelines for contractors, especially those involved in government projects. The analogy was drawn with engineers and architects who both have a code of ethics, which contractors do not. The aim was to place greater emphasis on self-regulation among contractors and consultants in ensuring ethical practice and continuous improvement. A government minister stated:

> Once implemented, the code of ethics and appraisal system will have to be adhered to by contractors registered with the construction Industry Development Board.
>
> (Vellu 2006)

Challenges

In the run-up to the Beijing Olympics, China's construction industry has been growing at an even more rapid rate than the 7% per annum that it has grown in the past ten years. Indeed, the annual consumption ratios of construction to national consumption are as follows: steel 25%; timber 40%; cement 70%; glass 70%; transportation 8%. Furthermore, the annual energy consumption of the construction industry is about 28% of the national energy consumption. Indeed, the annual total construction of two billion square metres of development is almost half of the global total of construction. Hence, a major challenge for China is to focus at a national and company level on developing green construction protocols and to focus specifically on energy saving. The government, in reflecting a growing awareness of CSR note that:

> Sustainable development of the building industry is closely related to improved quality of life for the people, the realization of the national energy strategy and resource conservation strategy, global climate change and sustainable development.
>
> (Baoxing 2008)

This is undoubtedly a great challenge to government, state and privately run organisations. Indeed, such a challenge also applies to the rapidly developing economies of Malaysia, India, Indonesia and Thailand to name but a few.

Generally, perhaps the greatest challenge facing Asian countries in implementing CSR is to go about the process in a culturally sensitive manner which both addresses the concerns of the West and also maintains the morals and values of the East. Hence, the task at hand is to devise corporate social responsibility with Asian characteristics. This means accepting, for the time being at least, Confucian and Buddhist concepts such as harmony, filial piety and consensus. For the West, there is a need to accept this diversity and to acknowledge the range of cultural and political values which exist in Asia. Asia is not homogeneous, and each country and city state has its own distinctive values and ways of doing business. There is no merit in striving for conformity when such practice will lead to a clash with local values. Hence, CSR as developed in Asia will be multifaceted and distinct, unique in each location and continuously changing as society develops.

The ability to meld a global economy with a set of ancient values and beliefs is not easily developed and this process will undoubtedly take time. Hence, the challenge is not only in adopting an appropriate and acceptable form of CSR, but also in driving this development forward at an appropriate pace. Essentially, time is not of the essence in this process: it is important that a uniquely Asian form of CSR develops which can meet the objectives of both East and West.

Conclusions

The concept of CSR has taken root in Asia, but one might be drawn to conclude that the driving force for it is rather different from that in the West.

A major factor that must be considered when looking at any business practice in Asia is culture. Many Asian countries have their roots in Buddhist or Confucian ideals, so the starting point in terms of responsibility is focused on collectivism and lack of confrontation. Also, power-distance, filial piety and respect for elders are commonly held Asian values. As such, there is an inbuilt necessity in Asian organiations to respect and care for workers as if they are part of an extended family. However, barriers such as *guang xi* also exist to truly open and fair competition. Hence, the concepts of CSR in Asia take on a very different dynamism. An underlying driver is the tension between rapid economic development and the imposition of sanctions and quotas through various trade organisations. Hence, the systems and approaches that are adopted in the West to drive CSR are inappropriate in the Asian context. Indeed, with the high power-distance and respect shown for those in authority, there is less of an expectation from the workforce in terms of equality and participation.

In countries such as China, the rule of law and procedures are a major driver for the development of CSR, along with the change in the economic system and the responsibilities that has brought and/or removed. However, the drivers in Hong Kong and Singapore are rather more similar to the drivers in the West. The drivers in Malaysia and the Philippines are different again. Hence, it is impossible to categorise the nature of the development of CSR in Asia except by looking at the individual culture and economy of each country or city state and then fitting the approach carefully to the culture and values of that location.

References

Anon. (2003) *Financial Times*, 18 July 2007.
Anon. (2004) *New Straits Times*, 11 December 2007.
Anon. (2005a) *Air China Inflight Magazine*, 6, vol. 127, p. 168.
Anon. (2005b) *South China Morning Post*, 21 February 2007.
Baoxing, Q. (2008) Quoted in, Press briefing on China's energy-efficient building and green building initiative speech by Qiu Baoxing, Vice Minister of Construction, http://english.gov.en/2008–02/28/content_904034.htm (accessed 22 June 2008).
BCI Asia (2007) Available at http://www.bciasia.com/news/news_view.cfm?news_id=100921340002652 (accessed 16 August 2007).
Chen, H. and Jia, S. (2003) Research on evolution of corporate social responsibility view: from the perspective of integrative social contracts. *China Industrial Economy*, 12: 85–92.
Cheng, D. (2005) Corporation social responsibility under SA8000. *Group Economy*, 172: 29–30.
Cheung, C.F. (2005) Firm issues proposal to develop green sites. *South China Morning Post*, 1 June 2007.
CIA (2007) *The World Factbook – Singapore*. Online. Available at https://www.cia.gov/library/publications/the-world-factbook/geos/sn.html (accessed 16 August 2007).

CSR Asia (2007) http://www.csr-asia.com/ (accessed 20 December 2007).

CSR Singapore (2007) Online. Available at http://www.csrsingapore.org/aboutus. php (accessed 16 August 2007).

Fellows, R.F., Liu, A.M.M., and Storey, D. (2004) Ethics in construction project briefing, *Science and Engineering Ethics*, 10: 2.

Hong Kong Trade Development Council (2005) *Corporate Social Responsibility and Implications for Hong Kong's Manufacturers and Exporters*. Online. Available at http://www.tdctrade.com/econforum/tdc/tdc050202.htm (accessed 17 August 2007).

Hong Kong Trade Development Council (2006) *Industry Focus-Building and Construction*. Online. Available at http://infrastructure.tdctrade.com/content.aspx? data = infrastructure_content_en&contentid = 173569&w_sid = 194&w_pid = 749&w_nid = 10738&w_cid = 2&w_idt = 1900-01-01 (accessed 17 August 2007).

Hua, H. (1985) Corporate's social responsibility: an interview with the Nanhua activator factory, *Outlook*, 38: 21–22.

Jensen, M.C. (1998) *Foundations of Organizational Strategy*, Cambridge, MA: Harvard University Press.

Ko, K. (2005) *South China Morning Post*, 8 May.

Kohler, A. (2005) *The Age*.

Lim, E., Spence, M. and Hausmann, R. (2006) *China and the Global Economy: Medium-term Issues and Options a Synthesis Report*. Center for International Development at Harvard University, Working Paper No. 126.

Lingard in Lingard, H.C. and Rowlinson, S.M. (2005) *Occupational Health and Safety in Construction Project Management*, United Kingdom: Taylor & Francis Books Ltd.

Lingard, H., Yip, B., Rowlinson, S. and Kvan, T. (2007). The experience of burnout among future construction professionals: A cross-national study. *Construction Management and Economics*, 25 (4), 345–357.

Luang, T.K. (2004) *New Straits Times*, 21 November.

Macau S.A.R. Statistics and Census Service (2007) *E-Publications, Construction*. Online. Available at http://www.dsec.gov.mo/index.asp?src=/english/html/ e_construction.html (accessed 17 August 2007).

Maslach, C., Jackson, S.E. and Leiter, M.P. (1996) *Maslach Burnout Inventory Manual* (3rd edn), Palo Alto, CA: Consulting Psychologists Press.

Ministry of Trade and Industry, Singapore (2007a) *Economic Survey of Singapore, Second Quarter 2007*. Online. Available at: http://app.mti.gov.sg/data/article/9302/ doc/ESS_2007Q2_Ch2.pdf (accessed 16 August 2007).

Ministry of Trade and Industry, Singapore (2007b) *Economic Survey of Singapore, Second Quarter 2007*. Online. Available at: http://app.mti.gov.sg/data/article/9302/ doc/ESS_2007Q2_Ch3.pdf (accessed 16 August 2007).

Mottershead, T. (2002) The evolution of the sustainable corporation – nonsense or good business sense? *Corporate Governance International*, 5 (4): 28–51.

National Income Section, Hong Kong Census and Statistics Department (2007) *Hong Kong Statistics Key Economic and Social Indicators*. Online. Available at http://www.censtatd.gov.hk/hong_kong_statistics/key_economic_and_social_indicators/index.jsp#domestic (accessed 17 August 2007).

Ning, L. (2000) An economical and sociological analysis of corporate social responsibility and its application in China, *Nanfang Jingji*, 6: 20–23.

Porter, M.E. (1990) *The Competitive Advantage of Nations*, New York: Free Press.

Shi, P. (2002) Social responsibility of corporates, *China Economist*, 10: 22–23.

Standards Australia (2003) AS 8003-2003: Corporate governance – Corporate social responsibility.

Vellu, S. (2006) Drastic measures in Malaysia to keep contractors on the job, Quoted in, *International Construction Review*, http://www.iconreview.org/news/view/977 (accessed 25 June 2008).

Wang, Z. (2005) SA 8000 and the construction of CSR in China. *Enterprise Management*, 8: 20–21.

Welford, R. (2004) *Awareness and Attitudes to Environmentally and Socially Responsible Investment in Hong Kong: Summary Report, Project Report 13, Corporate Environmental Governance Programme*, The Centre of Urban Planning and Environmental Management, The University of Hong Kong, June.

Welford, R. and Hills, J. (2004) *Business and Human Rights: A Study of Corporate Risk in Hong Kong, Project Report 15, Corporate Environmental Governance Programme*, The Centre of Urban Planning and Environmental Management, The University of Hong Kong, July.

Welford, R. and Mahtani, S. (2004) *Work Life Balance in Hong Kong Survey Results, Project Report 17, Corporate Environmental Governance Programme*, The Centre of Urban Planning and Environmental Management, The University of Hong Kong, September.

Wikipedia (2007a) *Languages of Singapore*. Online. Available at: http://en.wikipedia.org/wiki/Languages_of_Singapore (accessed 16 August 2007).

Wikipedia (2007b) *Culture of Singapore*. Online. Available at: http://en.wikipedia.org/wiki/Culture_of_Singapore (accessed 16 August 2007).

Williamson, O.E.E. (1967) *Economic Institutions of Capitalism: Firms, Markets, Relational Contracting*, New York: Free Press.

Yin, L. (2005) SA 8000 and CSR in China. *Enterprise Vitality*, 8: 38–39.

Yip, B., Rowlinson, S., Kvan, T. and Lingard, H. (2005) Job burnout within the Hong Kong construction industry: a cultural perspective, *Proceedings of the CIB W92/T23/W107 International Symposium on Procurement Systems*, 8–10 February, Las Vegas, Nevada, USA.

Yu, Q. (2005) China's CSR under SA 8000. *On Economic Problems*, 7: 33–35.

Yuan, H. and Pi, J. (2007) CSR practice in the United States. *China Economist*, 2: 93–94.

Zhai, W. and Liu, L. (2003) The social responsibility of corporates. *China Economist*, 10: 268.

Zhang, S. (1994) The social responsibility of modern corporates. *Reform of Economic System*, 3: 43–49.

Zhao, J. (2005) Corporate social accountability and SA 8000 certificate. *Commercial Research*, 331: 78–80.

Zhu, G. (2003) Corporate social responsibility: comparison between Western and China. *Science & Technology Progress and Policy*, 12b: 126–128.

Chapter 16

Corporate social responsibility in the Australian construction industry

Helen Lingard, Nick Blismas and Peter Stewart

> The rule of law has been supplanted in the building and construction industry by a culture of lawlessness.

> The relationships between parties in the building and construction industry are not conducive to an efficient, fair and innovative industry. There are many points of tension within this industry and many aspects of behaviour and culture have contributed to the breakdown in relationships between parties. This behaviour and culture encompasses commercial pragmatism, desire for uniformity in employment terms and conditions, excessive pursuit of self interest, in an industry characterised by conflict, high levels of lawlessness and a breakdown in rights.
>
> (Cole 2003)

Introduction

These opening quotes suggest an industry in disorder, which is unlikely to perform well in the area of corporate social responsibility (CSR). CSR has been defined as

> [a m]echanism for entities to voluntarily integrate social and environmental concerns into their operations and their interaction with their stakeholders, which are over and above the entity's legal responsibilities.
>
> (SA 2003b)

If, as Cole asserts, the Australian construction industry is characterised by lawlessness, commercial pragmatism and self-interest, it seems unlikely that discretionary attention to social and environmental concerns would be a key feature of the industry. Indeed, in the past, the Australian construction industry has paid little attention to its social-risk profile, despite the fact that the industry has had an enormous impact upon the country's physical and social landscape.

This chapter provides a broad overview of CSR in the Australian construction industry. The chapter develops the subject by discussing the general context of Australian history and society, and progresses to an overview of the Australian construction industry. The influence of key industry players on the industry's CSR performance is discussed, and a case example is presented to illustrate the role played by organised labour in policing the CSR performance of the industry. The chapter then briefly describes the CSR profile of Australian business in general, before turning to the ethical climate of the Australian construction industry. The chapter describes the findings of recent reports on the industry, most of which present a bleak picture, and presents an indicative case example to illustrate detrimental social impact of misconduct within the industry. The closing section of the chapter reports on initiatives under way in Australia to promote ethical conduct in the construction industry, pointing to the potential for improved CSR performance in the future.

The Australian context

Historical and political context

Australia's historical and political contexts play an important role in shaping the CSR performance of the Australian construction industry. Some of the defining features of Australia's history and political landscape are briefly described below.

The Commonwealth of Australia was formed in 1901 through the proclamation of the constitution for the federation of six states. The founders of the Commonwealth were progressive about human rights and democratic procedures. At the time of federation, Australian society was generally egalitarian, with no strongly defined aristocracy or ruling class (Australian Government 2005). The egalitarian ethos has prevailed in Australian society and is manifest in a strong organised labour movement, which has played a critical role in shaping employment practices and working conditions. The role of the labour movement in shaping the CSR performance of the construction industry is discussed in greater detail later in this chapter. The European population in Australia at the time of federation was 3.8 million of whom three-quarters had been born in Australia – the great majority being of English, Scottish or Irish descent (Australian Government 2005). The enactment in 1901 of the Immigration Restriction Act, ensured that immigrants would be primarily of European origin. This reinforced the link with Britain in many aspects of Australian life. Unsurprisingly, the structure and practices of the Australian construction industry have always closely resembled those of the United Kingdom and Ireland.

The six states and two territories of Australia are governed under a federal system. Power is distributed at a commonwealth, state, territory and local level. This structure has encouraged the development of specific state and territory laws, which have clearly influenced the approach taken by organisations operating in different states and territories. For example, although occupational health and safety (OH&S) legislation in Australia is modelled upon the British Robens-style Health and Safety at Work Act, each Australian state and territory has its own principal OH&S Act, subordinate regulations and codes of practice. The lack of national uniformity is evident in the recent inclusion of OH&S responsibilities for construction design professionals in OH&S legislation. In some jurisdictions (Western Australia, South Australia, Queensland and Victoria), specific OH&S obligations for designers of buildings and structures have been established in the OH&S legislation, although the scope and nature of these obligations vary (Bluff 2003). In the remaining states and territories, specific duties for designers are yet to be established. This situation is confusing and inequitable. The problem of fragmented OH&S regulatory framework was acknowledged by the Australian Labour Party (while still in opposition). As an election promise, the ALP promised to work with state and territory governments to pursue national harmonisation in OH&S legislation and standards within five years of winning office. The objective of this harmonisation is not to effect a Commonwealth 'takeover' of state and territory OH&S legislation but to 'achieve common standards across jurisdictions, to harmonise definitions, procedures and reporting requirements' (Gillard 2007).

Post-war politics in Australia have been dominated by two groups, namely the Labour Party (ALP) and the Liberal–National Coalition. A period of major change in Australian social and economic policy followed Labour's 1972 election victory, with a programme of reforms in health, education, foreign affairs, social services and industrial relations (Australian Government 2005). A further period of Labour government in the 1980s and mid-1990s entrenched many of these reforms. Between 1996 and 2007, the Coalition dominated federal government. During this time the state governments were held by Labour, introducing an interesting tension between federal and state leadership. Policies of the Coalition government (mostly directed at reforming the industrial environment) thrust the construction industry to the forefront of this tension. For example, the Coalition federal government made funding for large-scale infrastructure projects contingent upon state governments' agreement to sign a national code of conduct governing industrial relations in the construction industry. The construction code was designed to end 'union control' of building sites. Controversially, this code was signed by Victoria and South Australia but not by Queensland, Tasmania and Western Australia (*The Australian Financial Review*, 19 May 2005a). In 2007, the Liberal–National Coalition was ousted and a new Labour federal government was formed under the leadership of Kevin Rudd. The Rudd government has criticised the Coalition government's industrial policies which, they claim undermined teamwork and cooperation in workplaces and created an adversarial workplace culture.

It is against this backdrop of British-influenced ethics, a spirit of egalitarianism, industrial democracy and polarised state and federal governing structures, that the Australian construction industry has evolved its unique culture and work practices.

Geo-demographic characteristics of Australia

In geographic terms, the Australian continent is almost 7.7 million square kilometres in area. The land mass is comparable with the United States (excluding Alaska) and is one-and-a-half times greater than Europe (ABS 2005). In the past, the location of Australia contributed to a sense of isolation, however, in recent years, globalisation and advances in travel and electronic commerce have integrated Australian business with other economies. The growth of multinational companies, with systems and processes that traverse national boundaries, creates new issues with the potential to impact upon the social, as well as the financial, performance of firms.

A good example of the difficulties faced by multinational companies is the case of the Australian construction company Multiplex Construction, which won the prestigious Wembley Stadium Project in the United Kingdom. Media reports described how Multiplex ran into difficulty on the UK project, becoming embroiled in costly litigation with a specialist sub-contractor, Cleveland Bridge. It was alleged that Multiplex's 'hard ball' tactics with sub-contractors, which reflected conduct that was tolerated in Australia, were strongly resisted in the UK (*The Age* 2005a). Following the Latham Review of the British construction industry in 1994, the UK enacted legislation to protect sub-contractors from what was construed as 'exploitation' by principal contractors. It was widely reported that Multiplex Construction suffered a substantial financial loss on the Wembley Stadium Project. Moreover, around this time, the company's founder stepped down as executive chairman of the company and the Multiplex board was forced to suspend the company's securities because it could not quantify the extent of losses from the Wembley project. Further complicating the situation, the Australian Securities and Investments Commission also initiated an official investigation into the company's financial disclosures and a class action was brought by buyers of shares in the company around this time, placing Multiplex's corporate governance processes (at the time) under scrutiny (*The Age* 2005b). These actions could, in the eyes of some observers, have harmed the company's corporate image and standing within the Australian construction industry, as well as tarnishing its reputation in the UK. The case highlights how problems on one international project can have ramifications within a company's home country.

Demographic trends within Australia are also likely to impact upon the construction industry, both presenting challenges for CSR as well as making it imperative that the industry improve its image with respect to social and environmental performance. The population of Australia now exceeds 20 million, with a growth rate comparable with the overall world growth rate (ABS 2005). The main source of growth has been natural increase; however, net overseas migration is also a significant factor. The rate of overseas migration is an important factor for CSR, as immigrants come from countries and trading blocs of varying CSR maturity. Australia's population is also ageing rapidly. Figure 16.1 illustrates the predicted trend over the next four decades. This trend, as in other developed countries, is resulting in concern for the sustainability of the economy and health services. For construction also, this raises skills concerns.

The Australian construction industry is facing a critical shortage of skilled workers. It is estimated that if the construction industry is to replace its retiring workers and meet growth demands, between 40,000 and 50,000 new skilled workers will be needed in the next five years (*The Australian* 2005a). In response to the looming shortage, in April 2005, the then federal government announced that it would increase the skilled migration programme by 20,000 to 97,500 places. Under the programme, migrants with building industry-trade skills, such as bricklayers, cabinetmakers, carpenters, joiners, plasterers and plumbers, are given preferential immigration status. The programme also provided for a trade-skills training visa allowing eligible migrants to take up trade apprenticeships in regional areas of Australia. The increased reliance on migrant labour will present further challenges for the evolution of CSR in the construction industry, as the migrant labour force not only has language barriers, but often comes from developing countries with poor CSR records. In addition to attracting migrant workers, the construction industry needs to improve its ability to attract new workers from within Australia. Many Australian workers who enter trade apprenticeships fail to complete them, representing a significant loss to the industry. For example, between January and September 2004, 9,500 trainees withdrew from apprenticeships in building-related trades and only 6,430 completed apprenticeships. By comparison, in 2004, 5,640 workers

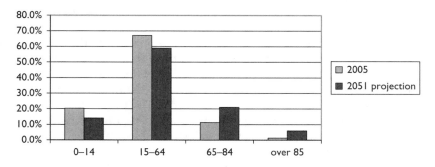

Figure 16.1 Australia's population trends projected to 2051 (ABS 2005).

were needed to replace those who left the industry and an additional 8,810 new workers were needed in building-related trades (*The Australian* 2005a). Increasingly, young Australian workers, in their employment decisions, are considering companies' environmental and social performance, and one way for the construction industry to address the labour shortage would be to improve its image with respect to CSR performance.

The Australian construction industry

The size of the industry

In 2002–2003, construction was the fourth-largest industry in Australia, contributing AU$45.9 billion (£18.4 billion) as at June 2003, or 6.3% of Australia's Gross Domestic Product (GDP) (ABS 2005). The income reported by construction businesses during 2002–2003 was $140.9 billion (£56.4 billion) as at June 2003. Of this, federal, state and local government organisations provided 11.5% of the total income, whereas 'householders and other organisations' accounted for 82.0%. Work in the residential sector accounted for 47.6% of income whereas non-residential construction and non-building construction accounted for 28.6% and 23.8% respectively. The dominance of the private sector presents a problem for the development of CSR initiatives within the industry. Although all construction clients can influence CSR performance by requiring that social and environmental criteria are addressed in the tendering process and by monitoring project performance in CSR, CSR is more closely aligned with the policy objectives of public sector clients. For example, the federal government has established a national Australian Government Building and Construction OH&S Accreditation Scheme. The Scheme is administered by the Office of the Federal safety Commissioner. The purpose of the scheme is to ensure that Australian governments, both state and federal, adopt strong leadership in promoting OH&S in the industry by requiring that all construction contractors engaged on public-sector construction projects have satisfactory OH&S management processes and performance. It establishes strict requirements for head contractors must be accredited in order to enter into contracts with government agencies. From 1 October 2007, the operation of the scheme was extended by lowering the threshold value of projects falling within the Scheme. The Scheme now applies to projects directly funded by the federal government of a value of AU$3 million or more. The Scheme also covers indirectly funded projects, which are those in which the federal government contributes to funding construction work procured by a third party agency, for example a state government agency, by means of a grant or funding agreement. The Scheme applies when the value of indirect funding is $5 million or more and at least 50% of the total project value or where commonwealth funding is $10 million or more, irrespective of the proportion of project value this makes up. The Scheme creates a strong business incentive for the construction industry to improve its OH&S management processes and it is difficult to imaging a similar system of control being implemented by private sector construction clients.It is difficult to imagine a similar system of control being implemented by private sector construction clients.

The Australian construction industry is characterised by a large number of very small businesses. Of all the people employed in construction, 36% work for subcontractors or 'own account workers', and, of these subcontractors, 94% employ fewer than five employees (Cole 2003). The predominance of small and medium enterprises (SMEs) in the Australian construction industry presents a challenge for CSR. These organisations are resistant to change and are unlikely to have the resources or knowledge to implement appropriate processes to manage CSR. Further, the focus of SMEs is often on business survival, to the detriment of social and environmental performance. Unlike large enterprises, which are continually scrutinised by investors and the public, the processes and performance of SMEs are largely 'unseen'. Consequently, the CSR performance of large construction companies is unlikely to reflect the tone of the whole industry.

In composition and character, the Australian construction industry still has significant vestiges of its British past. The professions are largely based on those of British systems, including the traditional roles of architect, quantity surveyor, engineer and contractor. Associated with these professions are the various traditional procurement routes with their concomitant problems. The adversarial contractual conditions, highlighted in the United Kingdom by the Latham (1994) and Egan (Department of Trade and Industry 1998) reports, are still very evident in the Australian industry and have a detrimental effect on CSR.

Industry groups

The industry is divided into various representative groups and industry associations, with no single organisation representative of any particular group within the industry. The major employer associations and groups include: Australian Constructors Association (ACA); Master Builders of Australia Inc. (MBA Inc.); Civil Contractors Federation (CCF); Housing Industry Association Limited (HIA); and Australian Industry Group (AIG). Membership within these various groups ranges from large construction organisations to single-person SMEs, and across diverse sectors including housing, commercial building and civil engineering projects. Cole (2003) suggests that 'the diversity of these groups provides a challenge for the industry to present cohesive input into policy debate', and the fragmentation of the industry prevents its participants from forging a unified and cohesive policy response to CSR issues. The industry organisations do not tend to benchmark their CSR performance, and are unlikely to be seriously monitoring or seeking to improve members' social and environmental performance.

Indeed, the Royal Commission went further and accused construction industry employers of unethical commercial pragmatism, stating:

> The rule of law has been supplanted in the building and construction industry by a culture of lawlessness. Participants in the industry instinctively succumb to the exercise of industrial muscle in the interests of commercial expediency and survival. They prefer to capitulate to unlawful or otherwise inappropriate demands or to allow unlawful or otherwise inappropriate practices to continue, rather than resort to the law to enforce their rights.
>
> (Cole 2003)

If true, this situation is likely to detract further from organisations' focus on issues of CSR.

The unions

One characteristic of the Australian construction industry, which distinguishes it from its international counterparts, is the relative strength of the trade unions. Trade unions were traditionally formed to represent the interests of workers and provide an effective balance to employer power; however, in Australia they have become dominant players in the political process. The peak body for the union movement, the Australian Council of Trade Unions (ACTU), is made up of 46 affiliated unions, representing 1.8 million workers (Cole 2003). The Construction, Forestry, Mining and Energy Union (CFMEU) is the largest and most active union in the construction industry. Others include the Communications, Electrical, Electronic, Energy, Information, Postal, Plumbing and Allied Services Union of Australia (CEPU); the Australian Workers' Union (AWU); and the Automotive, Food, Metals, Engineering, Printing and Kindred Industries Union, referred to as the Australian Manufacturing Workers Union (AMWU). Historically, the union movement has played a major role in the advancement of social responsibility within the construction industry, especially in relation to the improvement of working conditions and OH&S. One of the earliest achievements of the Australian building industry unions was the stonemasons' campaign for an eight-hour day (48-hour week) in the mid-1850s. Other early achievements include the following:

* 1875–1920: 44-hour week;
* 1897: first workers' compensation legislation in Australia;
* 1902–1980s: campaign for proper amenities on construction sites;
* 1902: the first building industry OH&S legislation, the New South Wales *Scaffolding and Lifts Act*.

More recent industrial campaigns have seen the introduction of the right for construction workers to full pay while off work as a result of an occupational injury (1971) and further improvements in working conditions and wages. Most recently, following an extensive campaign of industrial action by the unions, construction industry employers in the state of Victoria signed a new standard for wages and conditions in the industry. A ground-breaking element in this 'compact' was an agreement to introduce a 36-hour week for all construction workers by 2003. These achievements demonstrate the pivotal role played by the construction industry unions in securing decent working conditions for their members. However, the construction industry trade unions have also played a more altruistic role in ensuring the industry conducts itself in a socially responsible manner. This is illustrated by the case study that follows (Box 16.1).

The 'green bans' provide a good example of a construction union acting to protect the interests of the community and environment, even though this action went against the economic interest of its members.

Recently, the trade unions in the construction industry have been harshly criticised for corrupt and inappropriate practices and creating industrial havoc within the industry (see Cole, Royal Commission above). Whether this criticism is justified or not, the significant, positive impact of the union movement on the social and environmental performance of the Australian construction industry cannot be disputed.

Case study: 'green bans'

In 1971 the New South Wales branch of the Builders Labourers' Federation (BLF) commenced a campaign to stop the destruction of heritage buildings in Sydney. During the 1960s and 1970s speculative property development was rampant in Australian cities and many historic buildings were being demolished to make way for commercial office and high-rise residential buildings. The BLF, led in NSW by Jack Mundey, Bob Pringle and Joe Owens, argued that the workers' movement

> must engage in all industrial, political, social and moral struggles...that buildings which are required by the people should have priority over superfluous office buildings which benefit the get-rich-quick developers, insurance companies and banks..[and] that all construction work performed should be of a socially useful and ecologically benign manner.
>
> (cited in Burgmann 1993)

Having unsuccessfully lobbied the local council, the mayor, local state parliament member and state Premier, a group of residents in the suburb of Hunter's

(Continued)

Hill asked the BLF to help them save an area of open space known as Kelly's Bush, where the residential property developer A.V. Jennings proposed to build luxury homes.

The union asked the residents to call a public meeting at Hunter's Hill to demonstrate that there was community support for a union ban on the Kelly's Bush development. Over 600 people attended the meeting at which a ban was formally requested. The construction union members refused to work on the Kelly's Bush project. Initially, A.V. Jennings responded to the union ban by stating the company would build the proposed luxury homes using non-union labour. However, the union then threatened to withdraw labour on other A.V. Jennings' projects in Sydney. In particular, workers on an office block being built by A.V. Jennings in North Sydney threatened that the office block would never be completed, were the company to proceed with any development of the Kelly's Bush site.

The union action, which became known as a 'green ban', was successful and the Kelly's Bush open public reserve was saved. The action resulted in many more community groups seeking help from the BLF to protect heritage sites. By 1974, 42 green bans had been imposed preventing over AU$3,000 million of development projects and saving over 100 buildings considered by the National Trust to be worthy of preservation. Some areas protected by green bans are now prominent tourist areas in Sydney, including The Rocks, Woolloomooloo, Centennial Park and the Botanical Gardens. The green bans also prompted the NSW government to pass tighter planning laws. Although some criticised the green-ban campaign because it denied workers employment, the union argued that it wanted to build only 'socially useful' buildings.

(Burgmann and Burgmann 1998)

The professions

The final major group of bodies in the industry is that of the professions. These bodies represent the professions and protect the interests of their members. Like their British ancestors, they are the custodians of the ethical conduct of their members, often with legislative obligations to uphold such standards.

The main groups of industry participants have been portrayed, in the preceding sections, in a very simplistic manner, with some obvious omissions, such as clients and supplier groups. Obviously, all parties must play an active role in the promotion of CSR within the industry. For example, clients can play an important role in 'buying' socially and environmentally responsible construction projects. Similarly, construction designers and contractors can specify and purchase only those products and materials that are produced in a socially and environmentally responsible manner. Therefore, it is important to state that CSR must become a key performance indicator for all parties in the construction process.

CSR in Australia

CSR and ethics

Before discussing the CSR performance of Australian industry in general, and the construction industry in particular, it is important to clarify the relationship between CSR and ethics. Fisher (2004) points out that the terms are often used interchangeably, despite being conceptually different. One common view of the relationship between ethics and CSR is that ethics is concerned with the conduct of individuals, while CSR is concerned with the impact of organisations. According to Fisher (2004), this view is not adequate, because ethics is understood to guide the behaviour of both individuals and groups, and organisations are understood to have ethical responsibilities, as distinct from their legal and economic responsibilities. An alternative view is that social responsibility has several dimensions, one of which is ethics. This view is represented by Carroll's 'Pyramid of Corporate Social Responsibility' (Carroll 1991). This pyramid recognises that companies have four main types of responsibility to society and arranges these in a hierarchy. At the base of the pyramid is society's expectation that companies fulfil their economic responsibilities. Moving up the pyramid are legal, ethical and, finally, philanthropic responsibilities.

This view of the relationship between ethics and CSR is the most appropriate because it recognises that organisations have ethical responsibilities and places these in the context of an organisation's responsibilities to all of its stakeholders. Thus, in discussing the CSR performance of Australian business, we make reference to ethical conduct as an integral component of CSR.

CSR profile of Australian business

Although CSR has been a topic of discussion within Australia for a number of years, there has been some resistance from Australian business to the introduction of mandatory reporting of CSR (*The Australian* 2005a). A recent report revealed that the majority of Australians (80%) believe that companies put profit before people and engage in CSR as a goodwill exercise rather than integrating it into their everyday business. The same report concluded that 'the Australian business community is lagging behind the UK, USA and the EU in some key aspects of CSR' (The cited in State Chamber of Commerce (NSW), 2001: 5). Australian industry has been described by Paul Hohnen, former Australian diplomat and Director of Greenpeace, as having a 'culture of complacency' which has prevented it from being at the international forefront of CSR initiatives (*The Age* 2005b). Hohnen, who now runs an Amsterdam-based company called Sustainable Strategies, visited Australia in May 2005 to run a two-day workshop at the Australian Centre for Corporate Social Responsibility.

The list of high-profile companies attending the workshop, including BHP Billiton, BP Australia, Origin Energy, Transurban, the ANZ Bank and TXU (a leading utilities company), suggests that this culture of complacency is beginning to change and that Australian companies are becoming serious about CSR. In this respect, Australian companies need to 'catch up' with their global counterparts. For example, companies like General Electric (GE) and Nike have implemented wide-ranging social responsibility initiatives internationally. The increasing interest of Australian companies in CSR is likely to be a response to a growing belief that corporations should be held responsible and accountable for their actions. This is manifest in overt public sentiment, the publication of standards and guidance material, and the growth and performance of socially responsible investment (SRI) options on the Australian share market.

In Australia, public perception of the importance of CSR and ethical conduct has grown considerably in recent years. One comparative global study demonstrates that the Australian public had a relatively high awareness of issues related to CSR and strongly believed that companies should exceed the traditional role of profit-making and set very high ethical standards and help build a better society (*Millennium Poll* 1999). Also, in 2002, the Transparency International Bribe Payers Index reported that of the countries surveyed, Australia was rated the country with the least likelihood of the use of bribery for winning or retaining business. Further, the report found that Australia was one of those least likely to use any other corrupt means to gain or retain business, unlike other Western and industrialised countries, in which the likelihood was rated very high (e.g. United States, France, United Kingdom and Japan being the top four on the list).

The recent failure of large organisations in Australia, such as HIH insurance and One Tel, highlighted the need for good corporate governance and ethical conduct. In 2003, the Australian Stock Exchange (ASX) published principles of good governance and best practice recommendations to promote ethical conduct, transparency and accountability in publicly listed Australian companies. While these principles do not focus specifically on CSR, the fact that ethical responsibility is one component of CSR suggests the ASX principles would support the CSR performance of Australian business. The increased interest in CSR within Australia has also prompted the publishing of a formal framework for CSR in Australian Standard AS8003-2003. This Standard provides guidance for organisations intending to establish, implement and manage a CSR programme. It forms part of a suite of Australian Standards relating to corporate governance, which also includes Standards on corporate governance, fraud and corruption control and whistle-blower protection programmes. AS8003 also references Australian Standard AS3806-1998, which deals with principles for the development, implementation and maintenance of effective compliance programmes to ensure compliance with laws and regulations.

Ethical investment funds have also increased dramatically within Australia, indicating Australian shareholders' growing preference for companies that perform well in social responsibility. For example, an analysis conducted by Deni Greene Consulting showed that up to December 1999, managed ethical investment funds amounted to a few million Australian dollars. However, the research estimated the entire socially responsible investment (SRI) sector, including private portfolios as well as managed funds, to be worth AU$21.5 billion by June 2004 (*The Australian Financial Review* 2005b). Between 2003 and 2004, it is reported that SRI managed funds grew 41%, compared with growth in the mainstream market of 18% in the same period. There is also emerging evidence that ethical investment leads to better long-term results in the Australian share market. For example an AMP Capital study took 3,000 listed companies and assessed whether they had high or low CSR. After considering sector performance and size of the companies, the study found that responsible companies outperformed less responsible companies by 4.8% over four years and by 3% over ten years (*The Australian* 2005b). However, despite the growth and apparent success of SRI, it is estimated that the SRI managed-funds sector accounts for only 1% of the total Australian share market, whereas in the United Kingdom and Canada, SRI managed funds make up about 3% of the market (*The Australian Financial Review* 2005b).

The increasing awareness and commitment of Australian companies to CSR was highlighted in 2005 when Australian bank, Westpac, topped a list identifying companies in Australia and Britain with the greatest level of social responsibility. The bank scored a near-perfect score of 99.5 on the Corporate Responsibility Index (*The Age* 2005a). Given this backdrop, a progressive, socially responsible construction industry could be expected. However, a number of construction industry reports indicate that there is a 'disconnect' between the growing awareness of ethics and social responsibility in Australian business and the performance of the Australian construction industry. As the most recent Royal Commission into the industry's performance bluntly stated, 'At the heart of the findings is lawlessness' (Cole 2003: Part 1: 5). This disparity is discussed in more detail in the next section.

Ethical climate of the construction industry

> [T]here has been an unwillingness within industry leadership to recognise the long term advantages of structural and cultural change, accepting instead a short term project driven profit process. Pragmatism and self-interest have dominated.

> A culture of lawlessness pervades every level of the industry.
>
> (Cole 2003)

Ethical climate is defined as a set of shared perceptions of what is ethically correct behaviour and how ethical issues should be handled within organisations (Victor and Cullen 1987). Ethical climate has been measured and linked to unethical behaviour within organisations (Peterson 2002).

There have been few empirical studies examining the ethical climate of the Australian construction industry. However, the most noteworthy research into ethics in the industry examined the ethics of tendering and collusive practices. Ray *et al.* (1999) conducted an Australia-wide survey of construction industry participants who engage in tendering practices. The survey assessed respondents' perception of the appropriateness of a number of tendering practices and asked respondents to indicate the extent to which they had engaged in such practices. The practices in question were those considered to be of dubious ethical nature, such as:

- submitting cover prices (i.e. bids deliberately designed to lose a tender);
- loading tender prices with fees to compensate unsuccessful tenderers;
- bid shopping (i.e. the practice of taking subcontractors' or suppliers' quotes to their competitors);
- bid cutting (i.e. the practice of trying to drive down suppliers' and subcontractors' quotes);
- withdrawing bids before the formal acceptance of prices, for example, if a more lucrative opportunity arises.

The research revealed a disparity between the perception of industry participants and principles included in industry codes of tendering, suggesting that these codes are not aligned with industry perceptions of acceptable and unacceptable tendering practices. For example, respondents defended the right of tenderers to withdraw offers before the closing of tenders even though this could result in costly retendering, delays in project commencement and financial loss to construction clients. Similarly, the practice of bid cutting was believed by the majority of respondents (57%), particularly contractors (67%), to be a fair and practical method for winning tenders. Forcing suppliers and subcontractors to reduce their prices can cause them to make a loss and ultimately drive them out of business, further reducing competition in the industry. The study also found that the anti-competitive practice of submitting a 'cover price' was generally believed to be unethical but was reportedly widely used by respondents. The different perceptions of different players (e.g. clients, contractors and suppliers) indicated a lack of consensus in the industry concerning what is ethically permissible with regard to tendering.

A second survey of contractors' personnel engaged in estimating and tendering investigated industry attitudes towards collusive practices (Zarkada-Fraser and Skitmore 2000). This study revealed that, for the most part, respondents consider collusive practices to be unacceptable. However, for a significant minority of respondents, several forms of collusive practice were considered to be 'contingent' events in that the decision whether or not to engage in them was not clear-cut and would depend upon the circumstances. These included:

- submitting a cover price as part of a rotating low-bid-position scheme (25%);
- submitting a cover price as part of a geographical- or market-distribution scheme (40%);
- inflating tender price to compensate losing tenderers (18%);
- inflating tender price to cover undisclosed fees to trade associations or third parties (28%).

Some forms of collusive practice were more universally disapproved of. For example, accepting a bribe in order not to tender was the least acceptable practice, with only 7% of respondents indicating they might consider this. This empirical research indicates that there is some variation in what is considered acceptable conduct in the Australian construction industry and that a uniform ethical climate does not exist.

More recently, the industry's culture has been condemned in a number of independent government-initiated inquiries. The findings of these inquiries suggest that the industry is characterised by lawlessness, corruption and unsafe practices, suggesting that the ethical climate of the Australian construction industry is based more upon egoism (self-interest) than benevolence or principle. Some of the most significant inquiries and their main findings are summarised below.

Government commissions and inquiries

The Gyles Royal Commission

In 1991, a Royal Commission was established to review productivity in the building industry of New South Wales. The terms of reference of the commission, headed by Commissioner Gyles, were to make inquiry and report on practices or conduct that impacted upon productivity, in particular, the nature and extent of illegal activities occurring in the industry as well as to make recommendations as to how to increase productivity and/or deter illegal activities in relation to the industry. After conducting public hearings, formulating discussion papers and reviewing submissions from interested parties, Commissioner Gyles issued his report in 1992. The hearings revealed widespread occurrence of illegal activities, including the involvement of professional criminals, physical violence or threats of violence, theft of building materials, bribery, collusive practices and even bomb threats. Commissioner Gyles *et al.* (1992) drew the following conclusions in relation to employers within the industry:

> There is evidence of widespread lack of integrity and probity amongst the management of contractors and others involved in the industry... This is reflected, for example, in the offering of dishonest inducements to union officials, workers, council inspectors, Workcover [OH&S]

inspectors and those able to procure work; soliciting or accepting dishonest inducements from those seeking work; involvement in collusive and anti-competitive behaviour; the exploitation of the subsidy of a group apprentice scheme; the manufacture and use of false documents to cover up illicit or unlawful payments; the obtaining of free or discounted work or materials, either in fraud of the employer, or the revenue, or as a result of pressure on sub-contractors and suppliers; the engagement of persons of ill-repute to solve industrial or commercial problems by actual or threatened violence or other illegal means; continuing to trade while insolvent in fraud of creditors; and cutting corners in safety regulation legislation and industrial awards requirements.

He continued:

> Observance of the law and law enforcement in general play very little part in the industry. The law of the jungle prevails. The culture is pragmatic and unprincipled. The ethos is to catch and kill your own.
>
> (Gyles *et al.* 1992)

The Commission also reported widespread abuse of safety procedures and principles within the industry by both unions and employers. An example was the practice of inventing safety issues to justify the employer paying for time not worked to ensure industrial harmony. This practice represented a perversion of the principles underpinning the OH&S legislative framework adopted in New South Wales (and all other Australian jurisdictions). This framework establishes OH&S committees to facilitate genuine employer–employee consultation regarding safety issues. Commissioner Gyles concluded that the subversion of these principles presented a barrier to the industry's ability to improve its OH&S performance (Royal Commission into the Building and Construction Industry 2002).

Finally, Commissioner Gyles recognised that the lack of ethical culture in the construction industry would not only adversely impact upon the economic performance of the industry, but would also negatively impact upon the community in which the industry operates. He wrote:

> The effect of illegal activities upon the culture of the industry and upon the commercial and industrial morality of participants in it is, in the long run, greater than the direct economic consequences. Once it becomes acceptable to break, bend, evade or ignore the law and ethical responsibilities, there is no shortage of ways and means to do so. Those who pay and suffer the other consequences of disruption in the end are the public.
>
> (Gyles *et al.* 1992)

The Cole Royal Commission

Another controversial inquiry, into the building and construction industry, initiated by the Howard Coalition government, focused intensely upon corrupt and unlawful practices in the industry's industrial relations practices. On 26 March 2003, the Minister for Employment and Workplace Relations, the Hon. Tony Abbott, released the final report of the Cole Royal Commission into the Building and Construction Industry. The Report was in 23 volumes and covered a wide range of issues. However, Commissioner Cole was highly critical of the construction industry in the states of Victoria and Western Australia. He reported that 'the rule of law has long since ceased to have any significant application' in Victoria and 'Doing business in the industry in Western Australia means doing business in accordance with the prescriptions, and to the satisfaction of the CFMEU (Construction, Forestry, Mining and Energy Union)'. Commissioner Cole commented that, in New South Wales, respect for legal obligations had been replaced by 'commercial expediency and the application and submission to industrial pressure'. Overall, Commissioner Cole found that 23 union officials and eight employer representatives might have committed criminal offences, and the report identified 392 findings of unlawful conduct, mostly relating to the activities of union officials. The apparent one-sidedness of the findings led the Cole Royal Commission to be branded an anti-union 'political stunt' on the part of the Coalition government of the time. The report identified over 100 types of unlawful or inappropriate practices that occurred within the Australian construction industry. Specific issues that were identified in the findings included:

- unlawful strikes and threats of them;
- the use of threat, intimidation and coercion by the CFMEU to secure membership, ensure union-endorsed Enterprise Bargaining Agreements (EBAs) and to discourage the use of non-union labour;
- widespread disregard or breach of the Workplace Relations Act, 1996, including the provisions relating to right of entry, strike pay, enterprise bargaining and freedom of association;
- disregard of Australian Industrial Relations Commission (AIRC) and court orders;
- union officials attempting to regulate the industry by seeking 'donations' from contractors to fund the salaries of a union organiser;
- disregard of contractual obligations;
- avoidance and evasion of tax obligations;
- the payment of money by contractors and subcontractors to 'buy' industrial peace;
- the misuse of occupational health and safety as an industrial tool;
- the breach of strike pay provisions by contractors.

Commissioner Cole concluded that significant structural and cultural change is needed in the Australian construction industry to reestablish the rule of law. Consequently, he recommended that a new set of rules be established applying specifically to the building and construction industry. He proposed the enactment of new industry-specific legislation, The Building and Construction Improvement Act, to regulate industrial issues in the industry as well as the establishment of an Australian Building and Construction Commission with powers to monitor conduct within the industry and prosecute unlawful industrial action (Smith 2003). Commissioner Cole also recommended changes to enterprise-bargaining laws in the industry to ensure genuine enterprise-bargaining and eliminate the practice of 'pattern bargaining'. The report also recommended that the Commonwealth Government act as an 'exemplar' construction client and engage only contractors and subcontractors that adhere to an updated National Code of Practice for the Construction Industry and Commonwealth Implementation Guidelines. The industry's poor occupational health and safety performance was to be addressed through the appointment of a Commissioner for Health and Safety in the Building and Construction Industry, whose role would be to monitor OH&S and manage an OH&S pre-qualification system as a requisite for contractors and subcontractors to work on all projects on which the Commonwealth was a client or to which it provided funds.

The first attempt to pass the Building and Construction Improvement Bill 2003 failed as the Bill was overturned in the Australian parliamentary upper house, the Senate. However, following the Australian federal election of 2004, the Coalition government held the majority of seats in both the House of Representatives and the Senate. The Building and Construction Improvement Act received assent in September 2005. This Act was regarded, by many, as draconian. It significantly reduced the circumstances in which industrial action could be taken in the building and construction industry. Unlawful industrial action was prohibited by the Act and penalties increased. For example, an exemption whereby industrial action taken during the life of a certified agreement was protected under the Workplace Relations Act (created by Emwest v Automobile, Food, Metals, Engineering, Printing and Kindred Industries Union (2003)), was removed. The BCCI Act also placed restrictions on industrial action called on the basis of safety concerns and expanded the definition of industrial action to cover partial or conditional refusals to undertake work. The BCCI Act also gave the newly created construction industry industrial watchdog, the Australian Building and Construction Industry Commission, considerable powers of investigation. The ABCC Commissioner was given extensive powers to compel attendance at hearings, require people to testify and produce documents in the course of an investigation. If required to testify by the ABCC, a person's privilege against self-incrimination was partially restricted and penalties for failing to comply with the Commissioner's directions also included imprisonment for up to six months. Before it was passed, the International Labour

Organisation commented upon the Construction Industry Improvement Bill, reporting that the Bill directly contravened international obligations under ILO conventions including the Freedom of Association and Protection of the Right to Organise and the Right to Organise and Collective Bargain Convention – both ratified by Australia. Following the election of a Labour federal government in 2007, the industrial framework created for the building and construction sector is again the subject of political debate. The Rudd government has made a strong commitment to a 'Fair' industrial system but at the same time has vowed to ensure that businesses are protected from unlawful industrial action. The new government has announced its plan to maintain the Australian Building and Construction Commission until 31 January 2010 after which the work of the Commission will be taken over by a construction industry-specialist division of the new industrial watchdog, Fair Work Australia.

Public–Private Partnerships

In Australia, as in the United Kingdom, there is an increasing trend to privatise government enterprises and procure major construction projects and/or government services by means of Public–Private Partnerships. Such arrangements can bring benefits to both sectors. However, there are important differences between the two sectors in terms of their accountabilities, objectives and responsibilities, which can undermine the effectiveness of Public–Private Partnerships and have implications for CSR outcomes. The safety training and certification case study, presented below, is an example of a situation in which a difference in principle between a government agency and private sector contractors led to potentially-damaging corrupt practices in the Australian construction industry.

Case study: safety training and certification

In February 2003, the New South Wales Independent Commission Against Corruption (ICAC) commenced an investigation into aspects of the safety training and certification scheme in the state's construction industry. The investigation was instigated at the request of WorkCover NSW, which is responsible for accrediting assessors who certify the training and competency of construction industry personnel under the state's Occupational Health and Safety Act. The ICAC investigation targeted three areas of operation:

- the assessment and certification of machinery operators;
- induction of OH&S for construction work training and certification;
- the training and certification of crane operators working in proximity to overhead power lines.

(Continued)

The investigation revealed deliberate and widespread abuses of the competency assessment regulations by six accredited assessors. Several thousands of Notices of Satisfactory Assessment had been issued by assessors without following the required assessment procedures. For example, candidates were provided with answer sheets to exam-style questions, and obligatory practical tests were not performed. In some cases, Notices of Satisfactory Assessment were issued without the recipient having undergone any assessment whatsoever. Induction training certificates, 'green cards', were reportedly issued when no induction training had been given, providing recipients with the opportunity to work on construction sites all over Australia. One assessor reported that in the six months prior to the inquiry, 80%–100% of certificates he issued were 'bodgie' (*Sydney Morning Herald* 2003). It was estimated that more than 30,000 Notices were corruptly issued and the six assessors profited by around AU\$4 million. Further, the ICAC reports reasons to believe that the corrupt practices were not limited to the six assessors identified (Small 2004). Certification presents the potential for corrupt practices because it provides good employment prospects and rewards for people who are accredited, and therefore presents an opportunity for the making of corrupt profits for people in control of the accreditation process. Owing to the hazardous nature of construction work, such corruption of the OH&S accreditation process has the potential to cause serious injury to the recipients of false certificates, as well as other workers and members of the public. The potential for human, social and economic harm was considerable. Thousands of WorkCover certificates of competency have been cancelled as a result of the corruption, causing significant disruption to the industry.

Source: CFMEU 2004.

Fraud and corruption control

One of the recommendations of the ICAC inquiry, described above, was that the regulatory authority (WorkCover NSW) conduct a detailed fraud and corruption risk assessment of the certification system and develop a fraud-prevention plan. The Australian Institute of Criminology estimates that fraud costs the Australian economy AU\$3 billion each year, and the incidence of fraud in Australia is steadily increasing. Further, many Australian organisations are poorly equipped to detect and prevent fraud, which can take the form of:

- theft of plant or equipment by employees;
- false invoicing (e.g. claiming payments for work not undertaken or products/services not provided);
- theft of intellectual property;
- theft of information belonging to the organisation;

- tax evasion;
- misrepresenting information in financial statements;
- money laundering;
- misleading/deceptive conduct within the meaning of the Trade Practices Act.

Other corrupt practices include:

- collusive tendering (the act of multiple tenderers for a particular contract colluding in the preparation of their bids);
- payment or receipt of bribes;
- release of confidential information for other than proper business purposes;
- payment of donations to not-for-profit organisations for an improper purpose;
- facilitation payments, i.e. small one-off payments to secure prompt delivery of goods, services or harmonious industrial relations;
- conflicts of interest involving senior officers acting in self-interest rather than in the interest of the organisation;
- giving lavish gifts or entertainment to achieve an agreed objective.

The Australian Standard on Fraud and Corruption Control (Standards Australia 2003d) recommends that all organisations adopt a risk-management approach to fraud and corruption control. Thus, organisations should periodically conduct a comprehensive assessment of the risks of fraud and corruption within their business operations. In the case of construction, in which work is undertaken in complex multi-organisation project environments and where cash transactions and verbal agreements are still common-place, fraud and corruption risks are likely to be significant. Once risks have been identified and assessed according to their likelihood and the potential severity of their outcomes, a fraud and corruption control plan should be formulated. This should specify accountabilities for fraud and corruption control, internal control procedures to be implemented and monitoring and review arrangements to ensure that fraud and corruption risk controls are properly implemented and effective. Risk controls should include:

- fraud detection processes, such as post-transaction reviews, the use of computer systems and data mining to identify suspect transactions and for the analysis of financial reports;
- specification of line-management accountability for fraud and corruption control and incorporation into performance-measurement systems;
- anonymous internal and external reporting mechanisms, which are communicated to employees and subcontractors;
- pre-employment screening for new employees;
- a programme for the protection of whistle-blowers;

- an internal audit strategy addressing fraud and corruption risk; and
- appropriate insurance policies to transfer, where possible, the risk of fraud, corruption and theft.

Given the risk profile of the industry, all construction organisations, including government and private sector organisations, should adopt the Australian Standard on Fraud and Corruption Control.

Workforce

The construction industry's CSR performance not only affects workers within the industry, but it is also shaped by the workforce. This section describes working conditions within the Australian construction industry and suggests that relatively poor work conditions are a facet of the industry's social performance. However, at the same time, the industry's overall CSR performance is unlikely to improve as long as sub-standard conditions discourage the entry of trained, skilled workers.

The workforce in the Australian construction industry, when compared to most other industries, has a vastly inferior work environment. Construction work in Australia is unsafe, has poor conditions of employment and has no defined career path. Consequently, the image is inauspicious, rate of entry into the industry is low and skill levels are on the decline.

'[T]he safety record in the industry is poor' (Cole 2003). In the period between 1994 and 2000, 50 construction workers were killed each year as a result of their work. The construction industry fatality rate, at 10.4 per 100,000 persons, is similar to the national road toll fatality rate (McWilliams *et al.* 2001) and the rate of serious injury is 50% higher than the all-industries' average (Cole 2003). Reasons for this poor record are manifold and often complex, with aspects ranging from cultural inertia, technical ignorance, legislative inadequacy and political expediency all having an influence.

Employment contract conditions in the industry are likewise poor, and often unlawful. For instance, the proportion of the building and construction labour force who receive standard benefits in their main job is less than the national average for holiday leave, sick leave and long service leave (Table 16.1).

Table 16.1 Proportion of construction workers with benefits.

Type of benefit	Employees receiving benefit (%)	Average (all industries) (%)
Superannuation	89	90
Holiday leave	67	72
Sick leave	66	72
Long service leave	56	62

Source: ABS 2002 (Employee earnings, benefits and trade union membership, August 2001: Australia, Cat. no. 6310.0).

The poor conditions therefore discourage people from joining the industry and simultaneously attract illegal workers willing to work in unsatisfactory conditions. Field operations conducted by the Department of Immigration and Multicultural and Indigenous Affairs (DIMIA) indicated that the majority of illegal workers are often paid less than legal entitlements; some were not paying tax at all, or, if paying, not at the correct rate; and some were claiming social security benefits to which they were not entitled. Further, it is probable that illegal workers are inadequately trained, and thus pose a safety risk to themselves and others. In 2003–2004, 11% (373 persons) of illegal workers identified were employed by the construction industry (DIMIA 2005).

In addition to the unsafe and poor working environment, construction workers have no structured career path. The transient nature of construction projects means that workers move from site to site, between states with little recognition of experience gained on previous jobs (Curriculum 2005). In the late 1980s, only 20% of all employees had any formal training and fewer than 50% of workers paid at the trade rate had actually completed an apprenticeship (Curriculum 2005).

At a professional and managerial level also, work conditions in the Australian construction industry are undesirable. Hours are long and often irregular and the geographical dispersion of construction work means that employees are sometimes required to relocate on a frequent basis causing considerable disruption to family life. Recent research undertaken by Lingard and Francis (2004) revealed that the majority of managerial and professional employees in three large Australian construction organisations work more than 40 hours a week, with 17.8% reporting that they work 60 hours or more (see Figure 16.2). Work hours, which were significantly

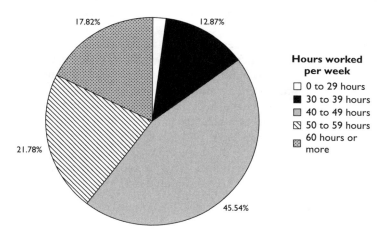

Figure 16.2 Hours worked per week by managers and professionals in the Australian construction industry (Francis and Lingard 2004).

Table 16.2 Cross-occupational comparison of mean burnout scores.

Burnout Dimensions	Construction Mean	Military Mean	Technology Mean	Management Mean	Nursing Mean
EE	2.76	2.05	2.65	2.55	2.98
CY	2.25	1.63	1.72	1.32	1.80
PE	4.29	4.60	4.54	4.73	4.41

EE – Emotional Exhaustion; CY – cynicism; PE – Professional Efficacy

higher in the private sector than in the public sector, were notably correlated with employees' perception that work interferes with family life. The research also revealed that site-based employees work longer hours and experience significantly greater work–family conflict than their counterparts in the head or regional offices of their companies (Lingard and Francis 2004). Work–family conflict has been consistently linked to negative outcomes for individuals and their families, including low levels of job and life satisfaction, damage to family relationships, substance-abuse and mental-health problems.

Australian 'white collar' construction employees were also found to have high levels of burnout, relative to international 'norms'. The mean burnout scores for the construction managers and professionals are presented in Table 16.2, along with the mean burnout scores from a range of different occupations presented by Maslach *et al.* (1996). At an individual level, burnout has been associated with the experience of psychological distress, anxiety, depression, reduced self-esteem and substance abuse. Research also consistently links burnout to lower levels of organisational effectiveness, job satisfaction and organisational commitment as well as higher levels of absenteeism and turnover.

The damaging impact of excessive work hours and the construction industry's adverse work conditions upon the well-being of employees and their families are socially unacceptable. The Australian research is now focusing upon developing, implementing and evaluating the effect of a variety of work–life balance initiatives in the Australian construction industry. These interventions, which include changing from a six-day week to a five-day week and providing employees with some flexibility in their work schedule and/or location, could yield dramatic improvements for construction industry employees and their families, as well as improving the overall performance and image of the construction industry.

Industry initiatives

As a response to the damning and highly publicised findings of public inquiries into conduct within the Australian building and construction industry, many construction industry bodies and large construction organisations in Australia have implemented codes of ethics or conduct, setting standards of behaviour expected of members or employees. An example code of ethics is reproduced in the box below. Codes of ethics or conduct should refer to specific issues of relevance to the organisation, rather than include broad, general statements of principle, so that employees are able to use the code to obtain practical guidance concerning ethical dilemmas, and managers can use the code when making important decisions. It is highly recommended that construction organisations develop codes of conduct or ethics. These codes should be communicated to all employees through inductions and ongoing training, and copies provided to external parties, including subcontractors and suppliers. The code of ethics or conduct should be publicised effectively within the organisation, e.g. displayed on noticeboards or available on the company Intranet, and compliance with the organisation's code of conduct should be regularly audited or verified (Standards Australia 2003a).

Australian Constructors Association Code of Ethics

Objectives

- To encourage its members to adopt the highest standards of skill, integrity and responsibility.
- To represent the views of members to government and other decision makers.
- To promote the construction industry and the important role played by construction contractors in the development of the Australian economy.
- To facilitate the exchange of information between members.
- To promote free, fair and healthy competition between members and throughout the industry.
- To encourage its members to adopt best practice principles and strive for increased productivity and enhanced skills through education and training.
- To promote improvement to those laws, regulations and subordinate legislation which impact on the construction industry.

The ACA approach to ethics

The Australian Constructors Association will conduct its meetings as forums for open, honest and frank discussion where members are encouraged to contribute in a constructive manner.

(Continued)

The Association will at all times put views to the Government and other decision makers that fairly represent the opinions of its members.

The Association will encourage its members to adopt the highest standards of ethical behaviour.

The members of the Association will, at regular intervals, jointly review and share ethical problems (and solutions generated) that have arisen in the course of members' business activity, with a view to creating a greater awareness of ethical issues, problems and solutions and for the encouragement of best practice.

The Association will encourage each member company to develop its own Code of Ethics. Whilst ACA will not prescribe the form or content of a company Code of Ethics, members will be encouraged to consider the inclusion of the following issues.

Issues

- *Honesty and integrity*: Members should undertake business dealings with honesty and integrity.
- *Confidentiality and intellectual property*: Members should respect intellectual property rights and maintain confidentiality in their dealings with each other and with clients, consultants, subcontractors and suppliers.
- *Tendering practices*: Members should not engage in practices such as collusion, hidden commissions, unsuccessful-tender fees or any other secret arrangement.
- *Payment practices*: Members should conduct relationships with consultants, subcontractors and suppliers in an ethical manner, and meet all financial obligations for payment.
- *Health and safety*: Members should provide a safe and healthy workplace for all employees.
- *The community and the environment*: Members should meet their obligations to the community arising from construction activity and to protect the physical environment.
- *Laws and regulations*: Members should abide by the spirit as well as the letter of all laws and regulations affecting the industry.
- *Respect*: Members should treat people with dignity and respect.
- *Moral courage*: Members should display the moral courage to stand up for what is right.

Source: Australian Constructors Association.

CSR indices

In Australia, several indices have been developed to assess organisations' CSR performance. While thus far, only the larger organisations with construction interests have been rated against these CSR indices, such rating systems present an ideal opportunity for construction organisations to evaluate and benchmark their CSR performance against industry or regional best practice, thereby improving their CSR performance. Two potentially useful CSR indices are The Corporate Responsibility Index and RepuTex.

The Corporate Responsibility Index

The Corporate Responsibility Index is a management tool designed to help organisations measure and benchmark their performance against best practice in CSR. The Corporate Social Responsibility Index was launched in Australia in February 2004. The Index is administered by the St James Ethics Centre, working in partnership with the *Sydney Morning Herald* and *The Age* newspapers and Ernst & Young accountants. The Index assesses organisations' CSR strategy and implementation process across the four key impact areas of community, workplace, marketplace and environment. In 2004, the overall average score of Australian companies assessed was 81.88%, an improvement on the 2003 average of 77%. Average scores for corporate values (100%) and purported principles around corporate responsibility (97.35%) suggest that Australian companies are starting to take CSR issues seriously. However, a closer analysis of the scores suggests a significant gap between rhetoric and practice, with the scores for values and principles being well above the average scores for integration (86.76%). On performance and impact, in which companies are asked to demonstrate what they are actually doing to implement CSR policies, the average score was lower again at 79.02% (*Sydney Morning Herald* 4 April, 2005). Thus far, no Australian construction contracting organisations have been assessed against the Corporate Social Responsibility Index, although some large suppliers of building products, such as Boral, have been assessed.

RepuTex

RepuTex is a private company based in Melbourne, Australia, specialising in the assessment and rating of organisations' social responsibility performance. RepuTex ratings represent an assessment of the extent to which an organisation is performing in a socially responsible manner and managing its social-risk exposures in the following four areas:

- corporate governance, including transparency, risk management and ethics;
- workplace practices, including occupational health and safety management, workplace culture and diversity;
- social impacts relating to products, services, policies and practices;
- environmental impacts of operations, policies, procedures, products and services.

RepuTex ratings are based on a model similar to credit ratings, and take account of past and current practices and future risk exposures relating to social responsibility. RepuTex analysts independently assess the effectiveness of organisations' management of social responsibility against a set of performance criteria, developed with stakeholder input. Organisation reports provide a basis for differentiating organisations in relation to their commitment to social responsibility and enable comparisons to be made between an organisation's performance and best practice in the industry sector in which the organisation operates. Several large Australian- and New Zealand-based companies with construction interests have been rated against the RepuTex criteria, including Leighton Holdings, Fletcher Building Ltd (New Zealand) and Lend Lease Corporation. Summary results of ratings are publicly available for stakeholders to view on the RepuTex website (http://www.reputex.com.au/).

Conclusions

In general, the Australian construction industry appears to be socially irresponsible. The severity of comments of recent Royal Commissions have very pointedly shown the industry to be corrupt and fraudulent; however, this view needs to be balanced against the many positive aspects of the industry. The industry is a major contributor to the economy, and services the country with the infrastructure that enables it to be a significant, albeit small, player on the world stage. The negative portrayal of the union movements likewise needs to be weighed against the positive social contributions made particularly in work conditions and towards environmental issues through resistance such as the 'green bans'.

Having painted a brief picture of CSR in the Australian construction industry, we feel that fundamental CSR change can come about only through general cultural and societal moral and ethical change. As in cases in other countries, the result of many initiatives can be merely 'whitewashing' in order to continue to compete within a new economy that values the notion of an organisation that 'cares'.

Note

1. Since this event the company has changed ownership and now operates as Brookfield Multiplex.

References

The Age (2005a) Taking care of business, 4 April: 13.
The Age (2005) Being responsible is suddenly serious, 17 May, 8.
The Australian (2005a), BCA against more corporate burdens, 17 October, 29.
The Australian (2005b), Feel good funds – ethical investments: Pick one that matches your views – nice buys come good ... Ethically, 13 April, TO1.

The Australian Financial Review (2005a) Sign IR code or we won't turn on the tap: Anderson, 19 May, 5.

The Australian Financial Review (2005b) Still a speck but SRI clout is on the rise, 21 April, 18.

Australian Bureau of Statistics (ABS) (2002) *Summary of Industry Performance 2000–2001, May 2002:* Australia, cat no. 8142.0.55.002, Canberra: ABS.

Australian Bureau of Statistics (ABS) (2005) *Year Book 2005*, cat no. 1301.0, Canberra: ABS.

Australian Government (2005) *Ancient Heritage, Modern Society*, Department of Foreign Affairs and Trade. Online. Available at: http://www.dfat.gov.au/aib/history. html (accessed 2 May 2005).

Bluff, L. (2003) *Regulating Safe Design and Planning Construction Works,* Working Paper 19, National Centre for Occupational Health and Safety Regulation, Canberra: Australian National University.

Burgmann, V. (1993), *Power and Protest: Movements for Social Change in Australian Society,* Sydney: Allen & Unwin.

Burgmann, M. and Burgmann, V. (1998), *Green Bans, Red Union: Environmental Activism and the New South Wales Builders Labourers Federation,* Sydney: UNSW Press.

Carroll, A.B. (1991), The pyramid of corporate social responsibility: toward the moral management of organizational stakeholders, Business Horizons, 34: 39–48.

CFMEU (2004) *ICAC Report into Safety Certification & Training in the NSW Construction Industry,* press release, July, Lidcombe, NSW: CFMEU.

Cole, T. (2003) *Final Report of the Royal Commission into the Building and Construction Industry,* Commonwealth of Australia, February.

Curriculum (2005) *About the Building and Construction Industry*, VET in Schools, Curriculum Corporation. Online. Available at: http://cms.curriculum.edu.au/the_cms/tools/new-display.asp?f=5280&seq=5249 (accessed 8 November 2007).

Department of Immigration and Multicultural and Indigenous Affairs (DIMIA) (2005) *Managing the Border: Immigration Compliance*, Australian Government. Online available at: http://www.immi.gov.au/illegals/mtb (accessed 22 June 2008).

Department of Trade and Industry (1998) *Rethinking Construction: The Report of the Construction Task Force*, London: HMSO.

Environics International Ltd (1999) *The Millennium Poll on Corporate Social Responsibility: Executive Briefing*, Canada: Environics International Ltd.

Fisher, J. (2004) Social responsibility and ethics: clarifying the concepts, *Journal of Business Ethics*, 52: 391–400.

Francis, V. and Lingard, H. (2004) *A Quantitative Study of Work-Life Experiences in the Public and Private Sectors of the Australian Construction Industry*, Construction Industry Institute Australia, September.

Gillard, J. (2007) *Workplace Health and Safety*, Election Fact Sheet, Australian Labour Party, Canberra, Australian Capital Territory.

Gyles, R.V., Yeldham, D.A., and Holland K.J. (1992), *Report of Hearings, Royal Commission into Productivity in the Building Industry in NSW Government of NSW*, Sydney, Australia.

Latham, M. (1994) *Constructing the Team: Final Report of the Government/Industry Review of Procurement and Contractual Arrangements in the UK Construction Industry*, London: HMSO.

Lingard, H. and Francis, V. (2004) The work-life experiences of office and site-based employees in the Australian construction industry, *Construction Management and Economics*, 22: 991–1002.

McWilliams, G., Rechnitzer, G., Deveson, N., Fox, B., Clayton, A., Larsson, T. and Cruickshank, L. (2001) *Reducing Serious Injury Risk in the Construction Industry, A Report for the Victorian Workcover Authority*, Melbourne, Australia: Monash University Accident Research Centre.

Maslach, C., Jackson, S.E. and Leiter, M.P. (1996) *Maslach Burnout Inventory Manual* (3rd edn), Palo Alto, CA: Consulting Psychologists Press.

Peterson, D.K. (2002) The relationship between unethical behaviour and the dimensions of the ethical climate questionnaire, *Journal of Business Ethics*, 41: 313–326.

Petrovic-Lazarevic, S. (2004) *Corporate Social Responsibility in the Building and Construction Industry*, Working paper 36/04, July, Melbourne: Monash University, Faculty of Business and Economics.

Ray, R.S., Hornibrook, J., Skitmore, M. and Zarkada-Fraser, A. (1999) Ethics in tendering: a survey of Australian opinion and practice, *Construction Management and Economics*, 17: 139–153.

Royal Commission into the Building and Construction Industry (2002) *Recent Reviews of the Building and Construction Industry*, Discussion Paper 9, Commonwealth of Australia, August.

Small, C. (2004) Corruption control and the pursuit of integrity: deserving confidence and trust of the public must be the goal, merely appearing to be worthy is not enough, presented to the *Corruption Prevention Network Annual Conference*, Sydney, 9 September.

Smith, G. (2003) *Cole Royal Commission Seminar – An Introduction to the Royal Commission*, 8 April, Melbourne: Clayton Utz.

Standards Australia (SA) (2003a) *Organizational Codes of Conduct*, AS 8002-2003.

Standards Australia (SA) (2003b) *Corporate Social Responsibility*, AS 8003-2003.

Standards Australia (SA) (2003d) *Fraud and Corruption Control*, AS 8001-2003.

The State Chamber of Commerce (NSW) (2001), *Taking the First Steps: An Overview, Corporate Social Responsibility in Australia*, February 2001.

Sydney Morning Herald (2003) Shortcuts a threat to safety, ICAC told, 26 August.

Sydney Morning Herald (2005) Taking care of business, 12 April.

Vee, C. and Skitmore, M. (2003) Professional ethics in the construction industry. *Journal of Engineering, Construction and Architectural Management*, 10 (2): 117–127.

Victor, B. and Cullen, J.B. (1987) A theory and measure of ethical climate in organizations, *Research in Corporate Social Performance and Policy*, 9: 51–71.

Zarkada-Fraser, A. and Skitmore, M. (2000) Decisions with moral content: collusion, *Construction Management and Economics*, 18: 101–111.

Index

Abbott, T. 367
Abercrombie, N. 28
ABN60 (firm) 275
ABRINQ Foundation (Brazil) 291
  Child Friendly Company Programm 291
ABS (Australia) 372
'absconders' 79–80
absence of robust infrastructure in
    developing countries 88–90
  six contributing factors 88–90
Accelerating change (2002) 43–5, 49
accountability 362
Ache, P. 150
Aceh province Indonesia 87, 341
Acres International Ltd. 152
Adams, C. 215
Adams, D. 17, 238, 240–1, 253, 255, 256
Adamson, D.M. 14–15
added ethical value 59–60
Adele, T. 316, 318
Adnams (Brewers) 231–2
Aerts, D. 199
affirmable business enterprises
    (S. Africa) 316
affirmative Action (S. Africa) 289, 307,
    316–17, 318
  corporate advancement 318
  outsourcing 318–19
  ownership 317–18
  small, medium and micro enterprises
      (SMMEs) 318–19
affordable homes 254
Afghanistan 144
Africa and corruption 149–50
African National Congress (ANC) 305
The Age 275–6, 296, 354, 361, 363, 377
ageing population 64
agency theory 327
Ahmed, R. 82
Air China 328
Alaska 354
Alexander, C. 216
Alexander, Douglas 39

Al-Khater, K. 312–13
alternative eco-building movement and CSR
    215, 221
  anti-establishment 218
  neglect by UK government 219
altruism, religion and solidarity 288–9
altruistic commitment 103
Amadigi, F. 18
American business in Hong Kong and
    China 342
American Council for Construction
    Education (ACCE) 292
  ethics education 292–3
Americans with Disabilities 289
American Society of Civil Engineers (ASCE)
    (2005, 2007) 144, 146
  'zero tolerance' for corruption 144
Amick, B.C. III 274
AMP Capital 363
Anglo-American world 27
Anglo-Saxon business culture 321
Ankers, Michael 226
annual improvement targets 42
annual number of natural disasters 85
  average death toll 85
  in developing countries 85
Anon. (2001) 121, 123, 137
anti-competitive activities 131, 166
anti-competitive behaviour 165–8, 175–6,
    182–3
Anti Corruption Code for Individuals in the
    Construction and Engineering Industry
    (England & Wales) 121
antiquated corruption statutes 130
Anti-Terrorism Crime and Security Act
    (2001) 144
ANZ Bank 362
apartheid 304–6, 309, 310, 314, 322
Apex Asphalt and Paving Co 182
application of anti-corruption codes 142
  in overseas developing countries 142
Aristotle 100
Armitt, J. 143, 145

arm's-length contractual relationships
    (ACR)   321
Armstrong, M.   55, 57
Armstrong, T.J.   262
arsenic to treat timber   229
    copper chrome arsenate (CCA)   229
Arup and Associates (Hong Kong) Ltd.
    342–3
Arup Group   107
    social responsibility programme   107
Arup international consultancy   87
    humanitarianism   86
    and RedR   87
Arup, Ove (Sir) 'The Key Speech'   13, 14
Arups   215
asbestos   275–7
Asia   327–48
    business practices   341
    challenges   347–8
    CSR Asia Weekly   346
    culture   343–5, 348
        power-distance   343, 348
        uncertainty avoidance   343
    employment practices   337–8
    environment and development   339–41
    ethics   345–6
        code for the construction industry   346
    freedom of association and human rights
        341–2
    health and safety   338–9
    investment   342
    philanthropy   342–3
        Hong Kong and China   342–3
    profitability   345
    sustainability and strategy   345
Asian Business Practices   347–8
Asia Pacific countries   329
Asia-Pacific Region   18, 334
aspirational nature of green building   224–5
    short and long-term costs   225
    supply and demand   225
    sustainable building   225
Associated Press (AP)   289
Association of British Insurers   274
Association of Community Technical Aid
    Centres (ACTAC)   217
    'Green-Building Fairs'   217
Association for Environment Conscious
    Building (AECB)   216, 217–18, 222
Atkin, Lord   275
Atkinson, G.   198, 202
attracting new graduates into construction
    64–5
    health and safety   65
    importance of CSR   65
    social and environmental issues   65
    workforce diversity   65
    work–life balance   65
Aupperle, K.E.   35

Ausol (Highways) Argentina   298
Australasia   158–159
Australia   219, 221, 351–78
    ageing population   355
    burnout   337
    demographic trends   354–5
    European population   352
    geo-demographics   354–6
    gross domestic product (GDP)   356–7
    history and society   352
    human rights   352
    immigration   355–6
    Immigration Restriction Act   352
    industry groups   357–8
    labour shortages   356
    law   353
    population   355
    small businesses   357
    social and environmental concerns   351–2
    young workers   356
The Australian   276, 363
Australian Associated Press   276
Australian Broadcasting Corporation   277
Australian Building and Construction
    Commission   368
Australian Building and Construction
    OH&S Accreditation Scheme   356
Australian building products   261
Australian Bureau of Statistics (ABS)
    35–6, 372
Australian Centre for CSR   361
Australian Commonwealth   352
Australian Constitution   352
Australian Construction Association
    (ACA)   357
Australian Construction Industry   18–19,
    262–3, 351–3, 356–60
    Civil Contractors Federation (CCF)   357
    and CSR   338
    designers   360
    employment practices   352
    ethics   352, 361
    health and safety standards   267
    history of political control   352–4
    Housing Industry Association Limited
        (HIA)   357
    Master Builders of Australia Inc.
        (MBA INC)   357
    organised labour   352
    procurement   356
    professionals   373–4
    small and medium enterprises SMEs
        (357–8)
Australian Constructors Association Code of
    Ethics   375–6
Australian Council of Trade Unions
    (ACTU)   358
    Automotive, Food, Metals, Engineering,
        Printing and Kindred Industries   358

Australian culture 353, 365
Australian Department of Employment and
    Workplace Relations 367
Australian Financial Review 263, 275,
    276–7, 353, 363
Australian Government 352–4, 356
Australian Government Commissions
    365–9
    bribes 365
    collusion 365
    fraud 365
    (The) Gyles Royal Commission 365
    illegal activities 365
    law 365
    productivity 365
Australian Industrial Relations Commission
    (AIRC) 367
Australian Institute of Criminology 370
Australian Labour Party 276, 353
Australian Security and Investment
    Commission 354
Australian Standards on Fraud and
    Corruption Control 371
    risk management standards
        (Australia) 371
Australian state government 277
Australian Stock Exchange (ASX) 362
Australia Standards 270, 338, 361, 375
    AS3806 362
    AS8003 362
Australia Standards (8003) (2003) 338, 362
Austria 215
avian influenza 93
Awarding sub-contracts 176
Axlerod, L.J. 201

B&Q 59–60, 66, 222
BAA Terminal 5 Heathrow 9
back-to-the-land movements 218
Baer, W. 177, 180
Baker, M. 237
Balabanis, G. (1998) 35, 55
Balanced Value 228
Ball, M. 242
Bandar Botanic project 340
Bangladesh 85
Bank of East Asia 341
Barbier, E.B. 198
Barker, K. 221, 241, 243, 250–2
Barker Review of housing (2006) 221
Barlow, J. 242
Barnes, June 226
Barratt Developments plc 238, 240, 244,
    248, 249, 251
Barratt Homes 225
Barrett, P.S. 206, 208
Barrios 86
    negative environmental impacts 86
Barry, M. 4, 12, 14

Barry, S. 123
Barthorpe, S. 4, 14
Bateson, G. 201
Bau-Biologie (2007) 219
Beazer (firm) 240
Beck, M. 32
Beijing 328
    Olympic Games 347
Belbington, K.J. 199
Belfast 217
Bellway (house builders) 244, 247–9, 254
benchmarks in specifications 83
benefits of eliminating corruption 130
Benjamin, Victor 217
Ben and Jerry's 59
Berger, G. 298
Berkeley (house builders) 244–5, 248–9,
    250, 254
Berrien, F.K. 26
'best value' 48
'Better Buildings, Better Lives'
    (May 2004) 217
better market information 58
Better Regulation Executive (BRE) 50
Beusch, A. 81
Beveridge settlement 24, 25, 28
Bhaskar, V. 198
Bhatti, M. 242
Bhoal 269
Bhopal 10
BHP Billiton (Australian Firm) 362
bid rotation 176
bids 364
    cutting 364
    shopping 364
    withdrawn 364
bid suppression 175
'Big Four' regulatory agencies (USA) 287
    Consumer Product Safety (CPSC) 287
    Environmental Protection Agency
        (EPA) 287
    Equal Employment Opportunity
        Commission (EEOC) 287
    Occupational Safety and Health
        Administration (OSHA) 287, 288–9
Bill of Rights 313
bio-composite materials 232
'Bioregional'/Bedzed project Survey
    (2007) 224
    iconic wind cowls 224
Black, C. 320
Black Economic Empowerment (BEE)
    (S. Africa) 306, 317–18
    Black entrepreneurs 317
blacklisting of trades unions 30, 31
Blair, Tony 50
Blake, Lapthorn Linnell 179
Blomstrom, R.L. 26–7
Bluff, L. (2003) 353

'Bob the Builder' 64–5
(The) Body Shop 54, 59, 66–7
Boge, Ulf 179
Boliden (Swedish firm) 184
Bolsa Escola (School Scholarships)
    Brazil 291
'bolshevism' 30
Bon, R. 73
Bonny, C. 56
'boom-to-bust economy' 11
Bossel, H. 196
Bosshard, P. 146
Bossink, B.A.G 194
Bottelier, P. 143–5
'bottom-line' profitability 38
Boureron, P.S. 206
Bousteads Consultants (UK) 230
Bovis Lend Lease 263
Bowen, H.R. 5, 25
Bowen, P. (2007) 4, 14, 143
Bowen, P.A. 316, 317
Bowley, M.E.A. 33
Boyle, G. 216
BP Australia (firm) 362
BPB (firm) 178
Braithwaite, J. 269
Brazil 288, 290–1, 329
    Paraná State 293
Brazilian construction industry 18
breadline 305
Brekke, K.A. 202
Bribe Payers Index (2006) 141, 147, 148
    and 'takers' 148
bribery 148, 156–7, 174, 186
'bribery and corruption' 121
bribery to obtain main contract award
    156–7
bribes and bribery 371
bribes and opportunity costs 129, 146–7
Briggs Cladding and Roofing Ltd. 185
Brilliant, Ashleigh 121
Bristow, M. 149
British Broadcasting Corporation
    (BBC) 251
    'Watchdog' 251
British class system 33
British Land 63
Brooks, S. 65
Broome, John 227
'brown envelopes' 121
Brouwers, H.J.H. 194
Brownfield sites 244, 247
Brown, Gordon (UK Prime Minister) 6,
    50, 226
Brundtland report 11, 194
BSRIA (2000) 55
Buddhism 347
Building 2007:33 10
'Building Businesses Not Barriers' 32

Building and Construction Improvement Act
    (Australia) 368
Building and Construction Improvement Bill
    (2003) 368–9
Building Regulations 248
Building Research Establishment (BRE)
    101, 226
Building Research and Information:
    Challenges Facing Construction
    Industries in Developing Countries 93
'build-it-yourself' and DIY 298, 300
built assets as parasitic 10
'bungs' 10
Burayidi, M. 198
Bureau of Labour Statistics (BLS)
    (USA) 296
Bureau for Market Research (South Africa)
    305
Burgmann, V. (1993) 359
Burke, C. 148
burnout due to overwork 337
Burns, T. 26
business case for community policy 103–5
business case for CSR 12, 35–6
'business case' dogma 35–6, 37–8, 44–9
    and accident reduction 42
    adversity to innovation 36
    power and ubiquity 37
    three foundations 37
'Business in the Community' (BITC) 3, 10,
    38–9, 57, 104, 113, 235, 240
    five key impact areas 57
business ethics 56–7, 59, 61
Business Ethics Online 237
'Business Integrity Management System'
    (BIMS) (FIDIC) 151–2
    ISO 9001, 2000 151
Business Schools 327
business and society 37–8, 98–100
business in society 54

CABE-housing audit 250
Cairns, J. 199
Caledonian Canal 9
Callaghan, James 29
Calland 1967 case 125
Callenbach, E. 202
Caliyurt, K. (2004) 61
Caleyron, N. 218, 224, 229
Campbell, A. 105
Campher, H. (2005) 85
Canada 99, 157, 221, 363
capacity and knowledge deficit 74
Cape Town International Convention Centre
    (CTICC) 317
capital–output ratios 78
carbon dioxide emissions 222, 231
carbon footprints 115
Carbon Trust 222

Carillion (and environmental performance) 110
Carmona, M. 253
Carpenter, Ralph 231–2
Carrion, D. 86
Carroll, A.B. (1979) 7, 34–639, 40–1, 98, 311, 361
  corporate social performance model 98
  'Pyramid of CSR' 361
Carson, W.G. 268
Cartels 16, 165–87
  air conditioning 179
  cement 179
  construction 180–3
  european dimension 176–9
  fines and penalties 183–5
  mechanisms to address 183
  plasterboard 178–9
  public sector 181
  refrigeration 179
  UK 180–3
cartels and monopolies in Hong Kong and Asia 341
  ready-mixed concrete in Hong Kong 341
Carter, R. 106
'Cash for Peerages Row' (early twentieth century) 121
Castle cement 231
Castle, S. 183
catalysts for unethical behaviour 143
Catholicism 297
Catton, W.R. 201
CDM regulations 44
CE marking 224
CEE Bankwatch network and Friends of the Earth Europe (2006) 9
cement and concrete 229
Cemex (Mexico) 301
  'Patrimonio Hoy' 301
certified timber 218
challenge to global capitalism 218
challenges of disaster management 85–8
  case study 87
challenges of poverty alleviation programmes 82–5
  access to health sanitation and education 82
  better housing 82
  communications 82
  roads 82
  water and power supplies 82
Chalmers, D. 16
Channel Tunnel Act 112
Channel Tunnel Rail Link (CTRL) 112
Chapter Seven (2007) 219
charitable donations 246
charity and charitable foundations 105–8
  charitable trusts 106–8
charity principle 106

Chartered Institute of Building (CIOB 2006, 2007) 14, 108, 122, 174
Cheadle, A. 274
Chernobyl 10
Cheung, C.F. (2005) 340
Cheung, S. 320
Cheung Kong Holdings 342
child labour 18, 54, 56, 81, 289–91
  and children's future 290
  children working illegally in the USA 290–1
  companies in the USA using child labour 290
  construction in New York 289–90
  grape harvesting in California 290
  USA profits from child labour 290
  vicious cycle of poverty and lack of education 291
child labour and construction 290
  brick making 290–1
    health risks to children 290
    hours worked 290
  inadequacy of protection against abuse 290
  international Legal Agreements 289
  isolation 290
  limitations on hours 290
  quarrying in Brazil 290
China 18, 81, 148, 221, 328–9, 330–1, 336–7, 344
  charity 336
  chemical factory example 336
  corporate manslaughter 337, 344
  cultural differences 336
  employment 336
  environment 336
  fake commodities 336
  Guang xi 343, 348
  human resources 336
  lack of social responsibility 336
  law 336
  loan difficulties 336
  low employee well-being 336
  production 336
  quality 336
  Social Accountability 8000 (SA 8000) 337
  speculatory investment 336
  stakeholders 336
  and Tanzania 148
  taxation 336
  tax evasion 336
  technological innovation 336
  voluntary activities 336
  western themes 336
China Light and Power (firm) 339
  exemplary OH&S record 368

China (People's Republic) 334–48
  construction industry 334–5
  culture 335
  economic overview 334–5
  family businesses 344
  GDP 335
  gross output value 335
  labour law 337
  population 335
  Yearly Macro-Economics Statistics
    (National) 335
Chinese construction industry 148–9
  anti-corruption measures 148
    punishment by execution 149
  and corruption 148
  expansion and construction 148–9
  reduction of corruption 149
  and value management (VM) 148–9
Chinese dialects 331
  Cantonese
    Yue and Min dialects 333
  Chiu-Chow 333
  Mandarin (Putonghua) 330, 331
  Shanghainese 331
Chinese language 331
Christopher, M. 172–3
Chung Ching 329
CIA (2007) 329
CIB-TG29 92–3
CIB-W107 92–3
CIRIA 102
Civil Rights Act (USA) 289
claiming benefits whilst working 122
Clarence Docks (Leeds) 254
Clarke, Greg 263
Clarke, S. 32
Clark, J.M. 25
Clarkson, M.B.E. 99
Clayton, A. 265
clean water 60
Cleveland Bridge 354
client-led tender requirements 56
clients 230–1
clients and CSR principles 61, 63, 67
clients and sustainability 218
climate change 73
CMS Cameron McKenna (2004) 177
Coates, B. 154
Cobb, J.B. 202
Cochran, P.L. 358–9
code of ethics 375
'Code of Practice for the Dissemination of
  Information on Major Infrastructure
  Projects' (Office of the Deputy Prime
  Minister-ODPM 1999) 112
'Code for Sustainable Buildings' 217
'Code for Sustainable Homes' 217, 221,
  226, 227
Coega Harbour 317

'cold war' 29
The Cole Royal Commission 367–9
Cole, T. 351, 357–9, 363, 367–9, 372
collective bargaining 295
Collier, P. 146
collusion 175
collusion and tenders 134
collusive tendering 371
combating corruption in construction 159
  ACCESS 159
Comben (firm) 240
commercialisation of sub-standard
  products 292
Commission of Inquiry in New South
  Wales 275
Commons Environment Audit Committee
  254
Community Development Venture Fund 42
community and the environment 376
community interaction and strategies 102–5
  commitment of resources 104
  industry/environment 102
  project strategy 102
Community Investment Tax Credit 42
community involvement 249
Community Land Trusts (2007) 220, 236
community perceptions of the construction
  sector 101–2
community politics 98
  disruption/expediting of construction
    projects 101
companies and communities 37–8
compensatory losing bids 365
Competition Act (1998) 130–1, 165,
  166–7, 175, 178, 180–1
Competition Appeal Tribunal 180
competition authority 177–8
Competition Commission 181
Competition Disqualification Order
  (CDO) 184
competition law 181
competitive advantage 328
competitive strategies 327
complete tax evasion by self-employed
  workers 122
compulsory competitive tendering
  (CCT) 48
Confederation of British Industry (CBI)
  31, 311
confidential information 371
confidentiality and intellectual property 376
conflicts of interest 371
Confucian values 338
consequences of corruption 128, 130
  costs and statistics 128
  personal ambivalence 128–9
  self-perpetuation 130
Conservative Government of 1979 28, 30–1

Considerate Constructors Scheme (CCS)
    101, 108–9, 114–16
    criticisms 109
    dealing with complaints 109
    eight key areas 108–9
    good neighbours 108
    non-compliance 109
Constructing Excellence (2004) 6, 12,
    14, 217
Constructing the Team 101, 108, 138
construction activity in developing countries
    73, 75, 78, 79
    and CSR 834
    and urban unemployment 74
construction in America and relationship
    with the community 297, 298
construction capability and challenges of
    development and reconstruction 78–9
Construction Clients Group (CCG) 217
construction companies and their long-term
    impact on communities 98, 102–3
construction contracts and corruption
    132–4
Construction Design and Measurement
    (CDM) regulations 63
construction firms and communities 46
construction in high-risk areas 85
construction industry 217
    carbon emissions 218
    CSR and tender returns 61
    difficulties with CSR due to
        transience 678
    landfill waste 218
    mistakes and successes 60
    resource consumption 218
Construction Industry Board Report 101
    1996 Report, 'Constructing a Better
        Image' 101
Construction Industry Council (CIC)
    108, 217
    Sustainable Development Committee 217
Construction Industry Development Board
    (CIDB) (S. Africa) 306, 314
construction industry as 'the most corrupt'
    128–30
    reasons why 128
construction industry 'norms' as
    criminal 121
    changing attitudes to corruption 124
Construction Industry Research and
    Information Association (CIRIA) 12
Construction Industry Training Board
    (CITB) 44
construction industry workforce as an
    asset 44
construction markets in developing countries
    78–9
Construction News Quality Awards (2005
    and 2006) 57

construction and poor business ethics
    292–3
construction process 246–7
construction procurement and
    sustainability 47
Construction Products Association 226
construction professionals and corruption
    121
Construction and Property Industry Charity
    for the Homeless (CRASH) 105–6
Construction Research and Innovation
    Strategy Panel (CRISP) 217
Construction Skills Certification Scheme
    (CSCS) 44
construction and society 14
Construction Task Force (CTF) 42, 264
construction and world-wide
    corruption 141
    and post-conflict reconstruction 146
consultation with interest groups 104
contentious nature of business ethics 40
contractors and the challenges of
    community/environmental relations
    115–16
    ten-point plan 115–16
contradictions between 'best practice' and
    anti-corruption measures 138
contribution of construction to unsustainable
    activity 191
Control Risks Group (2003) 125–6
Cooper, I. 10
copper tubing 184
Corner House (2000) 142
Cornwall Sustainable Building Trust
    (2007) 225
corporate accountability 266
corporate business/investment 73
corporate citizenship 100
corporate governance 362, 377
corporate governance and social
    responsibility 342
corporate level involvement 105
'corporate manslaughter' 33, 42
Corporate Responsibility Index 363
Corporate Social Responsibility (CSR)
    Academy 131
    and appropriate infrastructure
        development 81, 84, 88–9, 90–1
    and business behaviour 73, 81, 85
    (the) 'business case' in construction
        12–13
    Business Olympics 237
    and business strategy 54–68
        within construction 54
        outside construction 54
        raised brand awareness 56
        and staff turnover 56
    certification and auditable
        standards 5–6

Corporate Social Responsibility (CSR)
    (*Continued*)
and commitment to environmental
    responsibility 215, 254
competency framework 68
concept and its evolution 14
conceptualisation and reporting 5
and construction management
    research 4
corporate philanthropy or revenue
    opportunity? 7–8
against corruption 142
and 'cultural dilemmas' 81
and cultural/religious beliefs 18
dangers of PR polemics 61
as a defensive measure 54
definitions 4, 5, 25–6, 54–5, 62
    'fluffy concept for tree-huggers' 63
    need for a long-term investment view
        62–3
    as a risk-minimisation technique 56
    and the UK government 58
discourse 37
and economic performance 40
and ethics 191
forum 237
fresh perspectives in construction (since
    1993) 13–19
goals and principles 98–9, 108, 111
grounds for optimism 49
growing prominence in the construction
    industry 8–12
and health and safety 65–6
historical context, roots and development
    5, 24–50
holistic and multifaceted consideration 4
implementation by international
    construction firms 79, 94
introduction 1–19
and LHWP 155
links with corporate financial
    performance 8
management systems 245
and marketing 36
as 'a matter of principle' 40
and neoclassicism 210
outsourcing to quangos 39
parodies of the acronym 7
    'Companies Spouting Rubbish' 7
    'Complete Sidelining of Reality' 7
    'Corporate Slippery Rhetoric' 7
and 'people issues' 44
and procurement policy 42, 47–9
and profitability 24, 35–6, 48
'as the right thing to do' 38–9
and risk mitigation 37
second generation activities 83–4
and skills development 92–3
and social and political change 24

in a socio-economic context 6–8
    a case for localised conceptions?
        5–7
and strategic management 65
and sustainable development 195–7,
    197–9, 205
and the UK government 58
UK house building 235–5
    barriers in house-building 239–44
    land practice 255
    local employment 256
    speculative house building industry
        256
    training 150
voluntary focus 152
weakness of voluntary approaches 13
website (2005) 61
Csr.gov.uk 2007 6
welfare provision 13
corporate social responsiveness (strategies)
    98–9
corporate transparency 55
    and materiality 57–9
corporate values 377
Corporate Watch (2007) 5, 7
corrupt business practices 10
corruption and collusion 174–5
corruption in the developing world 121,
    125, 129
'corruption food chain' 122–36
corruption in government contracts in
    Central and South America 293
corruption and illegal practices 14, 16,
    124–8
corruption within international construction
    projects 79
corruption and international pressure 141,
    143–4
corruption introduction into infrastructure
    delivery 125–6
corruption law since 1906 125
corruption and poverty 148
corruption-types and methods 145–6
    criminal activities 145
    impact on other parties 146–7, 152–3
    and stages within the construction
        process 146–7
    'voluntary and coerced corruption' 145
corrupt practices 122
    'loans for knighthoods' scandal
        (2006) 137
    'sex for visas' scandal (2006) 137
corrupt practices 329
Cosa Nostra and construction in New
    York 130
'cosmopolitans' 79–80

Costain Construction 113–14
  'Community Chest' 114
  education, environment and
    regeneration 113
  work with community projects 113–14
Cost–benefit analysis of sourcing 111–12
cost of corruption 146–7
  cost overruns and corruption 147
  diversion of funds from essential
    projects 146
  perceptions of participants 147
  reduction of corruption in the Middle east
    through religion 159
Coull, Ian 225
Countryside Properties (firm) 254
The Courier-Mail 276
cover prices 364–5
cover pricing 10, 174
Cox, A. 320
crane operations 369
Crehan v. Inntrepreneur (Court of Appeal)
    2004 181
Creosote 229
Crest Nicholson (firm) 254
Crimes (Workplace Deaths and Serious
    Injuries) Bill 344
Croner 170
Crook, T. 243
Cropanzano, R. 266
Crosby Homes (firm) 254
Crosthwaite, D. 73
Crown Prosecution Services (England and
    Wales) 124
Crowther, D. 61
Crump, D. 224
Cullen, J.B. 362
Cullen, P. 127
Cullinan, B.P. Judge 155, 157
culture of complacency 361
culture of performance measurement 43
Curado, M 290
Curriculum 373
Curwell, S. 10
customer care 250–1
customer satisfaction 250–1
cynicism (about corruption) 135

Dahlman, C.J. 79
The Daily Telegraph 276
Dainty, A. 67
Daly, H.E. 201, 202
dangerous working environments 81
D'Arcy, J. 122
Darroch, F. 136, 157
Dasgupta, P. 198
David Wilson Homes (firm) 240, 251
Davis, J. 202
Davis, K. 25, 26–7
Dawson, S. 32

'dead men scam' 112–13
deaths 296
Debrah, Y.A. 150
decentralisation of procurement 47–9
deception 145
decimation of manufacturing 28
decline in training and in the apprenticeship
    system 31
defining corruption 124–8
  Chambers Dictionary 124
  cultural variations 124
  legal definition of corruption 125–6
  Transparency International (2003) 124
  US definition 127
De George, R.T. 313
demand and supply chains 170–1
Dembe, A.E. 274
Deni Greene Consulting (Australia) 363
Denison, Tony 106
Denmark 151–2
deontological view of the moral imperative
    of duty 100
Department for Business, Enterprise and
    Regulation 50
Department for Business, Enterprise and
    Regulatory Reform (BERR) (2007)
    224, 226
  as a replacement for the Department of
    Trade and Industry (DTI) 226
Department of the Environment, Transport
    and Regions (DETR) 42, 206
Department for Food and Rural Affairs
    DEFRA (2006) 47, 232
Department of Immigration and
    Multicultural and Indigenous Affairs
    (DIMIA Australia) 373
Department for International Development
    (DFID) 1997 82
Department of Labour (DOL) in the
    USA 289
Department of Transport and Public Works
    (S. Africa) 314–15
  accountability 315
  Department of Trade and Industry (DTI)
    3, 25, 37–41, 47, 49, 128, 166,
    236, 245
  Construction Statistics Annual 128
  DTI Corporate Social Responsibility:
    A Government Update (2004) 39
  DTI Corporate Social Responsibility
    Report (2002) 39–41
  equity 315
  ethics 315
  Five pillars of Procurement 315
  funding of CSR competency
    framework 57
    five attainment levels 57
  open competition 315
  value for money 315

Desai, (2003) 56
Design Quality Indicators 43
De-Veen, J.J. 82–3
Develay, D. 157
developing an ethical company 56
developing a community policy 103
developing a social conscience 56–7
Dewar, D. 306
Dietz, T. 199
direct action campaigning (1990s) 101
Direct Labour Organisations (DLOs) 30–1
    decline (1980–1995) 31
    reasons for 19th century foundation 37–8
Disability Discrimination Act 64
disaster events 85, 88
    high frequency with cumulative effect 85
    tropical cyclones, storm surges and
        floods 85
    low-frequency, high impact hazards 85
Disaster Resource Network (DRN) 86
    CSR related actions 86
    New Engineering Construction Initiative
        2006–2010 86–7
discrimination 288, 289
    colour 289
    disability, 288, 289
    gender 289
    nationality 289
    race 289
    religion 289
    veteran status 289
disorder 351
disorder and litter 101
DIY house decorators 224
DIY products 300
Doig, A 149
Donaldson, P. 89, 99
donor agencies 92
Donovan, C. 82
Dorcey, A.H.J. 199
Double Glazing Supplies Group plc
    (DGS) 184
Double Quick Supplyline Ltd. (DQS) 184
Dow Jones Sustainability Index (DJSI) 61
'downstream' impacts of construction
    decisions 215
Draft Corruption Bill (HM Government
    124–6
    extra-territorial effect 126–7
Drew, K. 153
Druker, P.F. 34
drying times of organic and toxic paints 229
Dryzek, J.S. 199
du Gay, P. 28
Dundas 175, 182–3
Dunfee, T. 135
Dunlap, R.E. 201
Dunn and Bradstreet rating 136
'duty of care principle' 100

Earle, A. 155–6
earthquakes and hurricanes 144, 146
earthquakes and landslides 85
earth's finite resources 192–3
East Thames Housing Association 226
ECD architects 222
Eckhardt, R.E. 266
eco-cabins 220
(The) Eco Centre South Tyneside 55
Ecocentric ethic 200
'EcoHomes' standards 244
ecological buildings and energy efficiency
    standards 216–17, 222
Ecological Design Association (EDA) 216
ecological development 341
'ecological' genuine environmentally
    responsible approaches to construction
    216, 224, 232
    building products 224
    range of products 224
ecological time frames 206
e-commerce for government
    procurement 293
economic development and sanctions 348
economic growth 74, 76
Economic League (1919–1994) 30, 31, 48
    construction industry majority 30
economic multiplier effect 111
economic and social costs of
    construction 143
The Economist (2006) 221
'EcoTech' 227
eco-towns 221
    Dontang (China) 221
Eco Trust (2007) 225
eco-villages in the countryside 219, 222
Ecuador 290
    Cuenca 290
    National Institute for Children and the
        Family (INNFA) 290–1
    Nueva Aurora 290
    Quito 290
Edinburgh 123
Edmonds, G.A. 82
education about the dangers of human
    immunodeficiency virus
    (HIV)/AIDS 83
educational differentials and information as
    power 297
Egan Report Rethinking Construction
    (1998) 42, 206
    domination of debate 42
    Egan compliancy 43
    Eganisation of construction 43, 45
    Eganites 42–346
Egan, Sir John 42, 46, 241, 264
Ehrlich, P.R. 193
Eigen, Peter 159
Einstein, Albert 327

Elizabeth, L.   215
Emerging Contractor Development
    Programme (ECDP) (DPW1998)   314
employee motivation   64
employee rights (South Africa)   305
employee screening   371
employees and human rights   63–4
employee social responsibility   41
employees as stakeholders   62, 66
employee welfare   58
employment   111
    local labour and training   111
employment contracts   372
Employment Standards Administration
    (ESA)   289
'encapsulators'   79–80
energy-efficient principles   222
Energy Savings Trust   226
Engage Report 2004   102
Engeli, G. (FIDIC)   151, 158
engineering and environmental ethics   4, 14
Engineering and Physical Sciences Research
    Council (EPSRC)   218, 224, 229
Engineering Without Frontiers (EWF)   83
Engineers Against Poverty (EAP)   75
Engineers Without Borders-UK
    (EWB-UK)   84
    experience for young engineers   84
English construction industry   10
English language   331, 333
English Partnerships   220
enlightened self interest   36–7, 39, 44–5,
    48, 107
Enron   287, 327
'Enronitis'   327
Entec (2000)   271
(The) Enterprise Act 2002   130–1, 134,
    180, 182
Enterprise Bargaining Agreement
    (Australia)   367
enterprise culture   24, 28, 30, 42, 48–9, 413
    in construction   30–4
    key policy dimensions   28
    undermining professionalism   24, 33–4
'enterprise discourse'   28, 29, 30, 31, 36–7
enterprise, education and training   111
'enterprise initiatives'   48
entrepreneurism   29
Environdec (2007)   230
Environics International Ltd.   369
environment agency   59
environmental appraisal   110
    three important factors   110
environmental audits   10
environmental benefits of the alternative
    building movement   216
environmental degradation   58
environmental-friendly housing   222
environmental-friendly paints   224

Environmental Impact Assessments
    (EIA)   158
environmental impact of construction
    100–1, 109–11
    Australia   378
    corporate versus local   111–12
    legislation   216
environmental issues   378
Environmental Management Systems (EMS)
    59, 264
Environmental Performance Indicators   43
Environmental Product Declarations
    (EPDs)   230
environmental projects
    six important activities   110–11
environmental reporting   59
environmental response   110
    four main issues   110
environmental and social objectives in
    procurement   83
environmental standards   228, 232
Environmental Supply Chain Improvement
    Plan   246–7
environment and the community   109–17
    record of the construction industry
        109–11
environment and construction   3, 7–9
Erlick, J.C.   294, 297
Ernst and Young (Accountants)   377
erosion of construction skills in the UK   111
'Essex Man'   31, 50
establishing an ethical business culture in
    South Africa   309–11
ethical behaviour and the built environment
    4, 14
'ethical climate'   18–19
ethical codes   132, 292–3
ethical codes of construction professionals
    142–4
    penalties for corrupt practices   134,
        144–5
    terms and characteristics   143
ethical dialogue   293
ethical dilemmas   143–4
ethical dimensions of CSR   99–100
    respect, integrity and empathy   106
ethical imperatives   297
ethical Investment   363
ethical and moral challenges of corruption
    143–4
ethical principles   218, 232
    as an impediment to commercial
        success   218
    of sustainable construction   220
ethical procurement of building materials
    11–12

ethical purchasing of green products and
    materials 228–31
  cost of petrochemicals 228–9
  problems of manufacturers/distributors
    229–30
  renewable sources 229
ethics 345–8
ethics (Australia) 354, 363–5
ethics and corporate governance (America)
    292–3
  Southern and Central lack 292
ethics, CSR and 'the bottom line' 67
ethics of house building 254
ethics *versus* tendering 364
Ethiopian famine 1974 85
  Northern Ethiopia 85
Ethiopian Road Authority 89–90
Europe 354
European Chamber of Commerce in China
    328
European Commission 10, 176, 177–9, 184
European Commission (2001) 242
European Commission Against Corruption
    (2003) 126
European Committee for Standardisation
    (CEN) 224
European Corporate Sustainability
    Framework 36
European Regulations 224
European Union Construction Products
    Directive (1988) 224
European Union (EU) 59, 335, 361
European Union (EU) Public Procurement
    Directives (2003) 136
Eurotunnel 60
Exeter (2007) 225
Expanded Public Works Programme (SA)
    306, 319
Export Credit Agencies (ECAs) 143, 152–4
  and corruption 152–4
Export Credit Guarantee Department UK
    (ECGD) 149, 152–4
  and corruption 152–4
  six main failures 154
extortion and bribery 127
Exworthy, M. 33

facilitation payments 371
Fadil, P.A. 277
'fair competition' 329
'Fairness for All' Government White Paper
    64
fair Trade 65
false invoicing 370
family participation in Latin American
    construction sites 290
  work done by children 290
famine and relief by roads 85
Fanning, Peter 225

Farmer, G. 10
fatalities in the construction industry 296
  US statistics 296
fatalities and injuries 262
faults in house building 251
federal contracting policy (1941) 289
Federal Government (USA) 289
Federal University of Paraná 300
Fellows, R. 345
Ferguson, H. 144
Field, B. 262
Field, S. 268
Finance, Design, Build, Operation and
    Transfer (FDBOT) 171
financial statements 371
Financial Times (July 2003) 178,
    181–2, 236, 327
Financial Times (July 2003) 327
financing infrastructure development 90–1
firms and the environment 202
  neoclassical and anthropocentric 200–2
  selfish ends 201
  'simple thinking' 201
Fisher, J. 361
Fisse, B. 268, 269
Flanagan, R. 171
Fletcher Buildings (firm) 378
flexibility and competitive advantage 31
'flexible economy' 28
flexible hours 64
flood defences 85
fluctuating demand cycles 31
Flynn, R. 33
Ford, C.R. 9
foreign direct investment (FDI) 89, 91
Forest Forum 230
Forest Stewardship Council (FSC) (2007)
    12, 223, 230–1
Forgery of Construction Skills Certificates
    Scheme (CSCS) 122
formaldehyde 229
formal and informal Job Market 294–5
  poor treatment of construction
    workers 295
Fortier, J. 200
Foundation World 176
Fournier, V. 33
Fox, E. 4, 14
fragility of infrastructure in developing
    countries 74, 75, 85–6
  and disasters 85
frameworks 67
France 277
Francis, V. 373–4
Fraser-Andrews, J. 239
Fraser Report (2004) 123
fraud and corruption within construction
    projects 126
fraud and corruption control 370–2

fraudulent invoices   121
Freeman, R.E.   27, 99, 100
free market economics   28
Free Trade Agreements (FTAs)   335
Freshfields Bruckhaus Deringer   61
Friedman, Milton   27, 29, 38, 46, 48, 54,
    60, 201, 264
Friend, A.M.   201
Friends of the Earth (FOE)   152, 218
FTSE   100, 235
FTSE4Good   61
future reduction of   38–9
    promotional materials on CSR   37

Gênesis Group Brazil   293
    transparency   293
    Wellness magazine   293
Gale, A.   320
Galenson, A.   82
Gambrill, B.   112
Gammon Construction Company Ltd.
    (Macau)   338
Ganesan, S.   308
GAP (clothes)   56
Gary, R.   199
Gattorna, J.L.   172
GCCP (2000)   47
Gekko, Gordon   35
General Electric (Australia)   362
Germany   215, 219, 223, 228–9
Ghosal (Professor)   327, 345
Gibb Africa   83
Gibb, K.   242
Gibbons, G.   54, 65
'gifting' to favoured subcontractors   123
'gifts'   371
Gillen, M.   262
Glackin, M.   122
Glasgow   123
Glass, J.   13
global anti-corruption initiatives   143
global construction industry   9
    and finite resources   9–10
global economies   28, 81
globalisation   28–9, 48, 62, 64
    and localisation   46
globalisation of construction   75
Global Reporting Initiative (GRI)   61
global responsibility   73
global sourcing of materials, plant and
    equipment   111
    environmental costs   111
global thinking   75
Glyn, A.   198
Goldstock   127, 130, 135
Goldzimer, A.   152–3
Goodchild, B.   253
Good Neighbour Scheme   278, 279
Goodpaster, K.   60

Gouldson, A.P.   202
Government encouragement of CSR   58
    ministers and an academy   58
The Grapes of Wrath   40
Great Depression   24, 25, 48
Greater London area   107
    south London   107
Great Pearl River Delta Region   334
'greed is good'   35
'Green bans' case study New South Wales
    359, 378
Greenbiz (2007)   61
'green-building'   17
Green-Building Digest   217
Green-Building Handbook   217
Green building materials   217
    green and round wood timber   220
    straw, mud and earth   220
    wattles and shingles   220
green construction   347
greenfield sites   253
'green lifestyles-development'   221–2
    increasing demand   221–2
    organic food production   222
Greenpeace (2007)   216, 218, 361
green products   224
    elimination of toxic and carcinogenic
        materials   224
    paints   224
Green, S.   14, 15
'greenwash'   17, 214, 227, 230, 232
Greenwood, M.R.   266
Griffiths, A.L.   154, 155, 158
Griffin, J.J.   35
Groak, S.   168
gross domestic product (GDP)   9, 73, 308
ground source heat pumps   222
ground works subcontractor and 'accounting
    irregularities'   123
Grove Village regeneration project
    Manchester   64
    first housing PFI contract   66
growing your own workforce   63
Growth, Employment and Redistribution
    Strategy (GEAR)   306, 308
Guangzhou   338
The Guardian   222, 269
Guatemala   293
    Acción Ciudadana   293
    transparency award in public
        procurement   293
Guberan, E.   262
Guerrera, F.   178
Guthrie, Peter   87
Gutiérrez, R.   288, 297, 298
Guy, S.   10
Gyi, D.E.   32
Gyproc, (firm)   178

Habeck, R. 274
Hagan, J. 60
Hall, E. 79
Hall, F. 79
Halford, S. 33
Hamann, R. 312, 314
Hamil, S. 87
Hampden Park, Scotland 123
Hampton, D. 9
Handy, C. 8, 64, 104, 108, 113
Harman, Sir John 217
Harper, B 216
Harper, M. 319
Harvard Business Review (HBR 2003) 8
Harvard Review of Latin America 297
Harvey D. 29
Harvey, M. 31
Haslam, R.A. 32
Hatch Energy 152
Haughey, Charles 128
Havana Declaration of the Group of 77 328
Hawley, S. 153–4
Haymills 232
HBG construction company 57
Heal, G.M. 202
health and safety 24, 32–3, 40, 44
    under reporting of injuries and
        fatalities 32
    and self-employment 32, 48
    unreliability of H&S statistics 32
health and safety (Australia) 376
health and safety in the building and
    construction Industries (Australia) 368
Health and Safety Commission (HSC)
    32, 273
health and safety in construction 11, 15
Health and Safety Executive 32, 278
    budget cuts 32
health and safety in the Hong Kong and Asia
    Pacific Construction industry 337–8
health and safety procedures 55, 66
health and safety programmes in Latin
    America 296
Health and Safety at Work Act (1974) 11
Health and Safety at Work Act
    (Australia) 353
'healthy' buildings 216
Hearne, B. 152
Heathrow Airport 9
    Terminal 5 123
Heidelberg-Cement (firm) 179
Heilman, B. 150
Heinberg, R. 219–20
    Power Down 219
Hemcore 222, 232
Hempcrete 231–2
    definition 231
    properties 231
    timber frame construction 231

Hemingway, C.A. 311
Hemp 231–2
Hencke, D. 30
Henriques, I. 265
Heritage sites (Australia) 360
Herridge, J. 3–5, 12
Hess, D. 135
Highlands Tower Condominium, Kuala
    Lumpur 85–6
    failure of design and construction
        systems 86
HIH isurance 362
Hildyard 156, 158
Hillebrandt, P. 11
Hill, R.C. 313
Hills, J. 328
Hilton, S. 54, 65
'hippies' 218
historically disadvantaged individuals (HDIs)
    (S. Africa) 315–18
    empowerment 315
    equity ownership 315
    management participation 315
    participation targets 315
    women 315–17
HIV/AIDS 305
HM Government (and anti-corruption
    legislation) (2003) 124, 125, 127
HMNB Devonport (upgrade) 123
HM Treasury 13
Ho, Edmund 334
Hoeffler, A. 146
Hofstede 343
Hogg-Johnson, S. 274
Hohnen, Paul 361
Holcim Corporation 291
Holcim Switzerland (firm) 179
holistic design *versus* 'cherry-picking' 223
Holistic and systemic framework for
    sustainability 205–6
    ascendancy 199–200
Hollingsworth, M. 30
Holton, R. 320
homelessness charities 106
Hong Kong 18, 327, 328, 330–3, 341
    Census and Statistics Department 331
    construction industry 330–3
        economic overview 330–1
        injured workers 342
        key characteristics 331–2
        labour 331–2
        non-value-adding subcontracting 342
        professionals 332
        stressful working conditions 337–9
        structure 331–2
    corporate social responsibility 332–3
    employment 337–8
    gross domestic product (GDP) 330–1
    Hong Kong Land 331

housing   323
languages   332
population   330–1
property crash late 1990s   338
small medium enterprises (SMEs)   332
as a Special Administrative Region of
   China   330–1
Swire, Sun Hung Kai and Hong Kong
   Land Consortium   340
   Tai Ho site into an ecology park   340
working hours   337
workloads   337–8
Hong Kong International Airport   273, 277
Hong Kong Trade Development Council
   (2005)   333
Honours (Prevention of Abuses) Act
   (1925)   121
Hooper, A.   238, 252
Hopkins, A.   268
Hopkins, M.   54–5, 66
Ho, Stanley   333
House Builders' Federation   251
house builders-regulatory environment   242
house building   17, 238–41
   changing the product   247–9
   key delivery responsibilities   245
   response to CSR   245
house building-culture and practice   250–4
house building-policy challenges   243–4
house building-production capacity   243
house-building profits   239
House of Commons (2004) Housing:
   Building a Sustainable Future   226
House of Commons Environmental Audit
   Committee   241–2, 243, 254
House of Lords (2005)   155
housing deficit in Brazil   288
   inadequate housing   288
housing problems   84
   affordability and sustainability   84
housing provision in the UK   38
housing and shelter   76
Howard Evans (Roofing) Ltd.   185
Howard, John (Australian Prime Minister)
   367
Howells, Kim   38
Hoy, Chris   249
Hua, H.   336
human beings and the environment   200
   ecological worldview   200
   neoclassical view and the anthropocentric
      ethic   200
human capital   64, 67
human dignity   322
Human Rights Act   64
Hunt, A.   274
Hunter, K.   13
hydropower and corruption   154

Ideal (firm)   240
Idowu, S   55
IG desiccant (chemical)   184
illegal buildings and settlements   220
illegal hardwoods   216, 231
illegal immigrants (USA)   295
   death rates for Mexican workers   296
   employment rituals   295
   low income   295
illegal payment mechanisms   122
Illegal workers   174, 373
image of construction in America
   (negative)   296
   the three Ds   297
   Wall Street Journal Almanac Poll of high
      school students   297
'The Image of the Industry'   108, 114
   improvement   108
image of women in construction   64–5
IMI group (firm)   183–4
impact of corrupt practices   141
impact of the industry on the public   115
imperatives for change   263
implementation of community policy   103–5
Implementation Manual on the Use of
   Targeted Procurement to Implement an
   Affirmative Procurement Policy
   (APP)   314
implementation of mitigation strategies   82
imports of ecological building materials   215
improper payments   371
inappropriate infrastructure programmes
   83, 85, 91
income tax losses   295
'incremental capital–output ratio' (ICOR)
   78, 88, 95
   high and low   78
   and skills   78
India   81, 85, 152, 347
   Maheshwar dam   152
Indian Ocean tsunami disaster
   (26th December 2004)   85–8
Indonesia   347
Indoor Air Quality (IAQs) standards
   216, 224
   toxic paints   216, 224
industry collaboration   278–9
industry (definition)   168
industry initiatives   374–6
industry structure   168–70
inequality in Latin America   297
inflated tender prices   365
influence of the alternative building
   movement   215–16
influences of the built environment   9
informal market and tax avoidance   292
Information Portal on Corruption in Africa
   (IPOC)   155
infrastructure   58

infrastructure constraint 73
infrastructure development 75
Inglehart, R. 197, 205
innocence 59
innovation 241
innovative forms of building 220
insider trading 344
instability of the construction
    industry 100
Institute of Chartered Accountants (New
    Zealand) (2002) 58
    market information and function 58
Institution of Civil Engineers
    (ICE) 83, 143
Instituto Ethos 292
insurance 372
'integrity manuals' and procurement
    procedures 151
interaction between social capital and
    ecological capital 200–2
inter-company corruption 123
interface of value chains 76
intergenerational equity 198
    elimination of poverty 198
internal audit strategy 372
International Council for Research and
    Innovation in Buildings 93–4
International Federation of Building and
    Wood Workers (2004) 9
International Federation of Consulting
    Engineers (FIDIC) 151–2, 155
    anti-corruption and business integrity
        151–2
    code of conduct 151
    consequences of corruption 151
    global figures 152
International Labour Organisation (ILO)
    41, 74, 262, 288, 290–1, 295
    child labour in construction and
        brick-making 291
    convention 100, (1951) 288
    convention 111, (1958) 288
    convention 135, 295–6
    convention 138 and Recommendation 146
        (1971) 290
    convention 87 (1948) 295
    Convention on the Elimination of the
        Worst Forms of Child Labour 289
    Declaration on Fundamental Principles
        and Rights at Work 1998 41
    International Programme on the
        Elimination of Child Labour
        (IPEC) 290
International Monetary Fund (IMF) 29,
    91, 158
international perspectives on the CSR
    debate 14
international pressure and construction 141

International Standards Organisation
    (ISO) 12
    Guidance on Social Responsibility
        (ISO 2600) 12
Internet 61
investment for capacity expansion in
    construction 73
    and employment generation 74
investment in construction (as a percentage
    of all investment) 73
investment and ethical indices 56
investors in People 44
    workforce diversity 44
Iraq 148
    war 41
Ireland 352
Irish Republic and planning corruption
    127–8
    1997 Flood Inquiry 128
'Iron Law of Responsibility' 25
Ison, S. 67
ISO 9000 292
Italy 146

James Hardie Industries Ltd. (JHIL) 261,
    271, 275–7
    Amaba and Amaca subsidiaries 275
    Asbestos trust fund scandal 275–7
James, P. 274
Japan 151–2, 277, 335, 362
Japanese construction industry and bid
    rigging 127, 130
    Dangó 127, 130
    'descent from heaven' 127
Jennings, A.V. (firm) 360
Jensen, M.C. 327
Jensen, M.E. 206
Jia, S. 336
job creation 76
Johannesburg Plan of Implementation 84
Johnson, G. 172
Johnson, H.L. 26
Johnstone, R. 260, 267–8, 277
Joint Contracts Tribunal (JCT) 133
    New engineering contract (NEC)
        provisions (edition 3) 133
    Core Clause 133
    Standard Building Contract with
        Quantities (SBCIQI) 2005 133
    and corruption 133
Joint Parliamentary Committee (on
    corruption) 125
joint-venture arrangements 80–1
Joint Ventures (JVs) (S. Africa) 320
    equity JVs 320
    foreign investment 320
    special purpose vehicle 320
    technology transfer 320
Jolly, J. 199

Jones, P. 4, 14
Jonker, J. 5
The Journal of Construction in Developing Countries 93
Jorg, N. 268
Jowitt, P. 83

Kaatz, E. 321
Kajimo-Shakantu, K. 316
Kalra, R. 56
Kang, B. 143
Kanter, R. 62
Kantian moral imperative 100, 108
Karekezi, P. 83
Karn, V. 253
Kast, F.E. 26, 33, 36
Karungu, P. 319
Kay, J. 169
Keat, R. 28
Kelly, J. 13
Kennerley, J.A. 112
Kenya 83
'kerb appeal' 253
Kessides, C. 89
Keynesian economics 25, 27, 29, 40, 48
'Keynesian multiplier' 74
key performance indicators (KPIs) 33, 43–4, 83, 245, 246, 272
  sustainability 245
Kim, S. 79, 81
Kimmins, S. 217
King, M. 142
Kletz, T. 270
Knauf (firm) 178
knowledge of construction specifications in South and Central America 292
knowledge movement 74
Kobe earthquake (1995) 130
Ko, K. 341
Korhonen, J. 271
Kota Kemuning project 340
KPMG survey (2005) 102
Kramer, M.R. 7, 8, 54
Kreps, T.J. 25
Kruse, D.L. 289

labour-based construction 93
labour and environmental practices 328
Labour Force Survey (Stats SA) 305, 322
Labour Government 1997 32
labour-only subcontracting 31
Labour-only subcontractors 122
Labour Party 29
labour violations 329
Lafarge (firm) 178
Laing Homes (firm) 248
Laing Partnership Housing 37
laissez-faire 25
Lall, S. 79, 92

'lame ducks' 28
Lammas (2007) 219
The Lancet 267
Land for People 219
Land Securities 236
land shortages 241
Langford, D. 169, 173
Langford, D.A. 30, 76, 79, 80, 82, 88–9
Langseth, P 149
large-scale infrastructure projects
  negative consequences 294
Larsson, T 262
Las Vegas 333
Latham, M. 11, 101, 108, 357
Latham Report (1994) 138, 321, 354
Latham Review Implementation Forum (RIF) 108
Latin American construction challenges 288
Latin American construction regulations 287, 288
  low level social responsibility 288
  weaknesses 288
Laurenson, J. 141
Law, Ben 220
  The Woodland House 220
Lawrence, P.R. 26
laws and regulations 376
Leão-Aguiar, L. 4, 14
'lean thinking' 42
legacy of architects and engineers 9
legal safeguards for workers 294
Legge, K. 28
Leighton Holdings (firm) 263
Leishman, C. 250
Lend lease Corporation 378
Lenssen, N. 191, 194
'lentils and sandals' 218–19
Lerer, L.B. 83, 158
Lesotho 154–7
  High Court 155
Lesotho Highlands Water Project (LHWP) 154–9
  Canadian involvement 157
  Chief Executive (CE) 155
  conflict of safety and progress 157
  contribution to GDP 155
  corruption and deceit 155
  costs and completion in 2020 154–5
  environmental action plan 157–8
  health and safety issues 157
  Katse dam and reservoir 155, 157
  seven company consortium 155
  Workforce and the spread of HIV/AIDS 158
Lesotho Highlands Water Project trials 136, 143
Lessard, D.R 9
Lewis, J. 142
Lewis, M. 146

Lhoist (2007) 231
Liberal–National Coalition Australia 344
Liedtka, J. 100
Life Cycle Analysis (LCA) tools 224, 230
'Lifting the Burden' 32
Li Ka Shing (Foundation) 342
Lim, E. 335
Lime mortars 229
Lime Technology (2007) 231–2
Limpopo province (S. Africa) 316–17
Lindley 1957 case 125
Lindsay, N. 262
Lingard, H. 17, 18, 274, 373–4
Lingard, H.C. 338–9, 344–5
Litigious society 56
Littlehampton 254
Liu, A. 4, 14
Liu, L. 336
Liu, G., 149
Living Villages 222
Lloyd George, David 121
local and international construction firms
    78–9
local and regional government 47
local stakeholders 83
London Business School 327
London Olympic Games (2012) 133
long-term needs 75–6, 87–8
    and disasters 86–8
Lorsch 26
low-cost housing on a self-help basis 84
low-impact development policies 220
low-income housing 298, 300
low wages 81
low-and zero-carbon development 220–1
Luang, Tay Kay 339, 340
Luo, J. 320
Luiz, J. 318–19
Lynch, M. 76, 83
Lynch, R. 122

Macalister, T. 254
Macao (now Macau) 18, 327, 331–9
    as a centre for the gaming industry 333
    Centre for Sustainable Development
        Strategies (CEEDS) 334
    changes in labour law 338
    corporate social responsibility (CSR) 334
    GDP 333
    government 334
    internationalisation of the economy 334
    languages 333
        Chinese dialects 333
        old Macanese language 333
        Portuguese 333
    population 333
    as a Special Administrative Region of
        China 330–1
McCarthyism 29–30

McCarthy, Richard 225–6
McCully, P. 154, 158
McGuire, J.W. 25–6
McIntosh, G. 274
McLean, C. 63
McNamara, C. 56
McWilliams, A. 35–6
McWilliams, G. 262, 372
Macau construction industry 333
    economic overview 333
    expenditure 333
    less stressful working conditions
        338, 339
Macauley, M.J. 126
Macau S.A.R. Statistics and Census (2007)
    333–4
macho culture 289
Maclagan, P.W. 311
MacPherson, S. 84, 92
Madi, P.M. 317
Mahtani, S. 337
Mahon, J.F. 35
Maignan, I. 312
    cross-cultural context 312
Maignan, I. 271
mainstream building needs 218–19
maintenance of rail infrastructure 65
Major Projects Association (2001, 2006)
    142, 273
'Make Poverty History' wristbands 63
Malay language 330
Malaysia 18, 85–6, 327, 346–7
    code of ethics for construction 346
    competition and cartels 339–40
    construction Industry Development
        Board 346
    Master Builders 346
Malaysia 339–41
    financial performance 340
    Gamuda Land 340
    'Garden City, Intelligent City' 339
    long view 339–40
    prosperity and construction 340
Male, S. 16
Male, S.P. 170
management of social issues by organisations
    98–9
managerialism 24, 42–3
    and codification of professional knowledge
        33–4
Manchester 217, 254
Manchester airport 101
Manchester Evening News property
    awards 66
Mandarin (Putonghua) language 331
man-induced disasters 85
    absence of CSR 85
    war and conflict 85
Mansfield, N 76, 80, 88

Maree, J.   311
marginalist reformers   201
Margolis, D.M.   8, 13, 61
Maria Auxiliadora   291
market (definition)   168
marketing and CSR rhetoric   62–3
marketing and green claims   230
market structures   168
Marks and Spencer (M&S)   60–1
   carbon neutrality   61
   five main issues   61
   eco-plan   60–1
   'Plan A'   60–1
Martineau, Johnson   182
Marxist/Leninist   36
Maslach, C.   37
Massie, R.K.   309–10
Matthews, J.   60
Matthews, P.   76, 83, 81
Mbigi, L.   311
Medical Research and Compensation
   Foundation (MRCF)   275–6
Médecins Sans Frontières (MSF)   342–3
Meghji, M.R.   16
Melbourne   377
Menzies, Walter   226
Mexican and Mercosur markets   287–8
Mexico   329
Michael, B.   149
middle-management skills   308–9
Midgley, M.   200
migrant workers   31
migration   64
Miles, D.   33, 80, 92
Millennium Development Goals (MDGs)
   77–8
   compliancy   76
   the eight goals   76–7
Miller (house builders)   244–5, 248–9
Miller, R.   9
Mindy, H.   101
Miners' Strike 1984   29
Minister for Employment and Workplace
   relations   367
Ministry of Defence (MOD)   248
Miralles, Enric   123
Mitrovic, D.   170
Mkapa, Benjamin (Tanzanian Prime
   Minister)   149–50
Mkono, N.E.   150
model of societal-ecological interaction
   192–3
modern management strategy   255
modern methods of construction   248–9
Monbiot, George   222, 269
money laundering   371
Monti, Mario   178
Moodley, K.   98–116
Moon, C.   56, 60

moral conscience   108
moral courage   376
moral education   159
'moral hazards'
   in disaster relief   88
   in infrastructure   91
Morin, E.   201
Mostakova-Possardt, E.   106
Mott Macdonald (2002)   9
movement of ecological/green lifestyles into
   the main stream   221–2
Movement for Innovation (M4I)   43, 278
Mueller Industries   184
multinational organisations   277
Multiplex (firm)   354
Mundey, Jack   359
Murray, J.   369
Murray, M.   15, 16, 43
myopic thinking   27

Naiker, K.   306
Napier, Robert   225–6
Naser, K.   312–13
National Audit Office (NAO) (1992, 2005)
   123, 153
national building strike (1972)   30, 48
National Code of Practice for the
   Construction Industry and
   Commonwealth Implementation
   Guidelines (Australia)   368
National Considerate Contractors Scheme
   (1994 and 1997)   108
National Construction Week   101
National Health Service (NHS)   182
   procurement   21, 170
National House Building Council
   (NHBC)   238
National Institute for Children and the
   Family   290
National Institute for Occupational Safety
   and Health (NIOSH) (USA)   296
National Lottery   41–2
National Non Food Crops Centre (NNFCC)
   (York) (2007)   232
National Small Business Act (S. Africa)
   318–19
National Traumatic Occupational Fatalities
   (NTOF) (USA)   296
National Trust (Australia)   360
National Union of Mineworkers (NUM)   29
National Vocational Qualifications (NVQ)
   122
'Natural Building'   215, 219
Natural Building Technologies (NBT)
   (2007)   222
natural insulation products   228–9
   sheep's wool *versus* fibreglass   229
Ndumbaro, L.   150
Neal, D.   268

Neale, R. 80
negative attitudes of workmen 101
neoclassical and ecological worldview matrix
    203–4
    Dependent social systems sustainability
        (DSS) 203–4
    Dominant product sustainability (DPS)
        203–4
    Ecosystem benefit sustainability (EBS)
        203, 205
    Ecosystem identity sustainability (EIS)
        197–8
    Ecosystem insurance sustainability (EIN)
        203, 205
    Global niche preservation (GNP) 203–4
    Global Product sustainability (GPS)
        203–4
    Human benefits sustainability (HBS)
        203–4
    Self-sufficient sustainability (SSS) 203–4
neoclassical economics 201–2
    resource substitution 202
neo-liberalism 27–8
Netherlands 275–6
New-Age mystical concepts of building
        design 219
    Feng shui 219
    geomancy 219
Newbury 101
Newby, T. 84
The New Civil Engineer 143–4
Newcombe, R. 321
New Hong Kong Airport at Chek Lap Kok
        142
new infrastructure 76
New Labour 29
    apparatchiks 4
    Government 1997 29, 41, 42
    managerialism 29
New South Wales 358, 359, 365
    Chamber of Commerce 361
    Commission Against Corruption 369–70
    government 276, 360
    Scaffolding and Lifts Act 358
    unlawful conduct 367
    strikes 367
    WorkCover (NSW) 368, 369
New Straits Times (Malaysia) (November
        2004) 59, 68, 339
New York State Organised Crime Task
        Force 127
New York Stock Exchange (NYSE) 292
    and business ethics 292–3
New York Times Magazine 264–5
Ngowi, A. 93
Nicaragua 291
Nicol, C. A. 238, 252
Nike (firm) 65, 271, 362
Nilsson, 34

nineteen eighties 34
nineteen fifties 25–6
nineteen nineties 35–6
nineteen seventies 27–8, 33, 48
    stagflation 35, 48
nineteen sixties 24–5, 33, 48
    social idealism 24, 26–7
Ning, L. 336
noise, pollution and dust 101
non-executive directors (S. Africa) 317, 318
non-governmental organisations (NGOs)
        65, 84, 92, 115, 124, 148, 152, 218, 291
    and development 84
    and disaster relief 85–8
    meeting infra structure needs 84
    'no-one got hurt so don't worry' 122
Norman, G. 171
Norton, B.G. 198–9
Notices of Satisfactory Assessment 370
Nottingham University 248–9
Nthako, S. 154, 155, 158
Ntsika Enterprise Promotion Agency
        (S. Africa) 319
Nurminen, M. 262

obligational contractual relationships
        (Asia) 321
Obrasnet Brazil 293
obstruction to roads and traffic 101
occupational health and safety (Australia)
        353, 367, 369
    performance 368
    workcover inspectors 365–6
occupational health and safety (OH&S)
        17–18, 261–83
    accountability 269
    corporate citizenship 271–2
    corporate social responsibility (CSR)
        264–5, 279–82
    costs 282
    culture 282–3
    economic theory 265
    education 281
    ergonomics 281–2
    fines 269
    law 267–70
        Australia and the UK 267
    management systems 272–3, 281
    morality 266–7, 281
    motivation 263–4
    organisational outline 269–70
    performance 262–3, 280, 282–3
    post Robens inspectors 267–8
    productivity 282
    standards 270–1
    supply chains 281
Occupational Safety and Health
        Administration (OSHAS) (2000)
        287–9

OECD and sustainability 197–8, 214
Oelschlaeger, M. 200, 206
Office of the Deputy Prime Minister
  (ODPM) (2003 and 2004) 42, 45–6,
  243
Office of Fair Trading (OFT) 10, 16, 134,
  136, 166–8, 180–6, 251
Office of Government Commerce 225
offshore banks and secrecy 155–6
Ofori, G. 72, 74–5, 83, 92, 93, 150
Olcayto, R. 115
Olympic Delivery Authority
  (ODA) 133
omission of insulation 223
'one planet living' campaign 221
  principles of 221
One Tel 362
open systems 26
open systems perspective 99
Operating and Financial Review (OFR)
  (2004) 58–9, 62, 67
organic farms 218
organisational interaction with the broader
  environment 26
Organisation for Economic Co-operation
  and Development (OECD) 41, 141–2,
  153, 155
  Anti-Bribery Convention 1997 142
  Convention 131
  Guidelines for Multinational
    Enterprises 41
Origin Energy (firm) 362
origins of the ecological building movement
  218–19
Outokumpu (Finnish firm) 184
outsourcing 31
Ove Arup Foundation 107
  learning in engineering and the built
    environment 107
Ove Arup Sir 107
Owens, Joe 359
Oxfam 87

Packham, G. 320
Papanek, V. 216
Parliamentary environment committee 226
partnering 173, 320–1
partnering between contractors, suppliers
  and sub-contractors 138
partnerships 67
Partners in Innovation Programme 228
Pauchant, T. 200
payment practices 376
payroll giving 41–2
Pearce, D. 9, 196, 198, 202
Pearce, D.W. 198, 202
Pearl, R. 369
Pearson, David 216
Peebles, L. 273

PEFC (2007) 230
penalties for corrupt parties 134
  accountability 130
  case studies 134
  confiscatory legislation 136
  tender blacklist for public sector works
    136
  'certificate of fitness to tender' 136–7
Peng, M. W. 320
people and development 83
people and profits 7–8
perception of corruption 147–8
Perham, Linda MP 57–8
perpetual poverty syndrome 88
Persimmon Homes 105, 238, 240, 244–5,
  248–9, 251
personal morality 143
Perspectives on Corporate Social
  Responsibility 4, 24
Peterson, D.K. 363–4
Petrovic-Lazarevic, S. 369
Petterson, T. 178
Pew Hispanic Center (PEW) USA 295
Pezzy, J. 202
philanthropy 3, 7–8, 15, 100, 103, 106–8
  cause related 8
  strategic 8
(The) Philippines 343, 348
Philippines' mini-hydro project 153
photovoltaic cells 223
Pidgeon, N.F. 270
Pieth, M. (FIDIC) 151, 158
'pirate developers' 86
planning criteria 225
planning restrictions 220
  rural areas 220
  twenty four hour farming and woodcraft
    220
Pollington T. 14–15
Poon, J. 4, 14
poor investment decisions for
  infrastructure 88
Port Elisabeth 317
Porteous, D. 306
Porter, M.E. 7, 8, 54, 100, 168–9, 170–1
post-carbon economy 219
post-conflict reconstruction 294
post-disaster recovery 15
post-disaster relief works 6
  high costs 86
post–peak oil era 219
Pouliquen, L.Y. 82
poverty alleviation, disaster management and
  construction works 75–85, 87, 93
poverty reduction 15
Prahalad, C.K. 301–2
Preece, C.N. 15, 115
preferential procurement plan (SA) 315–16

Preferential Procurement Policy Framework
   Act 322
Prescott, John 42
President's (USA) Council on Sustainable
   Development (1996) 196
Pressure, State, Response Model (PSR) 17,
   206–8
   CSR 210
   framework for change 209
   gap analysis 208–9
   stakeholders 208
   systemic nesting of scales over time and
      space 206, 207
   units of analysis 208
'prestigious projects' 91
Preston, L.E. 98–9
Pretty, David 225
previously disadvantaged persons
   (PDIs) 316
price fixing 16, 122, 170
Price, A. 15, 54–68
Price Waterhouse 321
Price Waterhouse Coopers (PWC) 155
Prince of Wales 38–9
principles of alternative building 215–16
Pringle, B. 359
Prior, G. 123
Private Finance Initiative (PFI) 13, 67, 91,
   170–1
   compatibility with CSR 13
   and infrastructure 91
   in Scotland 13
private financial investment in infrastructure
   90–1
proactive response to environmental
   problems 110
   six possibilities 110
Proceeds of Crime Act 136
Procurators Fiscal (Scotland) 124
Procurement Act (South Africa) 313–14
procurement routes 170–1
professional attitudes to sustainable building
   227–8
   accredited architectural training 227
   education 227
   integration of sustainable building
      practices 227
professionalism 33–4, 48, 345
   conflict with managerialism 33–4
   connection with social responsibility 33
   criticism from the left 33
   decline 33–4, 44
professional and managerial Staff 373–4
   hours worked 373
   working conditions 373
professionals 33
   associations 33
   definitions 33

firms and CSR in the construction
   sector 33
profit and business ethics 59–61
profits as a priority 34–5
Programa Das Nações Unidas Para O
   Desenvolvimento (PNUD) 288
Programa de Subsidío à Habitação de
   Interesse Social (Brazil) 298
   families building their own houses 298
project financing for infra structure
   development 88
project implementation failures 89–90
project performance 273
projects beneficial to society 9
projects and good business 65–6
projects not beneficial to society 9
   criteria for effective projects 9
   optimism bias 9
property developers 169
property development 345–6
   collision between CSR and ethics 346
property prices 220
   second and retirement homes 220
pro-poor infrastructure projects 78
Public Concern at Work 131–2, 136
public impression of sustainable buildings
   219–20
public information meetings 112
Public Interest Disclosure Act (1998, 1999)
   HMSO 13, 130–1m 183
Public–Private Partnership (PPP) (S. Africa)
   318, 320, 321
   prison initiatives in Malmesbury and
      Bloemflontein 320–1
Public–Private Partnerships (PPP) 48, 62,
   67, 91, 170–1, 322, 369
   engulfing illusions 9
   and infrastructure 91
Public Procurement Directive (Scotland)
   136
public sector confrontation 251–2
public sector procurement 13
   and sustainable development 13
public sector procurement reform and
      CSR in the SA construction industry
      313–16
   policies and characteristics 313–14
Pulliam, H.R. 206
Putnam, R. 64
Putrajaya Holdings Sdn Bhd (firm) 339

Qatar 312–13
quality of life at work 296–7
Quality Management System (QMS) 264
quality programmes in construction 296
'quasi-normality' of corruption and
   dishonesty in construction 121
Queensland, Australia 339, 353
Quito 86

racial inequality in Brazil 288
Rahman, A. 198
Rameezdeen, R. 4, 14
Ramsay, G. 84
Randles, L. 15, 54–68
rapid urbanisation and fast track
    construction 86
  inherent risks 86
'rational economic man' 27
Ray, R.S. 36
Rayman-Bacchus, L. 4, 7, 58
Reagan, Ronald 27–8
Reale, D. 154
Realpolitik 41
reasons for accidents 279
rebadging of construction vehicles 123
recognition of corrupt practices 144
Reconstruction and Development
    Programme (RDP) 305–6, 314
Reconstruction and Development
    Programme (DPW) 313–14
Redressing the inequalities of the past
    (SA) 314
Redrow plc (house builders) 244,
    247–8, 249
Reed, D.L. 27
Rees, J. 201
Rees, William 221
Reeves, R. 36
regeneration 111–12
  new generation schemes 111
  in urban areas 111
Registered Engineers for Disaster Relief
    (RedR) Engineers 86–8
  mission statement 87
regulation versus exhortation 227
regulatory relief 59
rehabilitation 274
Reichman, O.J. 206
relief after catastrophes 294
repair, maintenance and improvement (RMI)
    184
Reporting of Injuries, Diseases and
    Dangerous Occurrences Regulations
    1995 (RIDDOR) 11
reporting mechanisms 371
representation agreement (RA) 156
  'no duck–no dinner' clause 156
Republic of South Africa 306, 313
  The Broad Based Black Economic
    Empowerment Act 306
  The Construction Industry Development
    Board Act (CIDB) 307–9, 314, 317
  creating an enabling environment for
    reconstruction, growth and
    development in the construction
    industry 307, 313–14
  Department of Public Works (DPW) 307,
    313–14, 316–17

The Preferential Procurement Policy
    Framework Act 306, 322
RepuTex (CSR) 377–8
research priorities for construction in
    developing countries 93
respect 376
Respect for People initiative (2002) 42,
    43–4
  five consolidated themes 45
  minimal contribution to CSR 42–3
rethinking construction (2002) 42–9, 138,
    264, 277–8
  as an epitome of managerialism 42–3
Revitalising Health and Safety in
    Construction 278
'revolving door syndrome' 318
Rice, A.K. 26
Riely, S. 149
Rigby, N. 271
Rio Declaration at the Earth
    Summit 200
risks of investment in developing
    countries 91
Ritchie, J. 152
Riverson, J. 82
Robens Committee (1972) 267
Robens, J. 344
Roberts, J. 345
Robertson, C. 277
Robinson, S. 106
Roddick, Anita 59
Rogers, D. 122
Rogers, Peter 225–6
Roitter, M. 298
Roodman, D.M. 191, 194
roofing contractors 184
Roosevelt New Deal 25
Roosevelt, President USA 1941 289
  Executive Order 8802 289
Root, D. 316, 321
Rosen, N. 28
Rosenzweig, J.E. 26, 33, 36
Rossouw, G.J. 309, 310
Rosthnor, J. 36
Rough Sleepers Initiative 106
Rowan, J.J. 266–7
Rowland, V.R. 80
Rowlinson, S. 18
Royal Commission into the Building and
    Construction industry 262
  ethical responsibilities 365
  illegal activities 365
Royal Commission into the Building and
    Construction Industry (Australia)
    358–9, 363, 365–7
Ruf, B.M. 35–6
Rule of law (Australia) 351
rural decline 220
rural infrastructure projects 82

Russia 328
Rwelamila, P.D. (199, 2002) 312, 313
Rwigema, H. 319

safety record (Australia) 372
safety training (case study-Australia)
    369–70
safety training programmes 92
safe water supplies 76
Sadorsky, P. 265
St. James Ethics Centre 377
Sako, M. 321
Salaman, G. 28
sanitation 76
Santos A.D. 290
Santos (Dos) A. 18
Saravanamuthu, K. 270
Sassi, P. 4, 14
Saunders, A. 274
Saunders, M. 172–3
Save as You Earn 240
Scandinavia 158–9
Schein, E. M. 269–70
Scholes, K. 172
Schwartz, M. 199
Scottish Business in the
    Community 240
Scottish Ecological Design Association
    (SEDA) 217
Scottish Government 122, 123, 130
Scottish Government (formerly
    Executive) 47
Scottish housebuilders 242
Scottish Parliament building,
    Holyrood 123
Scudder, T. 83, 158
Secure and Sustainable Building Act (2004)
    225
    steering committee 225
security, health and safety 101
self-build in the countryside
    219–20
    cob (mud and straw walls) 219
    'long-haired self-builders in the
        countryside' 219
    planning and building control approval
        (or lack of) 219
    straw-bale walls 219
self-employment 31–2, 48
    growth and incentivisation 31, 48
Sen, A.K. 85
Serious Fraud Office 180
Seven One Four (714) certificate 31
Severe Acute Respiratory Syndrome epidemic
    (SARS) 329, 331
Sexton, M. 16
sexually transmitted diseases 83
Shakantu, W. 316, 317
Shanghai 338

Shantou province (China) 342
shareholders 15, 27, 55, 61–2, 68
Shaughnessy, H. 61
Shen, Q. 149
Sherwood, B. 181–2
Shi, P. 336
short-term needs 73–75
    and disasters 87
Shrivastava, P. 264
Sichuan Province 149
Siebert, H. 206
Siegal, D. 35–6
significance of construction in the American
    continent 287
Sillanpaa, M. 100
Silva, J.D. 76, 87
Simmonds, M. 13
Simmons, I.G. 199
Sin, D. 342
Singapore 18, 329, 348
    BCI Asia (2007) 330
    construction growth 330
    corporate social responsibility
        (CSR) 330
        National Tripartite Initiative (2004)
            330
        Singapore Compact 330
    financial 329
    GDP 329–30
    high technology industry 329
    languages and culture 330
        Marina Bay 330
        Ministry of Trade and Industry 330
    pharmaceuticals and medical technology
        329
Singapore construction industry 329, 330
Singleton, D. 93
Sinha, R. 146
site environment 56
Site Waste Management Plans (SWMP) 12
Skanska (Sweden) 12
skill and knowledge gap 92–3
skills development 102, 105, 111, 169, 239
skills shortage 63–4, 67
skills shortages in poor countries 88–90
skills for sustainable communities 42, 45–6
Skitmore, M. 364
Slough Estates 225
small to big corruption 144
small business classification (S. Africa)
    318–19
Small, C. 37
small and medium enterprises (SMEs)
    44, 122
    unjustified blame 44
small, medium and micro enterprises
    (SMMEs) (S. Africa) 315–16, 318, 319
    outsourcing 318–19
Smallwood, J.J. 261–82

Smith, Adam 25
Smith, C. 5, 8
Smith, G. 368
Snell, R.S. 29
Snider, J. 369
social accountability 8000 36
social capital 64
Social Contract years (1974–1979) 32
social-democratic consensus 24, 28
social and financial accountability 60
social and financial performance 8
social housing 36, 62
social impacts 377
'social licence to operate' 83
socially responsible investment (SRI) 362–3
social obligations 40
social responsibilities 56, 66, 275
Social Responsibilities of the BusinessMan
    5, 25
social responsibility 24–6, 214, 231,
    276–7, 363
  condemnation of 29–30
  DLOs 31
  four-part taxonomy 34–5
  links with profitability 35–64
  lack of evidence 35–64
  and three areas of business success 38–9
social responsibility and competitive
    advantage 287
social responsibility and design implications
    298, 299
  example of roof structure 298, 299
  pre-assembly in Brazil 298
social responsibility and education 297
social responsibility (of firms) 98–100
  dimensions 99
  slow response in construction 102
social responsibility and organisational
    structure 292–3
social responsibility and organisations 10
  across the decades 10–11
social responsibility in the USA 287
social responsibility and working at the base
    of the social pyramid 301
Societa Metallurgica Italiana (firm) 184
Society of Afghan Engineers 144
Society of Construction Law 130, 131
'socio-cultural system' 26
socio-ecological system interaction 193,
    194, 197
  built environment and construction
    industry contribution 193
  depletion of natural resources 194
  global situation and population growth
    194
'soft regulation' 35, 37, 39, 47, 49
Sole, Ephraim 155–7
solidarity of construction through
    community involvement 297, 298

Solow, R.M. 198, 202
solvents and additives 229
Somalia 148
Sommer, F. 102
South Africa 154, 161, 221, 264,
    311–22
  benefits 312
  business responsibilities 310
    community 310
    ethics 300, 301
  civil society 310
  commitment 311
  context 304
  corporate citizenship 312
  corporate rules 313
  corruption 305
  crime 305
  definition 311
  Department of Transport and Public
    Works 314–15
  development strategy
    304, 321
  disclosure 312–13
  diversity 313
  economic, legal, ethical and philanthropic
    responsibilities 311
  economic growth 305
  employees 312
  employment 305, 308–9, 313
  enabling strategy for the construction
    industry 314–15
  environment 313
  equal opportunities 307
  equity 313
  ethics 313
  governance 313
  gross domestic product (GDP) 308
  gross fixed capital formation (GFCD)
    308
  housing 304, 306
  income 305
  investors 312
  knowledge and skills 309
  labour 305
  local government reform 304
  media 310
  morals and morality 309
  municipalities 304
  National Treasury 306, 308,
    317, 320
  natural resources 304–5
  non-governmental organisations NGOs
    310
  parastatals (state owned companies) 306,
    308
    Eskom 306
    Telkom 306
    Transnet 306
  partnering 312

South Africa (*Continued*)
  philanthropy 312
  population 305
  post-apartheid 18, 304
  poverty 305
    reduction 307
  PPP 320
  Preferential Procurement Policy
    Framework Act 314
  proactive responsible view 311
  project managers 272
  public opinion 311
  quality of life 307
  Reconstruction and Development
    Programme 18
  redressing the inequalities of the
    past 314
  revenue services 316
  settlements 304, 306
  stability 307
  technology 309
  ubuntu 18
South African construction industry 143,
  304, 307
  processes and participants 304
  procurement 304, 307
South African government as a regulator
  306–7
South African Reporter (2004) 154–5
South America 18, 86, 298–9
South Australia 353
South/Central American construction as
  labour intensive 297
South China Morning Post (2005) 273,
  341–2
South Korea 79
South Summit (2000) 328
south west England 225
  voluntary sector sustainable
    buildings 225
Soviet Union 36
Spain 61
speculative housebuilding
  industry 237
speculative housing 251
spurious claims 121
Sri Lanka 86–7, 153
  Samanalawewa dam 153
stakeholder approach 26
stakeholder awareness 56
stakeholder models 99–100
  and ethics 100
  instrumental approaches 105
stakeholders 3, 5–7, 9, 10, 14, 15, 17, 27,
  57–8, 61–8, 114, 191, 195, 198–9, 203,
  208, 293
  broad and narrow definitions 27
  and ethical practices 3
  reports in the 1980's 5

stakeholders and the community 249
stakeholders' expectations 54–5
  external 55
  interaction 55
  internal 55
stakeholder theory 17–18, 24, 27, 48,
  99–100, 265
  aspects of the model 99
  definitions 99
  intrinsic-worth approach 108
  and social issues 99
Stalker, G.M. 26
Standard Assessment Procedure 247
standardisation and pre-assembly 43
  de-skilling consequences 43
Standards Australia 264
Stanhope plc 225–6
Stansbury, C. 121, 144, 157
Stansbury, Neill (2005, 2007) 121,
  144, 159
Staples, J. 123
state participation in infrastructure delivery
  89
Statistics South Africa (StatsSA) 305, 307
Stead, E.W. 202
Stead, J.G. 202
steel, glass and concrete 231
Steinbeck, John 40
Stern, P.C. 199
Stonehenge 218
Stoner, C. 100
Strategic Forum 42, 43–4
Strategic Forum for Construction 217
Strategic group 169
Strike pay 367
sub-Saharan Africa 74, 78, 82, 89–91, 95
  and ICOR 78
  Sahel Region and disasters 85
Suen, H. 4, 14
Suffolk 231
summer solstice 218
Sunter, C. 305
'super-builders' 257
suppliers and sub-contractors 271
supply chain 170–3
supply chain and CSR 10
supply chain management plan 246, 247
supply chain practices 54, 56, 62, 66–8
supra-system resilience 192
'survival of the fittest' 28
sustainability 6, 9–11, 14, 16, 17, 255
Sustainability Alliance of Professions 217
sustainability and alternative construction
  215, 218
sustainability as a devalued marketing
  term 214
Sustainability Forum 217
sustainability reports 102
Sustainable Building Association 216

Sustainable Building Code Steering Group
  (ODPM) 225, 226
sustainable building principles 222–4
  four main aims 223
Sustainable Building Task Group 225
Sustainable Communities (2004) 45–6,
  249, 250, 253–4
Sustainable Communities Plan 243–4
sustainable construction 54–5
sustainable development 11, 73, 75, 82,
  191–210, 236, 347
  constraints 196
  definitions 195–7
  development path 196
  differing objectives 196, 197
  focus and goals 195–6
  measures of performance 196
  principal elements
    endurable, appropriate progress 197
    equitable 197–8
    participatory 198
    socio-ecological system perspectives
      197
  sources of motivation and legitimacy 196
  worldviews 198–205
Sustainable Development Commission 226
Sustainable Homes Code standards 221
sustainable infrastructure 73, 89
'sustainable livelihoods business' 300
sustainable rural development 220
Sustainable Strategies (Amsterdam)
  (firm) 361
sweatshops 63
Sweden 230
Switzerland 215
Sydney (Australia) 359
  Botanical Gardens 360
  Centennial Park 360
  Hunter's Hill 360
  Kelly's Bush 360
  The Rocks 360
  Woolloomooloo 360
Sydney Morning Herald 37, 275–6, 370
systems 193–4
  ecological 193–4
  social 193–4
systems theory 27
  thinking 26–7

Tamil language 330
Tanzania 148–50
  bribe payments, 'grease or grit'? 149
  and China 148
  innovative anti-corruption measures
    (1995–96) 149–50
  'petty type' and 'grand type' corruption
    149–50

Tanzanian construction industry 148
  growth 1997–2005 150
  training requirements 150
targeted procurement (SA) 314
Tarmac Contractors 11
Tarmac in the Environment (First
  Report) 11
Tasmania 353
Tax evasion 336, 371
Taylorism 27
Taylor Woodrow Construction 228
Taylor Woodrow and Wimpey 240,
  244–5, 247–9
technological change 29
technology transfer 76–9, 92–4
Telford, Thomas 9
  roads in Scotland 9
  transgressions in social responsibility 3
tendering practices 376
Tepees 218–19
Thailand 328, 343
Thames Gateway area 221
Thatcher, Margaret 27–9
  government 32
  'there is no such thing as society' 27–9
Thatcher years 29, 65
Theft of plant, equipment, intellectual
  property, information 365, 370
Theodore, L. 4, 14
Thermoseal Supplies Ltd. (firm) 184–5
Thomas, P. 65
Thompson, M. 199
Thunstrom, M. 178
timber and forest management 230
timber and sustainable resources 60
timber from sustainable sources 12
time and cash-flow problems 89
time and cost-adverse effects 215
time–space compression 29
Timms, Stephen 39
Todeschini, F. 306
Tomlinson, R. 304
Tookey, J. 16
Towler, B. 55
Townsend, M. 320
toxic and carcinogenic chemicals 229
  EU legislation 229
Toyota Manufacturing System 42
Trade Practices Act 371
trades unions (in Australia) 35–60, 372
  achievements 358–9
  heritage buildings 359
  officials 367
    contractual obligations 368
trades unions rights 295–6
training (Australia) 373
training, education and research for
  sustainable construction 92–3
transition cost economics 334–5

Transparencia Brazil 294
  Rio de Janiero 294
  Santa Catarina Court of Audit 293
transparency 362
Transparency International (TI) 16, 132–5,
  137, 141–9, 155, 158–9, 165, 170–1,
  175, 186, 293–4, 362
  Bribe Payers Index 362
  Corruption Perception Index (CPI) 143,
    147–8
  economics and corruption (2004) 294
  five global priorities in the fight against
    corruption 144–5
  history, work and aims 144–5
  'Integrity in the Chinese Construction
    sector' 149
  Preventing Corruption on Construction
    Projects 159
  'Promoting Transparent Procurement and
    strengthening CSR' project 149
Transparency International (TI) (UK)
  121–2, 124, 128–30, 132, 141,
  143–8, 159
  examples of serious consequences 130
  four anti-corruption measures 145
  three objectives for the prevention of
    corruption in construction 145
transparency and political contributions
  294
transparency and social reporting 293–4
Transurban (firm) 362
Tremayne, C. 30
'trickle-down' economics 27, 28
Trinidad and Tobago's Piarco airport 142
  corrupt tendering 142
Trinidad and Tobago Transparency Institute
  (TTTI) 142
'Triple–Bottom-Line' case study 43–4
Triple-Bottom-Line (profitability,
  environmental and social responsibility)
  6, 10, 273–4, 282–3
  and small firms 10
  Environment, economy and sustainability
    55
Trist, E.L. 201
trust in business relationships 343
  China, Thailand and Hong Kong 343
trust between government and business
  10, 11
Tsunami (December 2004) 342–3
Turkey 146, 329
  Ilisu dam 152
Turner, B.A.
Turnkey projects 79
Turton, A. 155–6
Twyford Down (M3 Motorway) 11
TXU (utilities company) 362

Ubuntu (respect and human dignity) 304,
  311–13, 321, 322
UKae Ltd. (firm) 184–5
UK and the alternative construction
  movement 215–16
UK construction sector 24
UK as a flawed model about corruption for
  other countries 121–3
UK Government 112
UK Government and corruption 124, 137
  Laissez-faire approach to anti-corruption
    legislation 132
  White Paper (2000) 124
UK Government policies on sustainable
  building 226
UK Government Sustainable Building Task
  Group 217
UK house building-structure and
  organisation 238
UK measures against corruption 130–7
  legislation 131
UK Office of Fair Trading (OFT) 122
  OFT (2004a, 2005) 122
UN agencies 83
United Nations Declaration of 1974 84
'unauthorised workers' in the USA 295
UN Convention against Corruption 126
unemployment 28
unemployment in South Africa 307
unethical practices in the procurement
  process 16
unfairly bid tenders 10
unfair trading practices 54
unfeasibility of eliminating corruption 124
unfired earth bricks and blocks 229
unimaginative and repetitive design
  252–3
Union 358
  Australian Workers' Union (AWU) 358
  Communications, Electrical, Electronic,
    Energy, Information, Postal,
    Plumbing, and Allied Services Union
    of Australia (CEPU) 358
  community and the environment 359
  Construction, Forestry, Mining and
    Engineering Union (CFMEU) 358,
    367, 370
  improvements in OHS 358
  occupational injuries scheme 359
Union Carbide (firm) 277
'union control' (Australia) 353
unionisation in North America 296
  number and diversity of unions 296
unionisation in South and Central America
  296
  Construction Worker' Union of São Paulo
    296
Unison (Public Sector Trades Union) 13

United Kingdom (UK) 352,
   362–3, 369
   construction industry 354
United Nations (2004) 194
United Nations Children's Emergency Fund
   (UNICEF) 291
United Nations Conference on Environment
   and Development (1992) 200
United Nations Convention on the Rights of
   the Child 289
   Article 32, 289
United Nations (UN) 143
United States of America (USA) 228, 262,
   335, 361
   BREEAM 228
   CEEQUAL 228
   Green-Building Council 228
   Leadership in Energy and the Environment
      'LEED' standard 228
United Way 104
University of British Columbia (Canada)
   221
University of Hong Kong 342
University Sains Malaysia 93
UN Millennium Development
   Declaration 76
UN Millennium Development Goals
   (MDGs) 15
unplasticised polyvinyl chloride
   (uPVC) 229
UN World Conference on Disaster
   Reduction (2005) 86
UOP Ltd. (firm) 184–5
'upstream' impacts of construction
   decisions 215
urban bias of development 82, 88
Urban Splash (firm) 254
Usable Buildings Trust 223
USA construction industry
   regulations 287
   Clean Air Act 287
   Community Reinvestment Act (banking
      sector) 287
   Foreign Corrupt Practices Act 287
   Public Company Accounting Reform and
      Investor Protection Act 287
Usel, M. 262
utilities companies 60

valuing diversity in construction (America)
   288–9
   lack of in South and Central
      America 288
VanderKamp, J. 341, 346
van der Linde, C. 100
Van Marrewijk, M. 36
van Wyk 306, 308, 316, 319

Varney, D. 57–8
   mm02, 57
   Motivate Don't Legislate 57–8
Vee, C. 370
Venezuela 291
Vickers, Sir J. 181, 182
Victor, B. 362
Victoria State (Australia) 262, 344, 353,
   359, 367
Vietnam 328
Vietnamese boat people 87
Vinten, G. 13
'virtue ethics' 100
virtuous organisation 100
'Vivamos Mejor Foundation' 291
Vollmer, H.M. 33
Voltaire-'encourager les autres' 136
volume house builders and imported housing
   kits 223–4
Vreis, H.J.M. 199

Wall Street 35
Walsh, J.P. 8, 13, 61
Walters, D.W. 172
Ward, H. 5, 8
Warioba, Joseph 149
   Warioba Report 149–50
      six National priorities against corrupt
         practices 150
Warren, F. 250
Wartick, S.L. 98–9
Wassener, B. 179
waste management practices 55
waste minimisation 58, 246
water consumption 246
Water Integrity Network 154
Watermeyer, R. 76
Water Regulations Advisory Scheme 217
Wates Construction Group 107
Wates Foundation (and family) 107–8
   quality of life (improvement) 107
   racial equality 107
   social change rather than self-interest 108
   welfare of the young and disadvantaged
      107
'weak' to 'strong' sustainability continuum'
   202
Watkins, C. 238, 240–1
Weiss, E.B. 198
welfare rights 62
welfare state 28
Welford, R. 201, 328,
   337, 342
Welford, R.J. 202
Wellings, F. 238–40, 243
Wells, C. 268
Wells, J. 89
Welsh Assembly Government 47
Welsh valleys 218

Wembley Stadium 354
Westbury (house-builders) 244–5,
   248–9, 251
Western Australia (State) 353, 367
Western Cape Department of Economic
   Affairs, Agriculture and Tourism
   304–5
Western society and its attitude to
   corruption 121
West Midlands 122, 181, 182
Westminster Government 47
Westpac (firm) 363
Wheeler, D. 100
'whistle-blowers' 136–7, 182–3, 371
   leniency towards 185
'whistle-blowing' 12–13, 146
   and murder 146
Whitelegg, J. 155
Whiteside, A. 305
'whitewashing' 378
Wikipedia (2007b) 330
Wilcox, J.R. 4, 14
Wilkinson, P. 84
Wilkinson, S.J. 4, 14
William Leech (firm) 240
Williamson, Oliver E.E. 327
Wilson 175, 182–3, 184
Wilson Bowden plc 244–5, 246, 247–8
Wilson Connelly (firm) 251
Wilson, I. 313
Wimpey (firm) 24–9, 234–5, 238
Winch, G. 31
Wind patterns 102
'Winning with Integrity' 38–9
(The) Wintles in Shropshire 222
Winstanley, D. 100
Wolseley Group (2006) 222
   Leamington Spa 222
   'Sustainable Building Centre' 222
Women in construction 288, 289
Wong, P.S. 320
Wood, R.A. 35
Woodall, B. 127, 130
Wooley, T. 17, 217, 218, 219, 224, 225,
   228, 229
Woolfson, C. 32
workers' compensation 274–5
workers' health issues in the
   Caribbean 297
   adverse climate 297
work/family conflict 374
workforce (Australia) 372–4
   working conditions 372–3, 378
   working environment 372
   workplace practice 376
workforce dialogue and participation
   295–6

work–life Balance 374
Workplace Relations Act (Australia) 367
World Bank (1994) 73, 74, 89, 90
World Bank Group (2007) 75
World Bank (WB) 80–1, 90–2, 143–4, 145,
   157, 158
World Business Council for Sustainable
   Development (WBCSD) 55, 300–1
World Commission on Environment and
   Development (WCED) 192, 195
World Health Organisation
   (WHO) 229
world poverty 73
World Summit on Sustainable Development
   (2002) 84
World Trade Organisation (WTO)
   328, 335
worldviews and sustainable development
   198–200
   ecological 'umbrella' 199
   nature and role of world views
      198–200
   in decision-making 198
   neoclassical 'umbrella' 199–205
World War II 24, 25, 27
WorldWatch Institute 194
World Wildlife Fund (WWF) (2007) 218,
   220, 226, 255
   'Million Sustainable Homes' 218
   'one planet living campaign' 221

Yangtze River 221
Yin, L. 337
York, city of 105
York Minster 105
Yorkshire Copper Tube Ltd. 183–4
Young, J.W.S. 199
Yu, C. 216, 224
Yu, Q. 337

Zaghoul, R. 320
Zarkada-Fraser, A 364
Zawdie, G. and Langford, D. A. 76, 79,
   82, 88–9
Zawdie, G. 15
Zero carbon objective 220–1
Zero impact buildings 220
   off the grid 220
   renewable energy 220
   sewage treatment 220
Zhai, W. 336
Zhang, S. 336, 341
Zhu, G. 336, 341
Zou, P.X.W. 149
'Z squared' development 221